Legionella

Molecular Microbiology

Klaus Heuner and Michele Swanson

Caister Academic Press

Caister Academic Press
Norfolk, UK

www.caister.com

British Library Cataloguing-in-Publication Data
A catalogue record for this book is available from the British Library

ISBN: 978-1-904455-26-4

Printed and bound in Great Britain

Contents

Contributors

Christiane Albert-Weissenberger
Institut für Molekulare Infektionsbiologie
Universität Würzburg
Würzburg
Germany

christiane.albert@mail.uni-wuerzburg.de

Heike Bruhn
Institut für Molekulare Infektionsbiologie
Universität Würzburg
Würzburg
Germany

heike.bruhn@mail.uni-wuerzburg.de

Carmen Buchrieser
Unité de Génomique des Microorganismes Pathogènes
Institut Pasteur
Paris
France

cbuch@pasteur.fr

Christel Cazalet
Unité de Génomique des Microorganismes Pathogènes
Institut Pasteur
Paris
France

ccazalet@pasteur.fr

Audrey Chong
Department of Microbiology and Immunology
Dalhousie University
Nova Scotia
Canada

chonga@niaid.nih.gov

Nicholas P. Cianciotto
Department of Microbiology and Immunology
Northwestern University Medical School
Chicago, IL
USA

n-cianciotto@northwestern.edu

Paul H. Edelstein
Clinical Microbiology Laboratory
Department of Pathology and Laboratory Medicine
University of Pennsylvania School of Medicine
Hospital of the University of Pennsylvania
Philadelphia, PA
USA

phe@mail.med.upenn.edu

Gary Faulkner
Department of Microbiology and Immunology
Dalhousie University
Nova Scotia
Canada

gfaulkner@tupdean2.med.dal.ca

Maris V. Fonseca
Department of Microbiology and Immunology
University of Michigan Medical School
Ann Arbor, MI
USA

marisf@umich.edu

Rafael A. Garduño
Department of Microbiology and Immunology and
Department of Medicine
Division of Infectious Diseases
Dalhousie University
Nova Scotia
Canada

rgarduno@dal.ca

Thomas R. Hawn
Departmentt of Medicine
University of Washington School of Medicine
Seattle, WA
USA

thawn@u.washington.edu

Matthew Heidtman
Howard Hughes Medical Institute and Department
of Molecular Biology and Microbiology
Tufts University School of Medicine
Boston, MA
USA

matt.heidtman@tufts.edu

Klaus Heuner
Institut für Molekulare Infektionsbiologie
Universität Würzburg
Würzburg
Germany

klaus.heuner@mail.uni-wuerzburg.de

Ralph Isberg
Howard Hughes Medical Institute and Department
of Molecular Biology and Microbiology
Tufts University School of Medicine
Boston, MA
USA

ralph.isberg@tufts.edu

Carol Joseph
Respiratory Diseases Department
Health Protection Agency
Centre for Infections
London
UK

carol.joseph@hpa.org.uk

Matthieu Jules
Unité de Génomique des Microorganismes
Pathogènes
Institut Pasteur
Paris
France

mjules@pasteur.fr

Paul Christian Lück
National Reference Laboratory for Legionella
Institute of Medical Microbiology and Hygiene
Technical University Dresden
Dresden
Germany

christian.lueck@.tu-dresden.de

Tamara O'Connor
Howard Hughes Medical Institute and Department
of Molecular Biology and Microbiology
Tufts University School of Medicine
Boston, MA
USA

tamara.oconnor@tufts.edu

Katherine Ricketts
Respiratory Diseases Department
Health Protection Agency
Centre for Infections
London
UK

<email?>

John-Demian Sauer
Department of Molecular & Cell Biology
University of California
Berkeley, CA
USA

sauerj@berkely.edu

Michael Steinert
Institut für Mikrobiologie
Technische Universität Braunschweig
Braunschweig
Germany

m.steinert@tu-bs.de

Michele S. Swanson
Department of Microbiology and Immunology
University of Michigan Medical School
Ann Arbor, MI
USA

mswanson@umich.edu

Russell Vance
Department of Molecular and Cell Biology
University of California
Berkeley, CA
USA

rvance@berkeley.edu

Carr D. Vincent
Department of Molecular Microbiology
Washington University
St. Louis, MO
USA

cvincent@borsim.wurstl.edu

Joseph Vogel
Department of Molecular Microbiology
Washington University
St. Louis, MO
USA

jvogel@borcim.wustl.edu

Other Books of Interest

Preface

More than 30 years have passed since *Legionella pneumophila*, the causative agent of Legionnaires' disease, was identified as a new human pathogen. First recognized due to epidemic of pneumonia that followed the 1976 Legionnaires' convention in Philadelphia, USA, legionellosis is still a disease of medical and public interest. *Legionella* is commonly found in aquatic habitats, where its ability to survive and to multiply within different protozoa equips the bacterium to be transmissible and pathogenic to humans. On the other hand, these very traits also make *Legionella* a favoured model system to analyse the mechanisms by which bacteria survive, acquire nutrients, and replicate within macrophages and the lung.

With the application of modern molecular genetic and cell biological techniques, we have begun to understand the mechanisms used by *L. pneumophila* to multiply within protozoa and alveolar macrophages. Also, we have gained insight into the specific regulatory cascades that govern cell cycle-dependent differentiation as well as the general mechanisms of gene regulation in *L. pneumophila*. Thanks to the recent publication of the genome sequences of four *L. pneumophila* strains, it is now feasible to investigate the whole genome *in silico*, the transcriptome via micro arrays, and the proteome by two-dimensional gel electrophoresis. Thus, new questions can be asked and impressive amounts of data can be generated and evaluated. Furthermore, research in the fields of clinical features, diagnosis, treatment and epidemiology continues to generate new strategies for management and prevention of disease and more questions for basic scientists.

The topics covered by this volume range from the history of the identification of *Legionella* and clinical disease treatment, to the microbe's gene expression and secretion systems as well as its strategies for intracellular multiplication and nutrient acquisition. Since it was not possible for this book to cover all topics of interest, instead we sought to convey the current state of many of the most critical features of *Legionella*. Finally, we thank the authors and all who have generously contributed to the production of this volume.

Klaus Heuner and Michele Swanson

Abbreviations

AFLP	amplified fragment length polymorphism
BBC	British Broadcasting Corporation
BCYE	buffered charcoal yeast extract
bp	base pairs
CAP	community-acquired pneumonia
CDC	Centers for Disease Control
ECDC	European Centre for Disease Prevention and Control
ELISA	enzyme-linked immunosorbent assay
ER	endoplasmic reticulum
EWGLI	European Working Group on Legionella Infection
FISH	fluorescence *in situ* hybridization
IFN	interferon
IL	interleukin
LCV	*Legionella*-containing vacuole
LLAP	*Legionella*-like amoebal pathogens
LPS	lipopolysaccharide
Mab	monoclonal antibody
MIF	mature intracellular form
MRA	macrorestriction analysis
NT	nucleotides
ORF	open reading frame
PCR	polymerase chain reaction
ppGpp	guanosine 3′,5′-bispyrophosphate
Sg	serogroup
T2SS	type II secretion system
T4SS	type IV secretion system
TLR	Toll-like receptor
TNF	tumour necrosis factor
VBNC	viable but non-culturable

Legionnaires' Disease: History and Clinical Findings

1

Paul H. Edelstein

Abstract

The history of Legionnaires' disease began at least 33 years before the 1976 Philadelphia epidemic, when *Legionella micdadei* was isolated from human blood. Multiple isolations of several different *Legionella* spp. were made prior to 1976, and it was known by 1968 that tetracycline therapy prevented deaths in *L. pneumophila*-infected chicken embryos. The 1976 epidemic provided the scientific focus and resources necessary to determine that *L. pneumophila* caused epidemic pneumonia and to show that epidemics of Legionnaires' disease had occurred worldwide many years before 1976. Despite a surfeit of available resources and expertise, the effort to isolate the aetiological agent succeeded solely on the basis of one person's determination to solve a scientific problem and his willingness to re-examine his assumptions about prior laboratory results.

Pontiac fever, a disease of unknown aetiology, is a self-limiting and short duration febrile illness that has been associated with exposure to *L. pneumophila*. Because of non-specific clinical findings that overlap with other diseases, accurate diagnosis of Pontiac fever in non-outbreak settings is impossible. Legionnaires' disease can be diagnosed specifically through specialized laboratory tests, but not by clinical findings alone. This is because the clinical findings of Legionnaires' disease overlap with those of other more common causes of community-acquired pneumonia. Antimicrobial therapy of Legionnaires' disease requires the use of drugs that are active against intracellular *Legionella* spp., such as tetracyclines, macrolides, azalides and antibacterial quinolones.

Introduction

Legionnaires' disease is a type of pneumonia caused by *Legionella* spp., which are environmental Gram-negative bacteria. Pontiac fever is a febrile, non-pneumonic, short incubation period, non-fatal disease that is associated with exposure to water-borne aerosols containing *Legionella* spp. bacteria, but it is unknown if the disease is caused by the *Legionella* spp. bacteria. The vast majority of Legionnaires' disease cases are caused by *L. pneumophila*, and in particular *L. pneumophila* serogroup 1. A 1976 epidemic of pneumonia in American Legionnaires' attending a Philadelphia, Pennsylvania, congress was the first recognized occurrence of Legionnaires' disease, and it led to the naming of the disease and isolation and characterization of *L. pneumophila*.

Prompt and specific antibiotic treatment of Legionnaires' disease can markedly reduce the fatality rate of the disease, without which 15–80% of patients die, depending on underlying disease, host immunity, and duration and severity of illness before treatment. Unfortunately there are no clinical features that allow a clinicians to specifically diagnose the disease with sufficient accuracy, which places great emphasis on the need for empiric antibiotic therapy for the disease.

Historical aspects

Legionnaires' disease and Pontiac fever epidemics prior to 1976

SPAM City, 1957

The first known epidemic of Legionnaires' disease occurred in Austin, Minnesota, in the summer of 1957 (Osterholm *et al.*, 1983). Seventy-eight people in this small city of approximately 30,000 people developed pneumonia between June and August. A significant risk factor for illness was working in the city's meat packing plant, although 40% of cases did not work at the packing plant. Antibiotic therapy had no apparent benefit, but 97% of affected people recovered in 2–14 days after onset of illness. The source of the epidemic was never determined, and the pneumonia did not recur in subsequent years. The mystery of the cause of the epidemic was solved by investigation of survivors 22 years later. This showed that the survivors had significantly elevated antibodies to *L. pneumophila* in comparison to matched controls. Because both packing plant employees and non-employees became ill, it is most likely that the epidemic derived from a cooling tower. Of interest, Austin, Minnesota, is the headquarters of the Hormel Foods Corporation, and is known as 'SPAM City' after the name of the canned ham product made there since 1937.

St. Elizabeth's hospital, 1965

Another early Legionnaires' disease epidemic affected residents of a large psychiatric hospital in Washington, DC, St. Elizabeth's hospital (Thacker *et al.*, 1978). The hospital housed about 6000 patients in multiple buildings on a 350 acre (1.4 km^2) campus. In July to August 1965, at least 81 hospital patients developed pneumonia, 14 (17%) of whom died. Cephalosporin therapy was ineffective, but tetracycline plus streptomycin treatment appeared to have to some beneficial effect. Sleeping near open windows and having permission to walk on the hospital grounds were found to be the most significant risk factors for acquiring pneumonia. Soil excavation work for a lawn sprinkling system was causing very dusty conditions. A strong wind storm and summer rains occurred concurrent with the outbreak, which ended spontaneously, and contemporaneously with the cessation of the excavation work. A soil or dust borne pathogen was thought to be most likely. Extensive microbiological testing of lungs from dead patients and of the hospital environment, failed to reveal an aetiology for the epidemic. However, analysis of stored serum specimens in 1977 showed that 19 of 26 patients tested seroconverted to *L. pneumophila* serogroup 1. Subsequent Legionnaires' disease epidemics have rarely implicated soil excavation, although disruptions of potable water systems and contamination of building plumbing systems during construction have been alternative explanations (Mermel *et al.*, 1995). An interesting historical note is that the attempted assassin of US President Ronald Reagan is currently confined to this institution.

Pontiac fever, Pontiac, Michigan, 1968

In July 1968 visitors to, and employees of, the Pontiac, Michigan, Health Department developed a relatively mild self-limited febrile illness that was called Pontiac Fever (Glick *et al.*, 1978). Cases became ill over a period of several days, with at least 144 people affected. The attack rate was 95% in building occupants and 29% in building visitors. Being in the building was the sole risk factor for illness, with an increased risk the greater the duration of time spent in the building. After an incubation period of 5–66 hours, fever and other symptoms developed. The illness lasted for 2–5 days and was manifested mainly by fever, headache, myalgia and fatigue. Most cases recovered promptly without sequelae although a small fraction of disease victims required months to recover from assorted neuropsychiatric symptoms. No case was shown to have pneumonia. Re-entering the building after recovery from illness resulted in a lower attack rate, around 10–45%, and was associated with milder symptoms than during the first illness. Investigations were negative for allergic reactions to known environmental allergens, for known toxins and for viral infections. The cause of the illness was traced to a leak in the building's air duct that allowed water from the evaporative condensing system to enter the circulating air. Turning off the building air conditioning eliminated the risk of disease, while turning it back on caused disease.

Guinea pigs that were placed in the building during the outbreak developed pulmonary nodules (Kaufmann *et al.*, 1981). At the CDC, challenge of additional guinea pigs with aerosolized water from the evaporative condenser caused pneumonia in the animals, but not when the water was filtered or autoclaved. Extensive microbiological studies of the guinea pig lungs failed to demonstrate a convincing aetiological agent. The CDC employee performing the aerosol challenge studies became ill with what was later thought to be Legionnaires' disease (US Department of Health and Human Services–Public Health Service–Centers for Disease Control and Prevention and National Institutes of Health, 1993).

Legionella pneumophila serogroup 1was cultured from some of the frozen lungs of the sick guinea pigs after the aetiology of Legionnaires' disease was known (Kaufmann *et al.*, 1981). One of these isolates, Pontiac 1 is the type strain for the Pontiac monoclonal antibody type. In addition, serum from those with Pontiac fever contained antibodies to *L. pneumophila*.

Hotel outbreaks prior to 1976

A prelude to the 1976 Philadelphia Legionnaires' disease epidemic was an epidemic of pneumonia that occurred in Philadelphia in September 1974 (Terranova *et al.*, 1978). Approximately 1500 members of the Independent Order of Odd Fellows attended a convention at the Bellevue Stratford Hotel, the same hotel involved in the 1976 epidemic. The Independent Order of Odd Fellows are a fraternal order dating from eighteenth-century England (http://www.ioof.org/odd_fellows.htm). Twenty Odd Fellows developed pneumonia one to nine days after attending a September 16th meeting in main ballroom of the hotel; two of these people died of the pneumonia. The major risk factor for pneumonia was attending the 16 September meeting. Serological studies of survivors and matched controls in 1977 showed that 4 of 11 cases, but no controls, had elevated antibody titres to *L. pneumophila* serogroup 1. Of historical interest, there is speculation that some members of ancient Odd Fellows organizations were Roman Legionnaires stationed in Britain (http://www.oddfellows.co.uk/history/?page_id = 8).

Meanwhile in Spain an unrecognized epidemic of pneumonia had been occurring among British tourists to the Rio Park Hotel in Benidorm, a coastal resort town on the Costa Blanca (Grist *et al.*, 1979; Bartlett *et al.*, 1984). From 1973 to 1980, at least 150 British tourists acquired Legionnaires' disease at the hotel, possibly in several discrete epidemics. In 1973, 3 years before the Philadelphia epidemic, at least 89 hotel guests had febrile respiratory illnesses, three of whom died with pneumonia, one on the plane en route from Spain to the United Kingdom (Grist *et al.*, 1979). Serological testing of saved serum specimens, undertaken after the discovery of the cause of the 1976 Philadelphia outbreak, showed that these illnesses were due to Legionnaires' disease. A 1980 investigation showed that the source of the epidemic was the hotel's potable water system. The hotel had no air conditioning cooling tower and nearby cooling towers could not be implicated based on epidemiological findings. Engineering changes made to the plumbing system, chlorination, and maintenance of hot water at 50–60°C appeared to end the multiyear epidemic.

The 1976 Philadelphia epidemic

Legionnaires' disease was first recognized as a distinct entity during an epidemic of pneumonia that occurred in Philadelphia, in the summer of 1976. About 4,000 members of the Pennsylvania State American Legion, an organization of former military veterans, met in July for their annual meeting in the city, which lasted from 21 to 24 July (Fig. 1.1). The convention hotel, the Bellevue Stratford, considered the best hotel in the city, was a frequent convention site and the hotel where visiting dignitaries stayed. It also was the site of the 1974 Odd Fellows Legionnaires' disease epidemic.

July, 1976 was a special time in Philadelphia and the nation, as this was the bicentennial anniversary of the United States Declaration of Independence from Great Britain. Philadelphia was where the Declaration of Independence was written and first announced, and the nation's first capital. Celebrations and large parades were held in the city during the time of the American Legion convention. City officials were worried that outside protestors would create violent disruptions of planned bicentennial celebrations. In addition, the hot and humid Philadelphia summer was made even more unbearable by a garbage collector's strike that left high mounds of stinking garbage on city streets.

In the spring of 1976, the US Center for Disease Control warned about the possibility of pandemic swine influenza occurring in the

Figure 1.1 Cover of the convention programme book for the 1976 Pennsylvania Department of the American Legion Convention. The original colours are red, white and blue.

United States after two cases of the disease were diagnosed in army recruits at Fort Dix, New Jersey (Gaydos *et al.*, 2006; Sencer and Millar, 2006). A nationwide influenza vaccination campaign was advised, but this was stalled because of manufacturer's concerns about financial liability.

On Friday 30 July 30 1976, Dr Ernest Campbell, a Bloomsburg, Pennsylvania, physician, realized that he was treating three convention delegates for a febrile illness (Lattimer and Ormsbee, 1981). From Friday until Sunday, other health care providers became aware of clusters of pneumonia (Lattimer and Ormsbee, 1981; Thomas and Morgan-Witts, 1982), and at least one called the Philadelphia Department of Health on Sunday (Thomas and Morgan-Witts, 1982). Since the Pennsylvania State Public Health Department was closed on that day and through the weekend, these reports were made on Monday, 2 August. By that day 18 convention delegates or city visitors had died of acute pneumonia, and about 170 were sick with pneumonia. In retrospect, the first recorded death from the epidemic was on 27 July. Initial recognition of the epidemic was delayed because the convention delegates were widely disbursed after the convention ended, and because many lived in small towns. The Philadelphia epidemic was featured in national television and radio broadcasts on 2 August, with an emphasis on the unknown cause of the epidemic. Convention delegates last became ill on 3 August, with the last known cases of Legionnaires' disease being in non-convention delegates on August 16th. The epidemic was over by the time that the epidemiological investigation began in earnest.

The total number of Legionnaires' disease cases was 182, with 29 deaths. Both convention delegates (149 cases) and non-Legionnaires (33 cases) were affected by the disease, including some who never entered the convention hotel. Some members of a marching band and people who watched a parade outside the hotel became ill; this was initially termed Broad Street pneumonia to distinguish the illness from that suffered by the Legionnaires. One dramatic example is that of a marching band bus driver who became ill and died of the disease; he had spent only hours in Philadelphia, and most of the time in his bus. Great public concern and fear arose, such that funerals of those who died with the disease were not attended by the public because of concern of getting the disease from the corpse (Thomas and Morgan-Witts, 1982). Even before the public health investigations had been completed, Bellevue Stratford hotel guests cancelled their reservations and the hotel was left empty. The state governor stayed overnight in the hotel to emphasize its safety, but to no avail; the hotel was forced to close permanently on 18 November 1976. Reopened later under a new name and after major renovations, it is now known as the Park Hyatt Philadelphia at the Bellevue.

Epidemiological studies, led by David Fraser of the US Center for Disease Control (CDC), concluded that the focus of the disease was the Bellevue Stratford Hotel, that both American Legionnaires and non-convention delegates had acquired the disease, and that the disease was most probably spread by the air-borne route (Fraser *et al.*, 1977). However, the cause of the epidemic pneumonia was not determined until December 1976, some 5 months after the outbreak (McDade *et al.*, 1977). Tests for almost all known types of respiratory infections were negative, as were toxicological studies.

The inability of the CDC investigators to quickly determine the cause of the outbreak led to widespread, and often unfounded, speculations about the cause of the epidemic. Some suggested that the epidemic was a US Central Intelligence Agency germ or chemical warfare experiment gone bad. Others suggested that the epidemic was a hoax, designed to gin up support for the CDC and swine influenza vaccination. Some of this speculation was fuelled by the rapid US congressional subcommittee approval of a vaccine manufacturers liability protection act, on August 3rd, 1977, after congressional testimony that swine influenza could not be excluded as a cause of the epidemic (Lattimer and Ormsbee, 1981). Some thought that domestic terrorists had caused the illnesses through chemical or microbiological means. A leading Philadelphia pathologist and toxicologist, Dr William Sunderman, suggested that the epidemic was due to nickel carbonyl poisoning, a disease that he had studied for years. The New York Times excoriated the CDC for its failure to collect adequate specimens for toxicological studies (Anonymous, 1976). The possibility of nickel carbonyl poisoning was supported briefly by autopsy studies until it was realized that the autopsy instruments were the probable sources of the nickel that was detected; repeat testing and detailed analysis showed that nickel carbonyl poisoning was not the cause of the epidemic (Lattimer and Ormsbee, 1981). Regardless, Dr Sunderman and his son, also a toxicologist, maintained for years afterwards that nickel carbonyl poisoning was a strong possibility as the cause of the epidemic (Sunderman, 1977; Sunderman, 1978). Others speculations for an aetiology were toxic fumes from photocopy machines, air conditioning refrigerant decomposition producing phosgene gas, and nefarious foreign or extra-terrestrial forces.

Concern about the inability to identify the aetiology of the epidemic led to US Congressional hearings in November 1976, about a week after the permanent closure of the Bellevue Stratford Hotel. The hearings, representative of the frustration of many that no aetiology had been determined for the epidemic, were primarily political theatre. A month later Joseph McDade had determined that the aetiology of the outbreak was *L. pneumophila* (McDade *et al.*, 1977; Sharrar *et al.*, 1977). In April, 1977, the term Legionnaires' disease was first published by the Center for Disease Control as the official name of the epidemic disease, although it had been called that in the press as early as September 1976 (Anonymous, 1977; Anonymous, 1976; Lattimer and Ormsbee, 1981).

Discovery of the aetiological agent of Legionnaires' disease

McDade's achievement of determining that *L. pneumophila* was the cause of Legionnaires' disease was remarkable when so many others had failed for months to isolate the bacterium (McDade *et al.*, 1977). A rickettsiologist, he was assigned to determine if Legionnaires' disease was caused by *Coxiella burnetii*, the agent of Q fever (McDade, 2002). Using the technique of guinea pig inoculation followed by embryonated egg passage, he was unable at first to demonstrate an infectious agent affecting the embryos. The guinea pigs became ill after being injected with human lung homogenates, but passage into the embryos resulted in no embryo infections. Initial microscopic examination of guinea pig tissues stained with the Giménez stain, which stains rickettsia as well as many other bacteria, revealed just an occasional rod thought to be the result of contamination. However, when McDade re-examined guinea pig tissue slides he found clusters of rods, and convinced that these were real he repeated the hen's egg inoculations without adding the standard antimicrobials used for rickettsial isolation. By eliminating antimicrobial suppression he successfully infected embryos with *L. pneumophila*, discovering the aetiology of Legionnaires' disease. Axenic cultivation was achieved by Robert Weaver of CDC, using Mueller–Hinton agar supplemented with IsoVitaleX and haemoglobin, a source of needed L-cysteine and iron (Feeley *et al.*, 1978). Moreover, McDade was able to show that several patients with Legionnaires' disease had antibodies to the bacterial isolate, as did stored sera from the St. Elizabeth hospital and Pontiac epidemics (McDade *et al.*, 1977).

The Giménez stain used by McDade to visualize *L. pneumophila* in guinea pig and chicken embryos contains carbol fuchsin, which intensely stains Gram-negative rods, as well as rickettsia. Use of this stain was a key factor in McDade's discovery of *L. pneumophila* in guinea pig tissue, as was staining of fresh rather then fixed tissues. A number of conventional stains used to detect bacteria in tissues failed to detect *L. pneumophila* in the lungs of patients with Legionnaires' disease, despite the tremendous number of the bacteria in the lungs of fatal cases. In particular the usual Gram stain formula uses safranin as the counter stain, which barely stains the bacterium when taken from fresh growth on culture medium, let alone from infected tissue or sputum. In contrast, the Giménez stain is quite good at detecting the bacterium in culture, in infected macrophages, and in infected fresh and Formalin-fixed lung. However, Giménez stain does not stain the bacteria in paraffin-embedded lung, or in deparaffinized tissue, making it unlikely that use of the stain by pathologists investigating the disease would have been helpful, unless they had studied non-embedded tissues, something not ordinarily done by pathologists (Chandler *et al.*, 1977). The poor staining of *L. pneumophila* by conventional stains undoubtedly contributed to earlier failures to recognize the bacterium in the lungs of fatal cases.

The methods used by McDade to isolate and identify *L. pneumophila,* as well as to determine the link between the bacterium and human disease, were worked out by Walter Reed Army Institute of Research scientists over 30 years previously to detect non-conventional microbes causing human diseases. In fact, these prior efforts led to the first identification of several different *Legionella* spp., although at the time they were thought to be adventitial rickettsia-like agents.

First isolations of *Legionella* spp. bacteria

The first known isolation of a *Legionella* spp. bacterium was in 1943. Captain Hugh Tatlock, a military physician, was assigned by the Army to investigate an epidemic of a febrile rash illness at Fort Bragg, North Carolina, termed Fort Bragg fever (Daniels and Grennan, 1943). Many affected soldiers had bathed in a local creek. Tatlock injected the blood of ill soldiers into guinea pigs, and from one soldier's blood isolated what was thought to be a rickettsia-like agent organism from three of five inoculated guinea pigs (Tatlock, 1944). However, Tatlock could not demonstrate that serum from patients with Fort Bragg fever reacted with the bacterium, nor could the bacterium infect other animal species, including mice, monkeys, rats, or rabbits. He concluded that this bacterium was not the cause of Fort Bragg fever, and later expressed concerns that the bacterium was enzootic in the guinea pig colony (Tatlock, 1947). This rickettsia-like agent, which became known as the *Tatlock* strain, was identified in 1980 as *L. micdadei* (Hébert *et al.*, 1980a; Hébert *et al.*, 1980b). In 1944, Tatlock isolated a second microbial agent from the blood of soldiers with Fort Bragg fever, which when injected into human volunteers caused Fort Bragg fever (Tatlock, 1947). This second agent was later shown to be *Leptospira autumnalis* (Gochenour, Jr. *et al.*, 1952). In 1981 Tatlock reiterated that the *Tatlock* strain was not the cause of Fort Bragg fever, although he suggested that perhaps that isolate caused a disease that was not Fort Bragg fever (Tatlock, 1981; Tatlock, 1982). Since guinea pig isolation of *Legionella* spp. bacteria is insensitive for environmental samples, and presumably for clinical sources, it is difficult to understand how the *L. micdadei* isolate could have been a contaminant (Edelstein *et al.*, 1982). Whether it was enzootic is difficult to know, but Tatlock was unable to isolate a Tatlock-like bacterium from other guinea pigs in the same colony as those that yielded this bacterium (Tatlock, 1944). Whether some soldiers had both leptospirosis and Legionnaires' disease, from the same creek source, is impossible to know, but is an intriguing possibility. Apart from his investigative career, Dr Tatlock served as General Dwight Eisenhower's physician at Walter Reed Army Hospital, and later as a respected New Hampshire physician. He died at age 92 in 2005.

Guinea pig inoculation was used by the Rickettsial Laboratory at the Walter Reed Army Institute of Research in the 1940s and 1950s to diagnose human rickettsial infections. This led to the discovery of three rickettsia-like bacteria which were later shown to be *Legionella* spp.

bacteria. The first isolate, *Olda*, shown in 1979 to be *L. pneumophila*, was isolated in 1947 from the blood of a patient with a fever of unknown origin (McDade *et al.*, 1979). In 1959 Bozeman and colleagues at the same laboratory isolated a rickettsia-like agent from the lung of a Navy skin diver with fatal pneumonia (Bozeman *et al.*, 1968; Cordes *et al.*, 1979). The diver became ill in 1958, two days after undergoing scuba training in a Florida freshwater swimming pool, and died six days later of fatal pneumonia that failed to respond to penicillin and streptomycin therapy. This isolate, known as *Wiga*, was identified as *L. bozemanae* in 1979 (Cordes *et al.*, 1979; Brenner *et al.*, 1980). Bozeman and colleagues also reported the 1959 isolation of strain *Heba*, later shown to be *L. micdadei* (Hébert *et al.*, 1980b). *Heba* was isolated from the blood of a patient with pityriasis rosea; whether the patient had another disease is unknown, but likely, as *L. micdadei* is not a cause of pityriasis rosea (Chuh *et al.*, 2004). Tetracycline was shown by Bozeman and colleagues to be effective in preventing hen's egg embryo deaths when they were infected with *Wiga, Tatlock, Heba* or *Olda* (Bozeman *et al.*, 1968), the first evidence that *Legionella* infections could be potentially cured with tetracycline therapy.

Development of an artificial growth medium

Robert Weaver of CDC was able to grow *L. pneumophila* on an artificial medium from the infected chicken embryo yolk sacs that Joseph McDade gave him. Weaver did this by plating the yolk sacs to 18 different bacteriological media under different atmospheric conditions (McDade *et al.*, 1977), finding that only Mueller–Hinton agar supplemented with 1% haemoglobin and IsoVitaleX supported growth. Growth was optimal in 2.5% CO_2. At approximately the same time, Morris Dumoff, a clinical microbiologist in Flint, Michigan, had independently isolated *L. pneumophila* by plating autopsy lung and pleural fluid specimens onto a commercially made chocolate agar supplemented with haemoglobin and L-cysteine (Dumoff, 1979). Dumoff was the first person known to have isolated *L. pneumophila* from an artificial medium; he was honoured for this achievement by the naming of *L. dumoffii* (Brenner *et al.*, 1980). Growth of *L. pneumophila* on supplemented Mueller–Hinton and chocolate agars was very poor, and the lack of a good selective medium made continued use of guinea pig and embryonated hen's eggs inoculations important in some cases. Jim Feeley and George Gorman of CDC devised an improved non-selective medium, F–G medium(Feeley *et al.*, 1978). The formulation of F–G medium was based on an analysis of the composition of prior successful media, and of IsoVitaleX, a nutritional supplement devised to enhance the growth of *Haemophilus influenzae*. This analysis established that L-cysteine and iron were required for optimal bacterial growth. However, F–G medium was difficult to prepare and not very supportive of bacterial growth. The next iteration, buffered yeast extract charcoal medium, was superior to F–G medium as it was more supportive of *L. pneumophila* growth and much easier to prepare (Feeley *et al.*, 1979). Addition of α-ketoglutarate to the medium formulation enhanced the *L. pneumophila* growth rate, plating efficiency, and colony size; this compound is now part of most commercial formulations of buffered yeast extract medium (Edelstein and Finegold, 1979). However, there was still a problem in isolating the bacterium from contaminated clinical and environmental specimens, which required guinea pig inoculation with or without chicken embryo passage. The development of several different selective media resulted in the modern formulations of selective media for the clinical cultivation of *Legionella* spp (Edelstein and Finegold, 1979; Bopp *et al.*, 1981; Edelstein, 1981; Wadowsky and Yee, 1981), and eliminated the need for guinea pig inoculations (Edelstein *et al.*, 1982).

Discovery of the mode of spread of nosocomial Legionnaires' disease

During the period from 1976 to 1982 there were several multiyear epidemics of Legionnaires' disease in the US, the best example of which was at the Los Angeles Wadsworth Veterans Administration Hospital (Bock *et al.*, 1978; Haley *et al.*, 1979; Kirby *et al.*, 1980). A new hospital building was occupied in March 1977, and it soon became apparent that renal transplant patients were dying of nosocomial pneumonia. These pneumonias were confirmed to be Legionnaires' disease through a lengthy process involving sending serum and tissue specimens to the CDC. Once diagnostic testing became available on site, many cases of Legionnaires' disease were diagnosed in patients, hospital visitors, and some hospital staff workers. Because the environmental reservoir, and mode of spread of Legionnaires' disease, was unknown at the time, it was unclear what could be done to prevent further cases of the disease. The hospital cooling towers were chlorinated in November 1978 because the cooling tower contained *L. pneumophila* (Shands *et al.*, 1985). Cooling tower chlorination resulted in a temporal decline in cases from an average of 10 cases per month to five cases per month but no further. Air lock doors were installed in the hospital to prevent infiltration of contaminated air, and a HEPA-filtered nursing unit was bought, but never used, to protect immunocompromised patients. Isolation of the bacterium from environmental sources required guinea pig inoculation and was insensitive. It was not known at the time that *L. pneumophila* was common in building water systems, although the ubiquity of the bacterium in natural water sources had been shown by Morris and colleagues (Morris *et al.*, 1979). Several attempts were made to isolate *L. pneumophila* from nearby environmental sources that might be reservoirs and disseminators. For example, I was instructed to culture captured gophers to determine if they were a reservoir of the epidemic; a gopher hunter flushed the animals out of their tunnels, and their tissues cultured without avail. I was also assigned to culture a water culvert about 500 meters from the hospital, again without yield; these studies give a flavour of the desperation and ignorance of the time. A breakthrough was made in 1979 by John Tobin in Oxford (Tobin *et al.*, 1980). Using guinea pig inoculation to detect *L. pneumophila* in shower heads, Tobin reported that the source of a Legionnaires' disease outbreak in a renal transplant unit was caused by bacterial colonization of the hospital showers and potable water system. The big break at Wadsworth VA came in March 1979, when a huge rise in the numbers of nosocomial cases at Wadsworth VA occurred, up from an average of 4.5 cases per month to 25 cases in 1 month. It was discovered that a test of an emergency water pump failed immediately prior to the upswing in cases, resulting in a pressure shock and brown-black water throughout the plumbing system. This pressure shock was replicated a month later in an unoccupied hospital ward; a comparison of pre-shock and post-shock samples demonstrated a 35-fold increase in *L. pneumophila* concentration in the water, a convincing experiment which took several months to complete (Shands *et al.*, 1985). The epidemic was ended by hyperchlorination of the potable water system in September 1980, three and a half years after the start of the epidemic. By the end of the epidemic at least 250 inpatients, outpatients, visitors and employees had acquired Legionnaires' disease from the hospital water in the plumbing system.

Multiyear Legionnaires' disease nosocomial outbreaks still occur, although these are now due either to inadequate case surveillance or inability to disinfect plumbing systems, not ignorance about the reservoir and disseminators of the bacterium (Anonymous, 1997; Kool *et al.*, 1998; Lepine *et al.*, 1998).

Discovery of the role of amoebae in the environmental ecology of *Legionella* spp.

The ubiquitous presence of *Legionella* spp. bacteria in the environment was known by 1980, and by 1981 it was known that the bacteria were found in man-made aqueous environments (Morris *et al.*, 1979; Tobin *et al.*, 1981). However, how the *Legionella* spp. bacteria survived and grew in low-nutrient waters was unclear, in view of the fastidious nutritional requirements of the bacterium when cultivated on artificial media. Tim Rowbotham, a public health microbiologist in Leeds, made seminal observations on the growth of *L. pneumophila* in free-living amoebae (Rowbotham, 1980). These studies were performed almost exclusively based on the microscopic observations of the bacteria interacting with the amoebae, using the most basic equipment and facilities. Rowbotham hypothesized that inhalation of *Legionella*-infected amoebal cysts is the means of transmission of Legionnaires' disease. Based on these studies, several investigators determined that environmental *Legionella* spp. are dependent on the presence of amoebae for their growth and persistence, serving as an explanation for the growth of the bacteria in otherwise nutrient-poor water (Fields *et al.*, 1984; Wadowsky *et al.*, 1988; Fields *et al.*, 1989; Fields *et al.*, 2002). The formation of a consortium of microorganisms in a biofilm also helped to explain the erratic phenomena of some Legionnaires' disease epidemics, based on sloughing of the biofilm in response to physical or chemical changes.

Clinical features of Pontiac fever and Legionnaires' disease

Pontiac fever

Pontiac fever is a self-limited short duration febrile illness of unknown aetiology. It is unclear whether the illness results from non-*Legionella* spp. bacteria endotoxin inhalation, from the inhalation of live or dead *Legionella* spp. bacteria, or from inhalation of a toxic mix of *Legionella* and non-*Legionella* microbes and their toxins (Edelstein, 2007). There is neither an agreed-upon standard definition of Pontiac fever nor any specific laboratory test that can be used to diagnose the disease, although there is at least one proposal to standardize the definition (Tossa *et al.*, 2006; Edelstein, 2007). Most experts agree that a reasonable definition of epidemic Pontiac fever includes the following: (a) an epidemic febrile short duration illness without pneumonia or severe illness; (b) evidence of exposure to a microbe-containing aerosol, one component of which must be a *Legionella* spp. bacterium; (c) demonstration of an immune response to the *Legionella* spp. bacterium found in the aerosol source in some of the affected people AND evidence that a matched unexposed control population has a significantly lower antibody response to the same bacterium; d) lack of other credible explanations for the illness. Some would also add as a diagnostic test the presence of *L. pneumophila* serogroup 1 antigenuria in a number of the affected people, in the absence of documented pneumonia or prolonged illness, and with spontaneous recovery without antibiotic therapy (Burnsed *et al.*, 2007). It is unknown how to specifically diagnose single cases of Pontiac fever, and case reports of sporadic cases should be viewed with scepticism. The reason for this scepticism is that *Legionella* spp. are commonly found in environmental water without causing disease, and because up to 25–40% of healthy people may have detectable *Legionella* spp. antibody – especially against non-*L. pneumophila Legionella* spp. (Edelstein, 2006b). Thus, a high proportion of healthy people might meet laboratory diagnostic criteria for Pontiac fever. The requirement for an epidemic, and use of a matched control population, increases the positive predictive value of laboratory testing.

Whether Pontiac fever signs and symptoms can occur in the absence of exposure to *Legionella* spp. bacteria is unstudied; however based on the overlap of its clinical findings with a number of infectious disease, such as influenza and bacterial sepsis, this seems likely. Also where Pontiac fever ends and Legionnaires' disease begins is not defined, so in some cases people suspected of having Pontiac fever might really have mild Legionnaires' disease, and vice versa. Because of uncertainties about the pathogenesis and definition of Pontiac fever, the full clinical spectrum of the disease is unclear.

The incubation period after exposure to a bacteria-contaminated aerosol has a wide range, 4–120 hours, with variable median intervals ranging averaging around 1–2 days (Table 1.1). The sites of Pontiac fever outbreaks have included workplaces, hotels, recreational spas and restaurants, but disease can occur almost anywhere there is the possibility of encountering a bacterial aerosol. In some cases, exposed people may develop either Pontiac fever or Legionnaires' disease. Attack rates are very high, with more than 80–90% of such exposed people becoming ill. Fever, myalgia, headache and fatigue are the dominant symptoms (Table 1.2). Cough, dyspnoea, anorexia, arthralgia and abdominal pain occur less frequently. There is little information about physical examination findings in the first day of illness; examination 2–5 days after onset may show fever and tachypnoea, but little else. Pneumonia does not occur. In one outbreak pneumonia and hypoxaemia were documented, contrary to reported findings for all other outbreaks (Castor *et al.*, 2005). The authors reporting this outbreak speculated that the workers inhaled a very high concentration of endotoxin. Fatigue and non-focal neurological complaints have been reported to persist for up

Table 1.1 Incubation period, duration of illnesses and hospitalization rates for patients with Pontiac fever

Outbreak	*Legionella* spp.	No. cases	Incubation period (h)		Illness duration (days)		Hosp. (%)	Reference
			Median	Range	Median	Range		
Pontiac	*L. pneumophila*	144	33–40	5–66	3.0	1–6	0.7	(Glick *et al.*, 1978)
Vejers Strand	*L. pneumophila*, *L. micdadei*	15	43 (child) 70 (adults)	n.s.	3.6	1.7–4.5	0	(Lüttichau *et al.*, 1998)
Nashville	*L. anisa*	117	49	4–120	3.0	<1–8	0.8	(Jones *et al.*, 2003)
Minnesota	*L. pneumophila*	15	32	16–57	n.s.	1–3	87	(Castor *et al.*, 2005)
Lochgoilhead	*L. micdadei*	170	48	<24–216	3.0	1–14	0	(Goldberg *et al.*, 1989)
Rochester	*L. pneumophila*	14	23	4–38	n.s.	1–3	n.s.	(Mangione *et al.*, 1985)
Santa Clara	*L. anisa*	34	56	n.s.	2.0	1–8	0	(Fenstersheib *et al.*, 1990)
Windsor	*L. feelei*	317	46	n.s.	3.0	n.s.	n.s.	(Herwaldt *et al.*, 1984)
Oklahoma	*L. pneumophila*	101	n.s.	n.s.	4.0	1–9	2.9	(Burnsed *et al.*, 2007)

Table 1.2 Signs and symptoms of Pontiac fever, % present

	Pontiac	Oklahoma	Nashville	Lochgoilhead	Vermont	Windsor
Fever	85	90	100	65	85	91
Fatigue	97	90	93	72	n.s.	89
Headache	91	90	87	80	97	n.s.
Chills	91	87	92	50	85	92
Myalgia	95	78	93	60	74	93
Cough	57	65	44	35	32	n.s.
Dyspnoea	n.s.	52	n.s.	35	n.s.	28
Chest pain	45	n.s.	n.s.	n.s.	n.s.	36
Dizziness	54	n.s.	n.s.	n.s.	n.s.	68
Arthralgia	43	n.s.	n.s.	60	n.s.	n.s.
Vomiting	12	36	16	n.s.	15	15
Diarrhoea	15	28	30	n.s.	26	11
No. of cases	114	101	117	170	34	317
Reference	(Glick *et al.*, 1978)	(Burnsed *et al.*, 2007)	(Jones *et al.*, 2003)	(Goldberg *et al.*, 1989)	(Spitalny *et al.*, 1984)	(Herwaldt *et al.*, 1984)

n.s., not stated.

to several months in the minority of affected patients. Most people with Pontiac fever recover within 2–4 days, although some are sick for up to a week (Table 1.1). Illness severe enough to result in hospitalization is exceptional, so much so that a requirement for this should bring into question the diagnosis and the possibility of other diseases including Legionnaires' disease.

Legionnaires' disease

Definitions

Legionnaires' disease is a pneumonia caused by *Legionella* spp. that may or may not be associated with extrapulmonary infection. There are very rare case reports of *Legionella* spp. extrapulmonary infection in the absence of pneumonia, for which an appropriate term is either legionellosis or a description of the site of infection and infecting species, e.g. '*L. longbeachae* pyarthrosis'.

Clinical findings

Patients with Legionnaires' disease have pneumonia and in addition may have clinical findings suggestive of a systemic disease (Edelstein and Cianciotto, 2005). Symptoms and signs of the disease are often quite variable. The majority of patients have fever, which is usually one of the earliest signs of the illness. Accompanying the fever may be anorexia, myalgia, rigors, and headache. Chest pain, shortness of breath and cough may or may not be prominent findings. The cough may or may not be productive, and when it is productive the sputum can be bloody, purulent, or scant and mucoid. In some patients the absence of purulent sputum production, chest pain and cough may fool clinicians into discarding pneumonia as a possibility. When chest pain and haemoptysis are prominent the patient may be suspected of having a pulmonary infarction. Abdominal pain, diarrhoea, nausea and vomiting may occur as well, symptoms that have led to consideration of intra-abdominal infections and inflammatory conditions such as appendicitis, peritonitis, abscesses, inflammatory bowel disease and diverticulitis. Elderly and immunocompromised patients may not have fever or findings that localize to the lung. Confusion and memory loss are common presenting findings. Much less common are frank encephalopathy, focal neurological findings, seizures and meningitis.

As the untreated disease progresses, the major findings include fever or hypothermia, dense consolidation of the lung, and often respiratory failure. The majority (~70%) of non-immunocompromised patients without significant underlying diseases recover without specific therapy after being quite ill for 5–7 days. Death is usually the consequence of respiratory failure, often combined with sepsis syndrome.

Hyponatraemia, leucopenia or leucocytosis, thrombocytosis or thrombocytopenia, elevated serum creatinine kinase and elevated liver-associated tests are all non-specific laboratory findings that can be commonly observed. In the presence of severe respiratory failure, thrombocytopenia and evidence of disseminated intravascular coagulation is common. Chemical evidence of pancreatitis is an occasional finding, as is evidence of myocarditis.

Chest radiography almost always reveals alveolar filling infiltrates, often with consolidation (Kirby *et al.*, 1979; Tan *et al.*, 2000). Purely interstitial lung infiltrates are very uncommon although such infiltrates may be the only finding very early in the disease process. Cavitation of prior areas of consolidation occurs in up to 10% of immunocompromised patients. Pleural effusions are seen in about 40% of patients, often in patients with other causes for this such as congestive heart failure and renal failure. However, a small fraction of people may have a small pleural effusion without apparent chest infiltrates as the only radiographic manifestation of the disease.

Physical examination findings that are most common include pulmonary râles and other evidence of lung consolidation, tachypnoea and tachycardia. Confusion and memory loss are also relatively common findings.

Patients treated with specific antibiotic therapy usually improve promptly, with the systemic signs of infection such as confusion and sepsis clearing most rapidly. Up to a week may be required for patients to become completely afebrile, although the fever starts to decrease within 12 hours of the initiation of therapy.

Immunocompromised patients or patients with very advanced pneumonia may have either no response or a very slow response to specific therapy, with prolonged fever and respiratory failure. Fatality rates are the highest in patients treated late in the disease, with the fatality rate approaching 70% in those with severe respiratory failure on presentation, even with specific therapy.

Extrapulmonary manifestations of Legionnaires' disease can be of two types: organ or tissue disease without obvious infection, and metastatic or contiguous infection. Examples of the former type include sepsis syndrome, hepatitis, pancreatitis, myositis, myopericarditis, and encephalitis; the most common of which are sepsis syndrome, myositis and pancreatitis (Edelstein and Cianciotto, 2005; Edelstein, 2006a). Examples of the latter type include pleural empyema, purulent pericarditis, vascular graft infections, renal and brain abscesses, purulent peritonitis, bowel abscesses, pyarthrosis, and others; of these only pleural empyema has been reported in more than a handful of cases (McClelland *et al.*, 2004).

Not all of the clinical manifestations of Legionnaires' disease may be due to infection with *Legionella* spp., as up to 10% of patients with this disease have coinfections with other microbes. Documented coinfecting bacteria have included *Streptococcus pneumoniae, Haemophilus influenzae, Staphylococcus aureus, Escherichia coli, Aspergillus, Rhizopus,* cytomegalovirus, *Pneumocystis jirovecii* and many other organisms (Meyer *et al.*, 1980; Ruutu *et al.*, 1987; Marrie *et al.*, 1992; Edelstein and Cianciotto, 2005).

Legionella spp. infection in the absence of Legionnaires' disease is very uncommon (McClelland *et al.*, 2004). Most such cases involve direct inoculation of injured tissues with water containing *Legionella* spp. One example includes *L. pneumophila* and *L. dumoffii* sternal wound infections after bathing post-cardiac surgery patients with water colonized with *L. pneumophila* and *L. dumoffii* (Lowry *et al.*, 1991). Another example includes *L. pneumophila* wound infection after bathing a wound in *L. pneumophila*-colonized water (Brabender *et al.*, 1983).

Distinguishing Legionnaires' disease from other causes of pneumonia

Rapid diagnosis of Legionnaires' disease allows the use of therapy specific for the disease, and prompt notification of public health authorities that may curtail epidemics. Therefore, major efforts have been made to discover clinical findings specific for the disease. However, the presenting signs and symptoms of Legionnaires' disease are indistinguishable from those found in people with other common causes of community-acquired pneumonia, such as that due to *Streptococcus pneumoniae*. Although *Legionella*-specific laboratory testing is required to accurately diagnose the disease, these tests may be expensive, insensitive, unavailable or require many days to perform; details of these tests are found in Chapter 2. The similarity of the clinical findings of Legionnaires' disease to those of other pneumonias, and the deficits of specific diagnostic testing, led to interest in defining specific clinical findings that could be used to diagnose the disease quickly without resort to specific laboratory testing.

It was originally thought that Legionnaires' disease was a distinct and easily distinguishable clinical syndrome, characterized by rigors, absence of a productive cough, high fever with a disproportionately low pulse rate, headache, myalgia, anorexia, nausea and diarrhoea. In addition it was thought that several non-specific laboratory abnormalities were characteristic of the disease, including hyponatraemia, elevated liver-associated enzyme levels, hypophosphataemia, proteinuria and myoglobinuria, and elevated levels of muscle-associated enzymes (Tsai *et al.*, 1979; Kirby *et al.*, 1980; Lattimer and Ormsbee, 1981). Some thought that the radiographic presentation of the disease was also characteristic, with rapidly progressive multifocal consolidating infiltrates (Kirby *et al.*, 1979). However, a seminal study by Yu and colleagues showed that it was impossible to distinguish the presenting clinical findings of Legionnaires' disease from other common causes of community-acquired pneumonia (Yu *et al.*, 1982). This study was validated by several others, including studies of laboratory findings, radiography and nosocomial pneumonia (Sopena *et al.*, 1998; Mulazimoglu and Yu, 2001).

In each of the prospective studies comparing the clinical presentation of Legionnaires' disease with that of other causes of pneumonia some clinical or laboratory finding was found to be more common in patients with Legionnaires' disease. The most common ones were a greater frequency of hyponatraemia and of diarrhoea in patients with Legionnaires' disease. For example, in one study hyponatraemia was found in 44% of patients with Legionnaires' disease and in 14% of patients with other pneumonias (Yu *et al.*, 1982). However, there was enough overlap in these findings to make them unsuitable to distinguish unequivocally between Legionnaires' disease and other pneumonias (Edelstein, 2006a). In addition, some significant differences in laboratory abnormalities differed by relatively small amounts. In one study patients with Legionnaires' disease had mean serum sodium concentrations of 132.6 meq./L versus 135.7 meq./L for the patients with other pneumonias; both groups had hyponatraemia with a difference that can be observed from day to day in the same patient (Fernandez *et al.*, 2002).

Attempts at combining typical signs and symptoms of Legionnaires' disease into a diagnostic scoring system have failed. Two scoring systems, the Winthrop University criteria and the CBPIS criteria, were unable to accurately classify patients with and without Legionnaires' disease in prospective studies (Gupta *et al.*, 2001; Fernandez-Sable *et al.*, 2003). Even still, it appears as if there may be a clinically recognizable and distinct presentation of Legionnaires' disease which has yet to be captured by a clinical scoring system. In the study of the CBPIS, emergency department physicians were able to correctly suspect Legionnaires' disease in 64% of patients diagnosed with the disease, whereas they suspected the disease in only 4% of patients with pneumococcal pneumonia (Fernandez-Sabe *et al.*, 2003). In addition, specific therapy for Legionnaires' disease was given to 89% of those who turned out to have the disease, but only to 18% with pneumococcal pneumonia. While imperfect, experienced clinicians can accurately suspect Legionnaires' disease. Unfortunately, depending exclusively on clinical impressions would mean that at least 10% of Legionnaires' disease would not be treated for the disease, an unacceptable figure. Clinical diagnostic inaccuracy would likely be much higher if the physician assessors had less experience with Legionnaires' disease. So it is fair to say that many people with Legionnaires' disease have a distinct clinical syndrome, but that so far it is impossible to detect all patients with Legionnaires' disease using solely clinical and non-specific laboratory findings.

Analysis of inflammatory marker levels has been another approach to aid in the diagnosis of Legionnaires' disease. Despite the theoretical appeal of using markers that are most elevated in intracellular (neopterin) versus extracellular (procalcitonin) infections, measurement of relative levels of these two biomarkers is not useful for making a specific diagnosis of Legionnaires' disease (Prat *et al.*, 2006). This should not be surprising as procalcitonin is driven by inflammatory cytokines that *L. pneumophila* apparently induces to downregulate immune control of the bacterium by interferon gamma (Yoshizawa *et al.*, 2005). Depending on the stage of the infection, variable ratios of neopterin (interferon gamma driven) to procalcitonin would be expected. Whether there is a specific inflammatory signature of Legionnaires' disease has not been examined, but studies in macrophages show promise for this type of approach (Losick and Isberg, 2006; Stetson and Medzhitov, 2006). Acquiring microarray data on Legionnaires' disease patients during different stages of infection would help answer this question once daunting technical issues are addressed (Liu *et al.*, 2006). In the meanwhile preparation and storage of cDNA from the blood of such patients should be considered.

Treatment of Legionnaires' disease

Effective treatment of Legionnaires' disease requires the use of antimicrobial agents that are active against intracellular *Legionella* spp. Many different types of antimicrobial agents are active against the bacteria *in vitro*, but not necessarily *in vivo*, because of this need to inhibit or kill the intracellular bacterium. For example, both cefoxitin and gentamicin are very active against *L. pneumophila in vitro*, but are ineffective for treating Legionnaires' disease. Neither of these two drugs inhibits or kills a sufficient number of intracellular *L. pneumophila* to prevent intracellular growth of the bacterium. Examples of drug classes that are active for the treatment of Legionnaires' disease include the macrolides, azalides, ketolides, tetracyclines and antibacterial quinolones. Erythromycin, dirithromycin, clarithromycin, azithromycin, telithromycin, tetracycline, levofloxacin, and ciprofloxacin are all drugs that can be used to treat the disease. The most active drugs, used for severely ill or immunocompromised patients, are azithromycin and levofloxacin. Treatment duration is dependent on the drug given, the degree of immunosuppression, the presence of metastatic or contiguous infection, and the clinical course of the patient. Details about antimicrobial agent therapy of this disease can be found in Chapter 2 of this book and elsewhere (Edelstein, 1995; Roig and Rello, 2003; Sabria and Campins, 2003; Edelstein and Cianciotto, 2005). Recovery from the acute infection may be slow and plagued by fatigue, memory loss, and post-traumatic stress disorder, complications common to many types of community-acquired pneumonia (Marrie *et al.*, 2000; Lettinga *et al.*, 2002; Marrie *et al.*, 2004).

References

Anonymous (1976). 'Legionnaires' disease'. New York Times *125*, 44.

Anonymous (1977). Follow-up on Legionnaires' disease. Morb. Mortal. Wkly Rep. *26*, 111–112.

Anonymous (1997). Sustained transmission of nosocomial Legionnaires disease-- Arizona and Ohio. Morb. Mortal. Wkly Rep. *46*, 416–421.

Bartlett, C.L.R., Swann, R.A., Casal, J., Canada Royo, L., and Taylor, A.G. (1984). Recurrent Legionnaires disease from a hotel water system. In *Legionella:* Proceedings of the 2nd International Symposium, C. Thornsberry, A. Balows, J.C. Feeley, and W. Jakubowski, eds. (Washington, DC: American Society for Microbiology), pp. 237–239.

Bock, B.V., Kirby, B.D., Edelstein, P.H., George, W.L., Snyder, K.M., Owens, M.L., Hatayama, C.M., Haley, C.E., Lewis, R.P., Meyer, R.D., and Finegold, S.M. (1978). Legionnaires' disease in renal-transplant recipients. Lancet *1*, 410–413.

Bopp, C.A., Sumner, J.W., Morris, G.K., and Wells, J.G. (1981). Isolation of *Legionella* spp. from environmental water samples by low-pH treatment and use of a selective medium. J. Clin. Microbiol. *13*, 714–719.

Bozeman, F.M., Humphries, J.W., and Campbell, J.M. (1968). A new group of rickettsia-like agents recovered from guinea pigs. Acta Virol. *12*, 87–93.

Brabender, W., Hinthorn, D.R., Asher, M., Lindsey, N.J., and Liu, C. (1983). *Legionella pneumophila* wound infection. JAMA *250*, 3091–3092.

Brenner, D.J., Steigerwalt, A.G., Gorman, G.W., Weaver, R.E., Feeley, J.C., Cordes, L.G., Wilkinson, H.W., Patton, C.M., Thomason, B.M., and Lewallen Sasseville, K.R. (1980). *Legionella bozemanii* sp. nov. and *Legionella dumoffii* sp. nov.: classification of two additional species of *Legionella* associated with human pneumonia. Curr. Microbiol. *4*, 111–116.

Burnsed, L.J., Hicks, L.A., Smithee, L.M., Fields, B.S., Bradley, K.K., Pascoe, N., Richards, S.M., Mallonee, S., Littrell, L., Benson, R.F., and Moore, M.R. (2007). A large, travel-associated outbreak of legionellosis among hotel guests: utility of the urine antigen assay in confirming Pontiac fever. Clin. Infect. Dis. *44*, 222–228.

Castor, M.L., Wagstrom, E.A., Danila, R.N., Smith, K.E., Naimi, T.S., Besser, J.M., Peacock, K.A., Juni, B.A., Hunt, J.M., Bartkus, J.M., Kirkhorn, S.R., and Lynfield, R. (2005). An outbreak of Pontiac fever with respiratory distress among workers performing high-pressure cleaning at a sugar-beet processing plant. J. Infect. Dis. *191*, 1530–1537.

Chandler, F.W., Hicklin, M.D., and Blackmon, J.A. (1977). Demonstration of the agent of Legionnaires' disease in tissue. N. Engl. J. Med. *297*, 1218–1220.

Chuh, A., Chan, H., and Zawar, V. (2004). Pityriasis rosea--evidence for and against an infectious aetiology. Epidemiol. Infect. *132*, 381–390.

Cordes, L.G., Wilkinson, H.W., Gorman, G.W., Fikes, B.J., and Fraser, D.W. (1979). Atypical *Legionella*-like organisms: fastidious water-associated bacteria pathogenic for man. Lancet *ii*, 927–930.

Daniels, W.B., and Grennan, H.A. (1943). Pretibial fever. An obscure disease. JAMA *122*, 361–365.

Dumoff, M. (1979). Direct in-vitro isolation of the Legionnaires' disease bacterium in two fatal cases. Cultural and staining characteristics. Ann. Intern. Med. *90*, 694–696.

Edelstein, P.H. (1981). Improved semiselective medium for isolation of *Legionella pneumophila* from contaminated clinical and environmental specimens. J. Clin. Microbiol. *14*, 298–303.

Edelstein, P.H. (1995). Antimicrobial chemotherapy for legionnaires' disease: a review. Clin. Infect. Dis. *21*, S265-S276.

Edelstein, P.H. (2006a). Clinical features of Legionnaires' disease: a selective review. In *Legionella*: state of the art 30 years after its recognition, N.P. Cianciotto, Y. Abu Kwaik, P.H. Edelstein, B.S. Fields, D.F. Geary, T.G. Harrison, C.A. Joseph, R.M. Ratcliff, J.E. Stout, and M.S. Swanson, eds. (Washington, DC: ASM Press), pp. 3–7.

Edelstein, P.H. (2006b). Detection of antibodies to *Legionella*. In Manual of Molecular and Clinical Laboratory Immunology, B. Detrick, R.G. Hamilton, and J.D. Folds, eds. (Washington, DC: ASM Press), pp. 468–476.

Edelstein, P.H. (2007). Urine antigen tests positive for Pontiac fever: implications for diagnosis and pathogenesis. Clin. Infect. Dis. *44*, 229–231.

Edelstein, P.H., and Cianciotto, N.P. (2005). *Legionella*. In Principles and Practice of Infectious Diseases, G.L. Mandell, J.E. Bennett, and R. Dolin, eds.

(Philadelphia: Elsevier Churchill Livingstone), pp. 2711–2724.

Edelstein, P.H., and Finegold, S.M. (1979). Use of a semiselective medium to culture *Legionella pneumophila* from contaminated lung specimens. J. Clin. Microbiol. *10*, 141–143.

Edelstein, P.H., Snitzer, J.B., and Finegold, S.M. (1982). Isolation of *Legionella pneumophila* from hospital potable water specimens: comparison of direct plating with guinea pig inoculation. J. Clin. Microbiol. *15*, 1092–1096.

Feeley, J.C., Gibson, R.J., Gorman, G.W., Langford, N.C., Rasheed, J.K., Mackel, D.C., and Baine, W.B. (1979). Charcoal-yeast extract agar: primary isolation medium for *Legionella pneumophila*. J. Clin. Microbiol. *10*, 437–441.

Feeley, J.C., Gorman, G.W., Weaver, R.E., Mackel, D.C., and Smith, H.W. (1978). Primary isolation media for Legionnaires disease bacterium. J. Clin. Microbiol. *8*, 320–325.

Fenstersheib, M.D., Miller, M., Diggins, C., Liska, S., Detwiler, L., Werner, S.B., Lindquist, D., Thacker, W.L., and Benson, R.F. (1990). Outbreak of Pontiac fever due to *Legionella anisa*. Lancet *336*, 35–37.

Fernandez, J.A., Lopez, P., Orozco, D., and Merino, J. (2002). Clinical study of an outbreak of Legionnaire's disease in Alcoy, Southeastern Spain. Eur. J. Clin. Microbiol. Infect. Dis. *21*, 729–735.

Fernandez-Sabe, N., Roson, B., Carratala, J., Dorca, J., Manresa, F., and Gudiol, F. (2003). Clinical diagnosis of *Legionella* pneumonia revisited: evaluation of the Community-Based Pneumonia Incidence Study Group scoring system. Clin. Infect. Dis. *37*, 483–489.

Fields, B.S., Benson, R.F., and Besser, R.E. (2002). *Legionella* and Legionnaires' disease: 25 years of investigation. Clin. Microbiol. Rev. *15*, 506–526.

Fields, B.S., Sanden, G.N., Barbaree, J.M., Morrill, W.E., Wadowsky, R.M., White, E.H., and Feeley, J.C. (1989). Intracellular multiplication of *Legionella pneumophila* in amoebae isolated from hospital hot water tanks. Curr. Microbiol. *18*, 131–137.

Fields, B.S., Shotts, E.B., Jr., Feeley, J.C., Gorman, G.W., and Martin, W.T. (1984). Proliferation of *Legionella pneumophila* as an intracellular parasite of the ciliated protozoan *Tetrahymena pyriformis*. Appl. Environ. Microbiol. *47*, 467–471.

Fraser, D.W., Tsai, T.R., Orenstein, W., Parkin, W.E., Beecham, H.J., Sharrar, R.G., Harris, J., Mallison, G.F., Martin, S.M., McDade, J.E., Shepard, C.C., and Brachman, P.S. (1977). Legionnaires' disease: description of an epidemic of pneumonia. N. Engl. J. Med. *297*, 1189–1197.

Gaydos, J.C., Top, F.H., Jr., Hodder, R.A., and Russell, P.K. (2006). Swine influenza A outbreak, Fort Dix, New Jersey, 1976. Emerg. Infect. Dis. *12*, 23–28.

Glick, T.H., Gregg, M.B., Berman, B., Mallison, G., Rhodes, W.W., Jr., and Kassanoff, I. (1978). Pontiac fever. An epidemic of unknown etiology in a health department: I. Clinical and epidemiologic aspects. Am. J. Epidemiol. *107*, 149–160.

Gochenour, W.S., Jr., Smadel, J.E., Jackson, E.B., Evans, L.B., and Yager, R.H. (1952). Leptospiral etiology of Fort Bragg fever. Public Health Rep. *67*, 811–813.

Goldberg, D.J., Wrench, J.G., Collier, P.W., Emslie, J.A., Fallon, R.J., Forbes, G.I., McKay, T.M., Macpherson, A.C., Markwick, T.A., and Reid, D. (1989). Lochgoilhead fever: outbreak of non-pneumonic legionellosis due to *Legionella micdadei*. Lancet *1*, 316–318.

Grist, N.R., Reid, D., and Najera, R. (1979). Legionnaires' disease and the traveller. Ann. Intern. Med. *90*, 563–564.

Gupta, S.K., Imperiale, T.F., and Sarosi, G.A. (2001). Evaluation of the Winthrop-University Hospital criteria to identify *Legionella* pneumonia. Chest *120*, 1064–1071.

Haley, C.E., Cohen, M.L., Halter, J., and Meyer, R.D. (1979). Nosocomial Legionnaires' disease: a continuing common-source epidemic at Wadsworth Medical Center. Ann. Intern. Med. *90*, 583–586.

Hébert, G.A., Moss, C.W., McDougal, L.K., Bozeman, F.M., McKinney, R.M., and Brenner, D.J. (1980a). The rickettsia-like organisms TATLOCK (1943) and HEBA (1959): bacteria phenotypically similar to but genetically distinct from *Legionella pneumophila* and the WIGA bacterium. Ann. Intern. Med. *92*, 45–52.

Hébert, G.A., Steigerwalt, A.G., and Brenner, D.J. (1980b). *Legionella micdadei* species nova: classification of a third species of *Legionella* associated with human pneumonia. Curr. Microbiol. *3*, 255–257.

Herwaldt, L.A., Gorman, G.W., McGrath, T., Toma, S., Brake, B., Hightower, A.W., Jones, J., Reingold, A.L., Boxer, P.A., and Tang, P.W. (1984). A new *Legionella* species, *Legionella feeleii* species nova, causes Pontiac fever in an automobile plant. Ann. Intern. Med. *100*, 333–338.

Jones, T.F., Benson, R.F., Brown, E.W., Rowland, J.R., Crosier, S.C., and Schaffner, W. (2003). Epidemiologic investigation of a restaurant-associated outbreak of Pontiac fever. Clin. Infect. Dis. *37*, 1292–1297.

Kaufmann, A.F., McDade, J.E., Patton, C.M., Bennett, J.V., Skaliy, P., Feeley, J.C., Anderson, D.C., Potter, M.E., Newhouse, V.F., Gregg, M.B., and Brachman, P.S. (1981). Pontiac fever: isolation of the etiologic agent (*Legionella pneumophila*) and demonstration of its mode of transmission. Am. J. Epidemiol. *114*, 337–347.

Kirby, B.D., Peck, H., and Meyer, R.D. (1979). Radiographic features of Legionnaires' disease. Chest *76*, 562–565.

Kirby, B.D., Snyder, K.M., Meyer, R.D., and Finegold, S.M. (1980). Legionnaires' disease: report of sixty-five nosocomially acquired cases and review of the literature. Medicine (Baltimore) *59*, 188–205.

Kool, J.L., Fiore, A.E., Kioski, C.M., Brown, E.W., Benson, R.F., Pruckler, J.M., Glasby, C., Butler, J.C., Cage, G.D., Carpenter, J.C., Mandel, R.M., England, B., and Breiman, R.F. (1998). More than 10 years of unrecognized nosocomial transmission of legionnaires' disease among transplant patients. Infect. Control Hosp. Epidemiol. *19*, 898–904.

Lattimer, G.L., and Ormsbee, R.A. (1981). Legionnaires' disease. (New York: Dekker).

Lepine, L.A., Jernigan, D.B., Butler, J.C., Pruckler, J.M., Benson, R.F., Kim, G., Hadler, J.L., Cartter, M.L., and Fields, B.S. (1998). A recurrent outbreak of nosocomial legionnaires' disease detected by urinary antigen testing: evidence for long-term colonization of a hospital plumbing system. Infect. Control Hosp. Epidemiol. *19*, 905–910.

Lettinga, K.D., Verbon, A., Nieuwkerk, P.T., Jonkers, R.E., Gersons, B.P., Prins, J.M., and Speelman, P. (2002). Health-related quality of life and posttraumatic stress disorder among survivors of an outbreak of Legionnaires disease. Clin. Infect. Dis. *35*, 11–17.

Liu, M., Popper, S.J., Rubins, K.H., and Relman, D.A. (2006). Early days: genomics and human responses to infection. Curr. Opin. Microbiol. *9*, 312–319.

Losick, V.P., and Isberg, R.R. (2006). NF-kappaB translocation prevents host cell death after low-dose challenge by *Legionella pneumophila*. J. Exp. Med. *203*, 2177–2189.

Lowry, P.W., Blankenship, R.J., Gridley, W., Troup, N.J., and Tompkins, L.S. (1991). A cluster of legionella sternal-wound infections due to postoperative topical exposure to contaminated tap water. N. Engl. J. Med. *324*, 109–113.

Lüttichau, H.R., Vinther, C., Uldum, S.A., Møller, J., Faber, M., and Jensen, J.S. (1998). An outbreak of Pontiac fever among children following use of a whirlpool. Clin. Infect. Dis. *26*, 1374–1378.

Mangione, E.J., Remis, R.S., Tait, K.A., McGee, H.B., Gorman, G.W., Wentworth, B.B., Baron, P.A., Hightower, A.W., Barbaree, J.M., and Broome, C.V. (1985). An outbreak of Pontiac fever related to whirlpool use, Michigan 1982. JAMA *253*, 535–539.

Marrie, T.J., Beecroft, M.D., and Herman-Gnjidic, Z. (2004). Resolution of symptoms in patients with community-acquired pneumonia treated on an ambulatory basis. J. Infect. *49*, 302–309.

Marrie, T.J., Haldane, D., and Bezanson, G. (1992). Nosocomial Legionnaires' disease: clinical and radiographic patterns. Can. J. Infect. Dis. *3*, 253–260.

Marrie, T.J., Lau, C.Y., Wheeler, S.L., Wong, C.J., and Feagan, B.G. (2000). Predictors of symptom resolution in patients with community-acquired pneumonia. Clin. Infect. Dis. *31*, 1362–1367.

McClelland, M.R., Vaszar, L.T., and Kagawa, F.T. (2004). Pneumonia and osteomyelitis due to *Legionella longbeachae* in a woman with systemic lupus erythematosus. Clin. Infect. Dis. *38*, e102-e106.

McDade, J.E. (2002). Legionnaires' disease 25 years later: lessons learned. In *Legionella*, R. Marre, Y.A. Kwaik, C. Bartlett, N.P. Cianciotto, B.S. Fields, M. Frosch, J. Hacker, and P.C. Lück, eds. (Washington, DC: ASM Press), pp. 1–10.

McDade, J.E., Brenner, D.J., and Bozeman, F.M. (1979). Legionnaires' disease bacterium isolated in 1947. Ann. Intern. Med. *90*, 659–661.

McDade, J.E., Shepard, C.C., Fraser, D.W., Tsai, T.R., Redus, M.A., and Dowdle, W.R. (1977). Legionnaires' disease: isolation of a bacterium and demonstration of its role in other respiratory disease. N. Engl. J. Med. *297*, 1197–1203.

Mermel, L.A., Josephson, S.L., Giorgio, C.H., Dempsey, J., and Parenteau, S. (1995). Association of Legionnaires' disease with construction: contamination of potable water? Infect. Control Hosp. Epidemiol. *16*, 76–81.

Meyer, R.D., Edelstein, P.H., Kirby, B.D., Louie, M.H., Mulligan, M.E., Morgenstein, A.A., and Finegold, S.M. (1980). Legionnaires' disease: unusual clinical and laboratory features. Ann. Intern. Med. *93*, 240–243.

Morris, G.K., Patton, C.M., Feeley, J.C., Johnson, S.E., Gorman, G., Martin, W.T., Skaliy, P., Mallison, G.F., Politi, B.D., and Mackel, D.C. (1979). Isolation of the Legionnaires' disease bacterium from environmental samples. Ann. Intern. Med. *90*, 664–666.

Mulazimoglu, L., and Yu, V.L. (2001). Can Legionnaires disease be diagnosed by clinical criteria? A critical review. Chest *120*, 1049–1053.

Osterholm, M.T., Chin, T.D., Osborne, D.O., Dull, H.B., Dean, A.G., Fraser, D.W., Hayes, P.S., and Hall, W.N. (1983). A 1957 outbreak of Legionnaires' disease associated with a meat packing plant. Am. J. Epidemiol. *117*, 60–67.

Prat, C., Dominguez, J., Andreo, F., Blanco, S., Pallarés, A., Cuchillo, F., Ramil, C., Ruiz-Manzano, J., and Ausina, V. (2006). Procalcitonin and neopterin correlation with aetiology and severity of pneumonia. J. Infect. *52*, 169–177.

Roig, J., and Rello, J. (2003). Legionnaires' disease: a rational approach to therapy. J. Antimicrob. Chemother. *51*, 1119–1129.

Rowbotham, T.J. (1980). Preliminary report on the pathogenicity of *Legionella pneumophila* for freshwater and soil amoebae. J. Clin. Pathol. *33*, 1179–1183.

Ruutu, P., Valtonen, V., Elonen, E., Volin, L., Tukiainen, P., and Ruutu, T. (1987). Invasive pulmonary aspergillosis: a diagnostic and therapeutic problem. Clinical experience with eight haematologic patients. Scand. J. Infect. Dis. *19*, 569–575.

Sabria, M., and Campins, M. (2003). Legionnaires' disease: update on epidemiology and management options. Am. J. Resp. Med. *2*, 235–243.

Sencer, D.J., and Millar, J.D. (2006). Reflections on the 1976 swine flu vaccination program. Emerg. Infect. Dis. *12*, 29–33.

Shands, K.N., Ho, J.L., Meyer, R.D., Gorman, G.W., Edelstein, P.H., Mallison, G.F., Finegold, S.M., and Fraser, D.W. (1985). Potable water as a source of Legionnaires' disease. JAMA *253*, 1412–1416.

Sharrar, R.G., Streiff, E., and Parkin, W.E. (1977). Follow-up on respiratory illness – Philadelphia. Morb. Mortal. Wkly Rep. *26*, 9–11.

Sopena, N., Sabria-Leal, M., Pedro-Botet, M.L., Padilla, E., Domínguez, J.A., Morera, J., and Tudela, P. (1998). Comparative study of the clinical presentation of *Legionella* pneumonia and other community-acquired pneumonias. Chest *113*, 1195–1200.

Stetson, D.B., and Medzhitov, R. (2006). Recognition of cytosolic DNA activates an IRF3-dependent innate immune response. Immunity *24*, 93–103.

Sunderman, F.W. (1977). Perspectives on Legionnaires' Disease in relation to acute nickel carbonyl poisoning. The Henry M. Scharf Lecture on current affairs. Ann. Clin. Lab. Sci. *7*, 187–200.

Sunderman, F.W. (1978). The sophistry of the term 'Legionnaires' Disease'. Ann. Clin. Lab. Sci. *8*, 1–3.

Tan, M.J., Tan, J.S., Hamor, R.H., File, T.M.Jr., and Breiman, R.F. (2000). The radiologic manifestations of Legionnaire's (sic) disease. The Ohio Community-Based Pneumonia Incidence Study Group. Chest *117*, 398–403.

Tatlock, H. (1944). A Rickettsia-like organism recovered from guinea pigs. Proc. Soc. Exp. Biol. Med. *57*, 95–99.

Tatlock, H. (1947). Studies on a virus from a patient with Fort Bragg fever (pretibial fever). J. Clin. Invest. *26*, 287–297.

Tatlock, H. (1981). Rickettsia-like bacteria. Infect. Control. *2*, 254–255.

Tatlock, H. (1982). Clarification of the cause of Fort Bragg fever (pretibial fever) – January 1982. Rev. Infect. Dis. *4*, 157–158.

Terranova, W., Cohen, M.L., and Fraser, D.W. (1978). 1974 outbreak of Legionnaires' Disease diagnosed in 1977. Clinical and epidemiological features. Lancet *2*, 122–124.

Thacker, S.B., Bennett, J.V., Tsai, T.F., Fraser, D.W., McDade, J.E., Shepard, C.C., Williams, K.H., Jr., Stuart, W.H., Dull, H.B., and Eickhoff, T.C. (1978). An outbreak in 1965 of severe respiratory illness caused by the Legionnaires' disease bacterium. J. Infect. Dis. *138*, 512–519.

Thomas, G., and Morgan-Witts, M. (1982). Anatomy of an epidemic. (Garden City, NY: Doubleday).

Tobin, J.O., Beare, J., Dunnill, M.S., Fisher-Hoch, S., French, M., Mitchell, R.G., Morris, P.J., and Muers, M.F. (1980). Legionnaires' disease in a transplant unit: isolation of the causative agent from shower baths. Lancet *2*, 118–121.

Tobin, J.O., Swann, R.A., and Bartlett, C.L. (1981). Isolation of *Legionella pneumophila* from water systems: methods and preliminary results. Br. Med. J. (Clin. Res. Ed). *282*, 515–517.

Tossa, P., oge-Abarkan, M., Zmirou-Navier, D., Hartemann, P., and Mathieu, L. (2006). Pontiac fever: an operational definition for epidemiological studies. BMC Public Health *6:112, 2006.*, 112.

Tsai, T.F., Finn, D.R., Plikaytis, B.D., McCauley, W., Martin, S.M., and Fraser, D.W. (1979). Legionnaires' disease: clinical features of the epidemic in Philadelphia. Ann. Intern. Med. *90*, 509–517.

US Department of Health and Human Services-Public Health Service-Centers for Disease Control and Prevention and National Institutes of Health (1993). *Legionella pneumophila*; other *Legionella*-like agents. In Biosafety in Microbiological and Biomedical Laboratories, J.Y.Richmond and R.W.McKinney, eds. (Washington, DC: US Government Printing Office), pp. 92–93.

Wadowsky, R.M., Butler, L.J., Cook, M.K., Verma, S.M., Paul, M.A., Fields, B.S., Keleti, G., Sykora, J.L., and Yee, R.B. (1988). Growth-supporting activity for *Legionella pneumophila* in tap water cultures and implication of hartmannellid amoebae as growth factors. Appl. Environ. Microbiol. *54*, 2677–2682.

Wadowsky, R.M., and Yee, R.B. (1981). Glycine-containing selective medium for isolation of *Legionellaceae* from environmental specimens. Appl. Environ. Microbiol. *42*, 768–772.

Yoshizawa, S., Tateda, K., Matsumoto, T., Gondaira, F., Miyazaki, S., Standiford, T.J., and Yamaguchi, K. (2005). *Legionella pneumophila* evades gamma interferon-mediated growth suppression through interleukin-10 induction in bone marrow-derived macrophages. Infect. Immun. *73*, 2709–2717.

Yu, V.L., Kroboth, F.J., Shonnard, J., Brown, A., McDearman, S., and Magnussen, M. (1982). Legionnaires' disease: new clinical perspective from a prospective pneumonia study. Am. J. Med. *73*, 357–361.

Diagnostics and Clinical Disease Treatment

2

Paul Christian Lück

Abstract

The methods currently available to diagnose Legionnaires' disease are culture, urinary antigen detection, direct fluorescent antibody testing, detection of nucleic acid and detection of specific antibodies in serum samples. The advantages and limitations of each method are discussed. Presently, none of the diagnostic tests available offers the desired quality with respect to sensitivity and specificity. Culture should be obligatory, especially when hospitalized patients with underlying diseases are investigated. A positive culture is the prerequisite of molecular epidemiological investigations. Urinary antigen detection is a valuable tool in the majority of community-acquired cases when *L. pneumophila* serogroup 1 (sg 1) is the causative agent. In cases of nosocomial disease, when *Legionella pneumophila* serogroups other than sg 1 are frequent, this assay has limitations. The detection of nucleic acid is very promising, but needs further validation. The detection of antibodies in a patient's serum is of little use in the acute phase of the illness.

Several molecular subtyping techniques are in use to subtype *L. pneumophila* strains in epidemiological investigations. *Legionella pneumophila* is genetically very heterogeneous thus allowing an individual fingerprint of each strain. However, the majority of clinical cases are caused by a limited number of clones that cause disease worldwide.

The therapy of for Legionnaires' disease requires drugs that can access and are active intracellularly. Currently, fluorochinolones and macrolides are the most active agents.

Introduction

An outbreak of severe pneumonia among the participants of the American Legion convention in Philadelphia in 1976 led to the description of Legionnaires' disease and the isolation of *L. pneumophila* as the causative agent (McDade *et al.*, 1977; Chapter 1). Meanwhile it is well known that *Legionella* bacteria inhabit all kinds of freshwater reservoirs. The current list of validated names in the genus *Legionella* comprises 50 species with at least 73 serogroups (Euzéby, 2006). The most common species, *L. pneumophila,* contains strains that are classified as three subspecies (Table 2.1). Based on antigenic structures currently 16 serogroups are defined. Some *L. pneumophila* strains cannot be serotyped by using the established panel of monoclonal and polyclonal antibodies (Helbig *et al.*, 2007). Hence, further serogroups might be present. Some of the non-pneumophila *Legionella* species, e.g. *L. micdadei, L. dumoffii, L. bozemanae, L. longbeachae,* have been isolated from both human and environmental sources; others have been isolated only from the environment (Table 2.1). Infections by *L. longbeachae* are often associated with potting soil mixtures, especially in some areas of Australia. It is generally accepted that all species may cause disease, especially in immunocompromised persons.

Approximately 75% of *Legionella* infections are caused by *L. pneumophila* sg 1, whereas 20–30% are caused by other serogroups of *L. pneumophila,* and 5–10% by different *Legionella* species (McNally *et al.*, 2000; Helbig *et al.*, 2002; Yu *et al.*, 2002; Doleans *et al.*, 2004) (Fig. 2.1).

Table 2.1 Currently established *Legionella* species (serogroups)

Legionella species	Serogroup
L. adelaidensis	1
L. anisa (L, P)	1
L. beliardensis	1
L. birminghamensis (L)	1
L. bozemanae (L)	2
L. brunensis	1
L. busanensis	1
L. cherrii	1
L. cincinnatiensis (L)	1
L. drancourtii	1
L. drozanskii	1
L. dumoffii (L)	1
L. erythra	2
L. fairfieldensis	1
L. fallonii	1
L. feeleii (L,P)	2
L. geestiana	1
L. gormanii (L)	1
L. gratiana	1
L. gresilensis	1
L. haeckeliae (L)	2
L. israelensis (L)	1
L. jamestownsiensis	1
L. jeonii	1
L. jordanis (L)	1
L. lansingensis	1
L. londiniensis	2
L. longbeachae (L)	2
L. lytica (L)	1
L. maceachernii (L)	1
L. micdadei (L,P)	1
L. moravica	1
L. nautarum	1
L. oakridgensis	1
L. parisiensis	1
L. pneumophila (L,P) Subsp. *fraseri* Subsp. *pascullei* Subsp. pneumophila	16
L. quateirensis	1
L. quinlivanii	1
L. rowbothamii	1
L. rubrilucens	1
L. sainthelensi (L)	2
L. santicrucis	1
L. shakespearei	1
L. spiritensis	2
L. steigerwaltii	1
L. tauriniensis	1
L. tucsoniensis (L)	1
L. wadsworthii (L)	1
L. walthersii	1
L. worsliensis	1

(L), caused pneumonia.

(P), caused Pontiac fever.

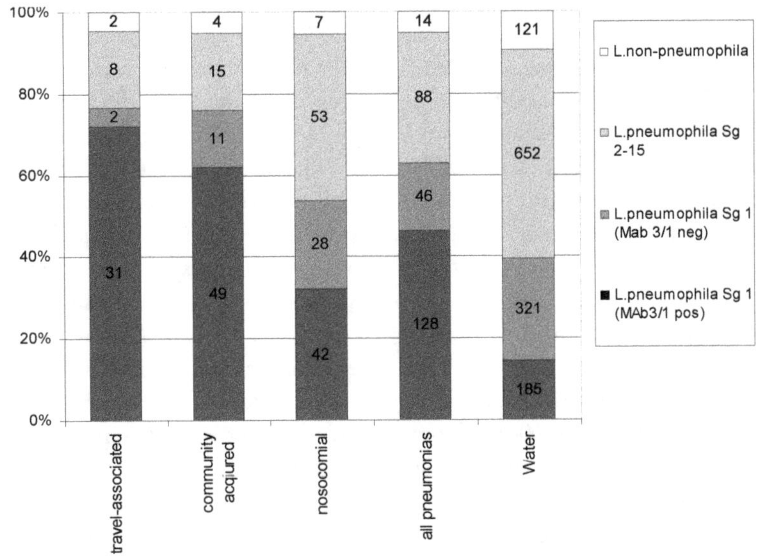

Figure 2.1 Prevalence of *Legionella* species and serogroups among clinical and environmental *Legionella* isolates 1986–2006, Dresden, Germany.

It is a well-known phenomenon that MAb 3-1-positive strains are more prevalent among clinical strains, especially in cases of community-acquired and travel-associated legionellosis. Therefore, it is believed that such strains exhibit a greater virulence. So far, this has not been defined on the molecular level. Recent studies suggest that a subset of clones, most of which react with MAb 3-1, are responsible for the majority of cases, whereas many *Legionella* strains of other serogroups are of lower virulence (Borchardt *et al.*, 2006; Harrison *et al.*, 2007). In contrast, the distribution of strains isolated from nosocomial cases reflects the distribution in the environment (Doleans *et al.*, 2004; Fig. 2.1).

The distribution of *Legionella* species and serogroups has great influence on the sensitivity of the diagnostic methods and therefore on the reported prevalence of legionellosis. This is reflected in the current case definitions that are used in epidemiological investigations (see Chapter 3).

Prevalence of legionellosis among respiratory infections

Legionellosis occurs most frequently as sporadic cases, either nosocomial or community-acquired. Travel-associated cases were collected and reported from a European-wide surveillance system (Chapter 3).

Depending on local factors, the age of the patients and risk factors, the proportion of legionellosis among pneumonia cases has been reported to range from 1% – 22% (Blanquer *et al.*, 1991; Fields *et al.*, 2002; Fine *et al.*, 2004; Sopena *et al.*, 2005). In an ongoing study on the aetiology of community-acquired pneumonia, we found in 112 out of 2556 patients (4.2%) urinary antigen and/or *Legionella* DNA in respiratory or serum samples (Lück *et al.*, 2006). If one accepts that these data represent the true incidence of *Legionella* pneumonia, then about 10,000 to 20,000 cases of *Legionella* community-acquired pneumonia occur in Germany each year. Based on these estimates the annually incidence in the United States would be 25,000 to 50,000 cases.

Laboratory diagnosis

Culture of *Legionella* spp.

Culture is still the 'gold standard' among the diagnostic methods for *Legionella* infections. The medium necessary for the cultivation of legionellae is buffered charcoal yeast extract (BCYE) agar supplemented with antibiotics. However, some *Legionella* strains might be susceptible to the antibiotics in selective media. Therefore, antibiotic-free agar should be used as an additional culture medium for material with no or low content of oral flora (Stout *et al.*, 2003). Since quality control of media designed for the recovery of *Legionella* spp. is difficult to achieve in routine laboratories, the use of ready-to-use plates is recommended (Lück *et al.*, 2004). For quality control of in-house preparations, standard *Legionella* suspension are commercially available (e.g. Health Protection Agency, London).

Since legionellae are environmental, aquatic organisms which do not colonize humans and cannot be isolated from healthy persons, the specificity of culture is estimated to be 100% (Stout *et al.*, 2003). 'False-positive' culture results may occur if clinical samples are contaminated with water containing legionellae, although reports of such cases are very rare (Lightfoot, *et al.* 1991).

The sensitivity of culture for the diagnosis of Legionnaires' disease has been estimated to be in the range of 11–65% by retrospective studies usually performed in reference laboratories (Hayden *et al.*, 2001; DenBoer *et al.*, 2004; Lindsay *et al.*, 2004). In a retrospective study conducted in our laboratory we found a sensitivity of 80%. In contrast, the percentage of positive cultures tends to be lower if studies are performed prospectively in routine laboratories (Blazquez *et al.*, 2005; Lück *et al.*, 2006). Bronchoalveolar lavage fluid, bronchial aspirates, lung biopsies, post-mortem tissue specimens and sputum are particularly suitable for culture, whereas pleural fluid is less suitable (Edelstein, 2000; Fields *et al.*, 2002; DenBoer *et al.*, 2004). *Legionella* has been cultured from extrapulmonary sites (Stout *et al.*, 2003) only seldom, in most cases due to systemic spread of the *Legionella* bacteria. However, the frequency of extrapulmonary legionellosis has not been studied extensively.

Legionella colonies usually form within 3–5 days. Suspected colonies are subcultures on BCYE Agar and cystein-free BCYE agar or Columbia blood agar. Colonies that grow on the former but not on the latter are presumptive *Legionella* species. One of the limiting factors for cultivation of *Legionella spp.* seems to be the experience of the laboratory staff (Edelstein, 2000). In some instances, *Legionella* colonies have unusual morphology that might be not recognized (Stout *et al.*, 2003) (Fig. 2.2).

Identification of *Legionella* species

The most important technique for the identification of *Legionella* in the clinical laboratory is the serological characterization of isolated strains. A fluorescein-conjugated monoclonal antibody (MAb) which recognizes an outer membrane protein of *L. pneumophila* is commercially available. This species-specific MAb detects all serogroups and can be used for rapid identification of *L. pneumophila* from clinical samples or environmental specimens (Helbig *et al.*, 2007). As mentioned before, this species can be divided into at least 16 serogroups. The serogroup-specificity is based on the chemical composition of the lipopolysaccharide, and the division into serogroups is based on the reactivity with polyclonal antisera and monoclonal antibodies. A few serogroups, mainly sg 1 of *L. pneumophila*, can be divided into MAb subtypes. This differentiation is useful for epidemiological investigations (Joly *et al.*, 1986, Helbig *et al.*, 2002). Polyclonal antisera, either fluorescein conjugated or coupled to latex beads, are commercially available for many, but not all, *Legionella* species.

Of practical relevance is the differentiation of *L. pneumophila* sg 1 strains from all other serogroups of this species. For this purpose, latex-coupled antibodies can be obtained from different suppliers. In general, all polyclonal antisera can cross-react with a variety of other bacteria. Such cross-reactions present few problems for the identification of colonies isolated on artificial media because the results can be verified by their morphology and requirement for cysteine.

Recently an immunochromatographic assay that utilizes monoclonal antibodies was developed and evaluated. This method is suitable to confirm the majority of *Legionella* species isolated from clinical and environmental samples (Helbig *et al.*, 2006).

Fluorescently labelled oligonucleotide probes can also used to verify that a given isolate belongs to the genus *Legionella* (Buchbinder *et al.*, 2004). This test correctly identified 43 of the described species (C. Lück, unpublished).

By using polyclonal antisera, FISH or the immunochromatographic test, it is difficult or even impossible to identify isolated strains to the species level since many species and serogroups

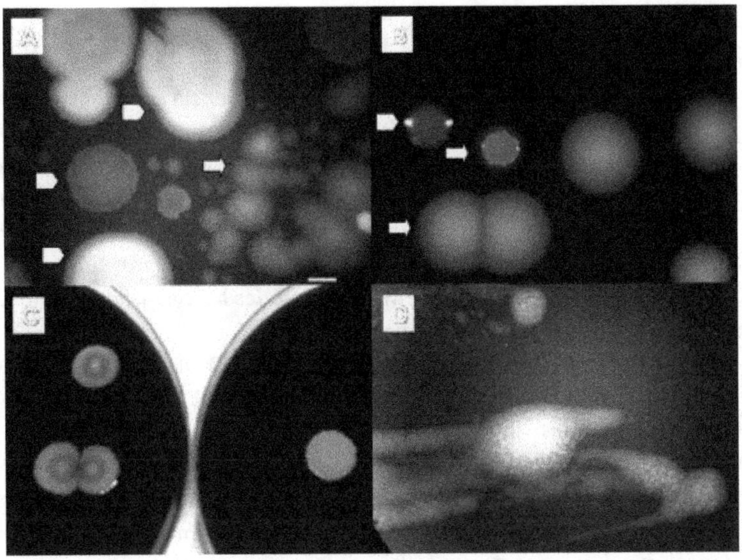

Figure 2.2 Colony appearance of *Legionella* and non-*Legionella* bacteria on BCYE agar. (A) *Legionella pneumophila* serogroup 1 cultured from a bronchoalveolar fluid of a patient with community-acquired pneumonia, *Legionella* colonies (→) appear with the typical ground-glass structure. Non-*Legionella* bacteria (▷) can be easily distinguished. (B) Typical and atypical colony appearance of *Legionella* (→) and non-*Legionella* bacteria (▷) isolated from a hot water sample (C) Atypical colony appearance of *Legionella* isolated cultured from a bronchoalveolar fluid of a patient. (D) *Legionella*-like colony appearance of a *Pseudomonas* spp. strain isolated from a bronchoalveolar fluid of a patient.

show considerable cross-reactions or cannot be distinguished by these assays. Therefore, the most accurate identification of *Legionella* isolates to the species level can be obtained by sequence analysis of the 16S rRNA genes (Rubin *et al.*, 2005) or the macrophage infectivity potentiator (*mip*) gene (Ratcliffe *et al.*, 1998). A web-based tool for the identification based on their *mip* gene sequence is available from the European Working Group on *Legionella* Infections (www.ewgli.org). Likewise, sequences from other genes, like *gyr*A, *rpn*B and *rpo*B, are also suitable for differentiation of *Legionella* species (Fedderson *et al.*, 2000; Ko *et al.*, 2002; Rubin *et al.*, 2005).

Detection of *Legionella* antigen in urine

An antigen excreted with urine has been characterized as heat-stable, resistant to enzymatic cleavage, and of about 10 kDa molecular weight. These characteristics are typical for lipopolysaccharide components. *L. pneumophila* is now divided into at least 16 serogroups and several monoclonal subgroups (Helbig et a., 2007). Indeed, all assays for the detection of *L. pneumophila* urinary antigens show sufficient recognition of the antigens which are homologous to the serogroup/monoclonal subgroup used as immunogen for preparation of the antisera, i.e. *L. pneumophila* sg 1 (Dominguez *et al.*, 1998; Helbig *et al.*, 2001; Guerrero *et al.*, 2004).

Presently, several ELISAs are commercially available (e. g. Binax, Biotest, Barthels). The specificities of all assays, which were mostly evaluated by testing urine specimens from patients with urinary tract infections or pneumonias caused by other pathogens, have been reported to be > 99.9% (Dominguez *et al.*, 1998; Helbig *et al.*, 2001; Helbig *et al.*, 2003; DenBoer *et al.*, 2004).

With respect to the sensitivity for *L. pneumophila* sg 1 infections, all assays are able to detect more than 92% of cases identified by culture (Helbig *et al.*, 2001). When culture-proven cases caused by other serogroups and legionellosis diagnosed by serology were included, the reported sensitivities dropped to 75–90%. This decrease reflects that all assays available have sensitivities for non-serogroup 1 infections of less than 35% (Helbig *et al.*, 2001). The Biotest ELISA seems to be slightly more sensitive in diagnosing legionellosis caused by serogroups 2 to 15 (Horn, 2001).

Recently, it has been reported by several investigators that the sensitivity of the urinary antigen correlates with the severity of illness (Yzerman *et al.*, 2002; Blanques *et al.*, 2005; Lück *et al.*, 2006) and the presence of underlying diseases (Sopena *et al.*, 2002). This seems to be plausible since the 'antigen load' is higher in severe cases and the processing of the antigen might depend on the immune state. The use of concentrated urine samples increases the sensitivity without decreasing the specificity (Dominguez *et al.*, 1998; Yzerman *et al.*, 2002; Guerrero *et al.*, 2004). Since a concentration step is time and labour intensive, we resort to this approach when the urine assay gives borderline results (Horn, 2001).

Besides these, a rapid immunochromatographic assay (BinaxNow) has been on the market for several years. Using this assay it is possible to detect antigenuria within a very short time, and no laboratory equipment is required. The BinaxNow assay is slightly less sensitive than the ELISA formats (Guerrero *et al.*, 2004). Recently, other immunochromatographic assays were released to the market (SAS *Legionella* test, Rapid U Test, SD Bioline). Evaluation of these newer tests revealed specificity values below 100% (Diederen *et al.*, 2006a; Diederen *et al.*, 2006b; C. Lück, unpublished). Keeping in mind the low prevalence of legionellosis (2–5%), a lower specificity might be problematic for a diagnostic assay. Furthermore, the sensitivities for all newer assays are lower than for the BinaxNow. Thus, at present, these newer tests must be used with great caution.

The advantages of urinary antigen detection are striking: specimens are easy to obtain and can be investigated repeatedly; antigenuria is detectable very early and, therefore, often provides the first evidence of *Legionella* infection. Furthermore, the test is very rapid, and it has a very high specificity. In addition, positive urinary antigen results can also be obtained in cases of non-pneumonic, Pontiac fever-like illness (Lück *et al.*, 1992; Lüttichau *et al.*, 1998; Burnsed *et al.*, 2007). In most cases, the antigenuria ends after 10 to 14 days. Despite the initiation of appropriate treatment, antigenuria may persist for some weeks or months, but this persistence of antigenuria does not reflect a failure of treatment. It is significantly associated with immunosuppressive therapy (Sopena *et al.*, 2002; Koide *et al.*, 2004). Despite the great advantages of the urinary antigen detection, one should keep in mind that a negative urinary antigen result never excludes a *Legionella* infection.

Detection of *Legionella* by direct fluorescent antibody testing

Direct fluorescent antibody (DFA) testing of respiratory specimens is a rapid method for the detection of *Legionella* antigen in respiratory samples. A monoclonal, fluorescein-conjugated antibody against an outer membrane protein of *L. pneumophila* is commercially available (Fig. 2.3). This reagent is highly specific but gives occasional false-positive results with *Staphylococcus aureus*, probably due to non-specific binding to protein A. The test detects only *L. pneumophila*. Since 90–95% of all cases of legionellosis are caused by this species, this limitation seems acceptable. The sensitivity of DFA testing has been reported to range from 27% to 70% (Hayden *et al.*, 2001; Lindsay *et al.*, 2004, C. Lück, unpublished). However, sensitivity depends mainly on the type of specimen used, the technical equipment (e.g. a cytospin centrifuge for concentration of diluted specimens), and the experience of the laboratory staff. In contrast to the urinary antigen detection which is mainly sg 1-specific and therefore has a high sensitivity for community-acquired cases, the sensitivity of the DFA test is not different for cases of community-acquired, travel-associated or nosocomial legionellosis (C. Lück, unpublished). The protein antigen detected by this assay is not degraded after fixation with formalin; hence, the test allows the detection of the aetiological agent in formalin-fixed lung tissue, which is not possible with the other methods available. An important issue is the prevention of false positive results of DFA testing due to contact of clinical specimens with water containing *Legionella*, due to contaminated buffers or legionellae that have washed off of positive control slides.

For the non-pneumophila species that are rare causes of legionellosis, no equivalent test exists. The use of polyclonal antisera for detection

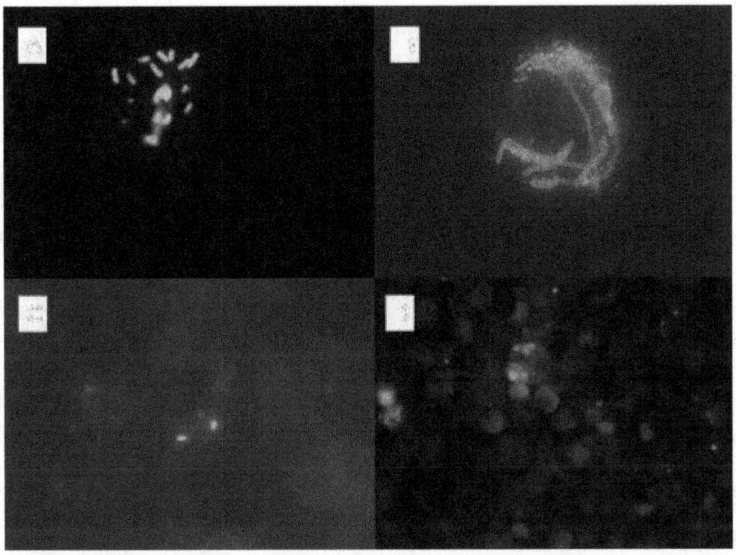

Figure 2.3 Detection of *Legionella* by FITC-labelled monoclonal antibodies. (A) Typical *Legionella* bacteria in a bronchoalveolar fluid of an immunosuppressed patient with community-acquired pneumonia. (B) Filamentous intracellular *Legionella* in a bronchoalveolar fluid of an immunosuppressed patient with nosocomial pneumonia. (C) Single coccoid rods detected in a culture-positive bronchoalveolar fluid of a patient with travel-associated pneumonia. (D) Detection of intra- and extracellular *Legionella* in a formalin-fixed lung section of a lethal case of community-acquired pneumonia. Detection by FITC labelled anti-lipopolysaccharide monoclonal antibody. Note extracellular amorphous antigen.

in clinical samples is limited by cross-reactions to other Gram-negative bacteria.

Detection of *Legionella* by fluorescence *in situ* hybridization (FISH)

Fluorescently labelled oligonucleotide probes can also used to detect Legionellae in clinical samples (Hayden *et al.*, 2001; Buchbinder *et al.*, 2004). However, FISH is technically demanding and requires an experienced laboratory staff.

Detection of *Legionella* nucleic acids in clinical samples

Over the last years, detection of nucleic acid has been more frequently used to identify *Legionella* in clinical samples. Depending on the primers used, the polymerase chain reaction (PCR) assays detect either *L. pneumophila*, or several, or all, of the published species of the genus *Legionella* (DenBoer *et al.*, 2004).

Recently, extensive technical advances have been made for the direct (real-time) monitoring of fragments generated by PCR. Since *Legionella* is not considered to be part of the normal human flora, the presence or absence of *Legionella* DNA in specimens is the main clinical criterion, rather than the quantity of the pathogen. Therefore, the advantages and limitations of real-time PCR for quantification purposes are not the subject of discussion in this article. Beyond quantification, several advantages make real-time PCR superior to conventional PCR, even for non-quantitative applications. First, the post PCR processing steps like gel electrophoresis, hybridization or sequencing of PCR products, will become obsolete since PCR products can be detected immediately and specifically. This results in a significant acceleration of the diagnostic procedure and an improved specificity of the reaction. In addition, real-time PCR abolishes the risk of contamination that occurs during processing of the PCR products.

Detection of *Legionella* nucleic acids in respiratory samples

The first application of PCR for respiratory specimens was reported in 1992 by Jaulhac *et al.* (1992). By investigating the performance of PCR in bronchoalveolar lavage (BAL) fluid specimens, the investigators established the principle that PCR is suitable for the detection of *Legionella* DNA in clinical samples.

The currently available data allow a preliminary estimation of sensitivity and specificity values for PCR relative to other methods. The high sensitivity for the detection of *Legionella* DNA in respiratory samples demonstrated by several studies suggests that PCR may exceed culture in its ability to detect *Legionella* in the above mentioned specimens (Cloud, *et al.* 2000; Hayden *et al.*, 2001; Rantakokko-Jalava *et al.*, 2001; Reischl *et al.*, 2003; DenBoer *et al.*, 2004; Koide *et al.*, 2004). For respiratory specimens an external quality assurance (EQA) system has been operating in Germany since 2004 (www.instand.de). This EQA scheme is open to all laboratories and should be use as a part of laboratory quality assurance systems. The results of this EQA scheme revealed certain problems in terms of sensitivity and specificity that occur in routine laboratories, but these are likely to be overcome in the future.

Detection of *Legionella* nucleic acids in urine

The excretion of DNA fragments in the urine has been described for several bacterial pathogens suggesting the suitability of urine PCR for the detection of *Legionella* DNA. Maiwald and colleagues (Maiwald *et al.*, 1995) demonstrated the practicability of PCR from urine samples and revealed positive results not only in 29 of 37 samples from infected guinea pigs but also in samples from 11 patients with pneumonia, nine of whom were tested positive by other methods (Maiwald *et al.*, 1995). Whereas the sensitivity of PCR seems not to be superior to the detection of *Legionella* antigens in urine, the possibility of using primer systems with specificity broader than the serogroup level appears to be advantageous for diagnostic purposes (Helbig *et al.*, 1999). In general, the reported sensitivity varies from 17% to 72% (Murdoch *et al.*, 1996; Helbig *et al.*, 1999, Matsiota-Bernard *et al.*, 2000, Lück *et al.*, 2006). Reasons for this remarkable variation are not yet known.

Detection of *Legionella* nucleic acid in serum samples

The detection of *Legionella* DNA in serum has also been described (Lindsay *et al.*, 1994; Murdoch *et al.*, 1996; Matsiota-Bernard *et al.*, 2000; Diederen *et al.*, 2007; Lück *et al.*, 2006). *Legionella* DNA is detectable in all acute phase and convalescent phase sera from patients with confirmed legionellosis, but not in the sera from over 100 patients without evidence of legionellosis. This approach appears rather promising since collection of this kind of clinical samples is inexpensive and simple to perform. As for urine samples, the sensitivity of the DNA detection assays in sera varies remarkably from 29% to 82% (Lindsay *et al.*, 1994; Murdoch *et al.*, 1996; Matsiota-Bernard *et al.*, 2000; Diederen *et al.*, 2007; Lück *et al.*, 2006). No DNA was detected in serum samples from patients with pneumonia due to other organisms.

Antibody detection in human sera

Among the methods for antibody detection, only the indirect fluorescent antibody (IFA) test has been evaluated and standardized (Stout *et al.*, 2003). It is the test format distributed by most of the commercial suppliers. ELISA tests are also available from different suppliers but are of lower specificities (Boshuizen *et al.*, 2003; Malan *et al.*, 2003; Rojas *et al.*, 2005). This might be acceptable in epidemic situations but needs to be used with great caution when single cases are investigated.

The sensitivity of serology is generally limited by the time required to develop a detectable antibody response during the course of the infection. Approximately 25–40% of patients seroconvert within 1 week of the onset of symptoms; about 10–15% of patients seroconvert as late as 6–9 weeks after onset of the illness; and approximately 20–30% of patients do not develop significantly elevated antibody titers, even after prolonged observation. This variation limits the overall sensitivity of serology to 70–80% (Stout *et al.*, 2003). On the other hand, the specificity of seroconversion (fourfold titre rise) using *L. pneumophila* sg 1 antigen in the IFA test has been reported to be approximately 99%. In general, it is recommended that *Legionella* serology be performed with polyvalent conjugates detecting IgG, IgM and IgA antibodies, since the immune response in the course of *Legionella* infections varies with respect to different immunoglobulin classes. IgM-specific IFA tests have a limited specificity and are not in use.

It must be emphasized that the exact sensitivity and specificity of serology has only been thoroughly evaluated for IFA tests using *L. pneumophila* sg 1 antigen. Sensitivity and specificity of antibody detection by using other serogroups of *L. pneumophila* and of non-*pneumophila* legionellae are not precisely defined. Probably, the values are considerably lower than those for serogroup 1. However, the detection of antibodies can be used to identify infections by novel species if appropriate serum samples are available (McNally *et al.*, 2000). Infections with other serogroups of *L. pneumophila* can be detected with serogroup 1 antigen in some, but not all, cases. Therefore, many laboratories perform *Legionella* serology with a number of antigens. Pools containing different antigens are commonly being used for screening purposes, but positive reactions must be confirmed by monovalent antigens, as non-specific reactions occur frequently with pool antigens.

The background frequency of elevated anti-*Legionella* titres in the normal population has been found to vary from 1% to 36% (Stout *et al.*, 2003). This limitation should be considered when single elevated titres are used as diagnostic criteria.

Epidemiological subtyping of isolated *Legionella* strains

So far, a positive culture is the only method that allows the comparison of patient and environmental *Legionella* strains necessary to confirm or exclude a given environmental reservoir as the source of the infection.

Epidemiological investigations of legionellosis are often complicated by the ubiquity of legionellae in nature. Because the incubation period varies from 2 to 10 days, the length of stay in a hotel, private accommodation, public building or hospital before onset of clinical signs does not establish with certainty where the infection has been acquired. Recognized sources of Legionnaires' disease confirmed in epidemiological investigations are: warm and cold water supplies (shower, taps) in private homes, hospitals, hotels, public buildings, cruise ships, cooling towers with 'water based cooling', whirlpools, thermal springs, moistener/respirators, decorative fountains, humidifiers for food display cabinets, car washers.

Often different species, serogroups, and monoclonal subtypes of *Legionella* are isolated from a given environmental source (Lück *et al.*, 1998; Visca *et al.*, 1999; Beyrer *et al.*, 2006). In contrast to this, simultaneous infection with multiple *Legionella* strains seems to be a rare event (Horbach *et al.*, 1988; Lück *et al.*, 1998; Buchbinder *et al.*, 2004). Such reports again underline the differences in virulence of different strains as illustrated in Table 2.2. In each case provided, the causative strain was in the minority among environmental isolates.

In some cases, no corresponding environmental isolate could be found, though all suspected water sources were investigated (Jonas *et al.*, 2000). In such cases, it might be that the patient acquired the infectious strain from other environmental sources that were not investigated, e.g. outside the hotel or hospital, during overnight stays in the nearby private accommodation, etc. On the other hand, the causative strain might not have been isolated and subsequently typed because the source of infection was decontaminated or the bacterial population in the environment had changed and/or the causative strain had been overlooked.

All the above mentioned problems lead to discussions as to the number of colonies which should be typed after primary isolation and to the preferable typing method(s). At present, we serotype at least six colonies from environmental samples. Further, a combination of antigenic and genomic typing systems applied as a step-by-step procedure is recommended for the identification of *Legionella* strains that cause the infection (Table 2.3).

Monoclonal antibody typing

Subtyping of Legionella *pneumophila* strains by using monoclonal antibodies (MAb) was the first technique used in epidemiological studies and is now a well established method (Joly *et al.*, 1986). As major advantages, it is technically simple and quick to perform. The reactivity patterns are stable, although changes have been observed due to point mutations and deletions in genes involved in the synthesis of lipopolysaccharide (LPS)

Table 2.2 Number of colonies typed during epidemiological investigations in relation to the 'causative' strain

Case/cluster	Patient's isolate	Environmental isolates	Serogroup–MAb–subtype	Identity of patients- and environmental strains*
A	1	11	Lp 1–Bellingham	Yes
		8	Lp 6–Chicago	No
B	1	8	Lp 10	Yes
	0	4	Lp 6–Dresden	No
		2	Lp 14	No
C	4**	24	Lp 1–Bellingham	Yes
	2**	6	Lp 12 strain A	Yes
	2**	4	Lp 12 strain B	Yes
		3	Lp 1–Camperdown	nt
		3	Lp 6	No
D	1	1	Lp 1–Philadelphia	Yes
		24	Lp 1–Philadelphia	No
		3	Lp 1–Olda	No
		4	Lp 3	No
		2	Lp 6–Dresden	No
		2	Lp 12	No
E	1	12	Lp 5–Cambridge	Yes
		6	Lp 1–Knoxville	No
		3	Lp12	No
		35	Lp 6–Dresden	No

*Defined by using monoclonal antibodies and macrorestriction analysis and/or sequence-based typing.
**Infection caused by three different strains.
nt, not tested.

(Zou *et al.*, 1999; Lück *et al.*, 2001; Bernander *et al.*, 2003). In general, genetic variation is of minor relevance, since such events occur at a relative low frequency under natural conditions.

The major disadvantages of MAb typing are the time-consuming and expensive establishment, maintenance and quality assurance of the hybridoma cell lines producing MAbs. Furthermore, the antigenic diversity, or number of subtypes, is limited. Thus, the index of discrimination, an important characteristic which reflects the ability of a typing system to recognize different strains as different, is in the range of 0.8. Nevertheless, MAb subtyping is applicable to *L. pneumophila* sg 1, which can be divided into at least 12 subtypes (Joly *et al.*, 1986; Helbig *et al.*, 2002). When the MAb type of the clinical isolate does not match that of environmental strain(s), these reservoirs can with high probability is excluded as the source of the infection. At times, the use of subgroup-specific and cross-reacting antibodies has allowed the subtyping of strains belonging to serogroups 2–15 (Helbig *et al.*, 2007).

All MAb subtypes are named according to a reference strain that shows this particular reactivity pattern. This kind of data can easily be exchanged between laboratories. In a multicentre evaluation of typing methods for the epidemiological typing of *L. pneumophila* serogroup 1, the epidemiological concordance was very close to 1 (Fry *et al.*, 1999).

Table 2.3 Step-by-step typing of *Legionella pneumophila* strains in epidemiological studies

First line typing – screening
Defining the species by antibody typing (*mip* gene sequencing)
Serotyping by using polyclonal to distinguish sg 1 from all other serogroups
Monoclonal antibody (MAb) subtyping: Subtyping for *L. pneumophila* sg 1; determination of the serogroup for all other sg)
Secondary typing – confirmation
Genetic fingerprinting: PFGE, AFLP
Sequence based typing

Macrorestriction analysis (MRA) by pulsed-field gel electrophoresis

Subtyping of *Legionella pneumophila* strains by MRA is an excellent tool for subtyping *Legionella* species and was considered to be the gold standard for more than 15 years (Lück *et al.*, 1998; Fry *et al.*, 1999; Jonas *et al.*, 2000). Macrorestriction patterns can be analysed both visually and by computer-aided methods. The index of discrimination is greater than 0.95, the value which is required for a good subtyping system. However, the intra- and inter-laboratory standardization and exchange of data are difficult (Fry *et al.*, 1999). Nowadays, sequences based typing (SBT) has replaced MRA.

Amplified fragment length polymorphism (AFLP) typing

This technique uses a combination of DNA restriction by asgn endonuclease and amplification by the polymerase chain reaction. It was the first typing system standardized within the European Working Group on *Legionella* Infection (EWGLI) (Fry *et al.*, 1999, Jonas *et al.*, 2000; Fry *et al.*, 2002;). However, its index of discrimination was lower than for MRA, and the inter-laboratory exchange was impaired by the subjective methods of evaluation.

Sequence-based typing

Sequence based typing (SBT) is a variant of multilocus sequence typing that employs variations from multiple chromosomal locations, or genes. Currently, the European SBT panel includes six *L. pneumophila* genes: *flaA*, *pilE*, *asd*, *mip*, *mompS*, *pro*A. Thus, an SBT type comprises a string of the individual allele numbers of each of these genes separated by commas. The major advantages of SBT are stability of the marker, good discriminatory power if appropriate loci are selected, and flexibility, since additional gene loci can be investigated if necessary. Data are readily exchanged among laboratories either as sequence data or as designated alleles. Furthermore, SBT reduces the need to transport live bacteria, since nucleotide sequence determination from PCR products can be achieved from killed-cell suspensions, purified DNA, or clinical material. While SBT is particularly suited to long-term and global epidemiology, as it identifies a variation which is accumulating slowly within a population, the data can also be used to investigate single cases or outbreaks (Gaia *et al.*, 2005). Currently, the allocation of the allele formula can be done using the EWGLI website (http://www.ewgli.org).

Antibiotic therapy

In vitro susceptibility of *Legionella*

In vitro susceptibility testing can be performed on artificial media to screen for active agents. It must be considered that components of the media (e.g. charcoal) might influence the *in vitro* susceptibility data. Macrolides, chinolones, ketolide, quinupristin/dalfopristin, doxycyclin, imipenem, rifampicin and tigecyclin are *Legionella* factors that are active *in vitro* (Stout *et al.*, 1998; Nielsen *et al.*, 2000a; Hammerschlag *et al.*, 2001; Edelstein *et al.*, 2003; Stout *et al.*, 2005).

Resistance to all clinically relevant substances can be induced in laboratory experiments (Dowling *et al.*, 1985; Nielsen *et al.*, 2000b; Jonas *et al.*, 2003). However, so far, there is no evidence that resistance occurs in clinical situations. Case reports describing clinical failure and/or prolonged clinical illness were never related to the development of resistance to antimicrobials (Rudin *et al.*, 1984; Kurz *et al.*, 1988; Tan *et al.*, 2001; Gläser *et al.*, 2005)

Intracellular activity of antimicrobial agents against *Legionella*

As legionellae are intracellular pathogens, L. pneumophila has been cultivated *in vitro* in a number of macrophage-like cell lines. In this way, the intracellular activity of antimicrobials can be assessed (Stout *et al.*, 1998; Jonas *et al.*, 2003; Stout *et al.*, 2005). Generally, these results show a good correlation with animal experiments. In earlier studies it was shown that although bacterial growth is inhibited by erythromycin and rifampicin, it recurs when the drugs are removed from the cells (Edelstein, 1998). This is in contrast to newer fluoroquinolones and macrolides, which kill intracellular *Legionella* bacteria and do not permit bacterial regrowth. Azithromycin is the most active macrolide, and it has a much higher activity than erythromycin against intracellular *L. pneumophila* (Edelstein *et al.*, 1991, Fitzgeorge *et al.*, 1993; Jonas *et al.*, 2003).

Clinical experience in the treatment of legionellosis

The clinical experience of azithromycin in the treatment of Legionnaires' disease is known to be safe and efficacious (Edelstein, 1995; Plouffe *et al.*, 2003). The vast majority of patients who receive monotherapy with intravenous azithromycin for 2–7 days, followed by oral azithromycin, are cured.

The efficacy of levofloxacin was reported in a study analysing six clinical trials encompassing a total of 1997 patients with community-acquired pneumonia. More than 90% of mild-to-moderate and severe cases of *Legionella* infection were cured; no deaths were reported (Yu *et al.*, 2004). Recently, three observational studies comparing levofloxacin vs. macrolides (not Azithromycin) in the treatment of legionnaires' disease have been published (Blázquez-Garrido *et al.*, 2005; Mykietiuk *et al.*, 2005; Sabria *et al.*, 2005). There were no significant differences in clinical outcomes among the groups of patients with mild or moderate pneumonia; but, in patients with severe pneumonia, levofloxacin was slightly more effective. The combination of rifampicin and levofloxacin provides no additional benefit but does increase the rate of side effects.

In summary, as compared with older macrolides in the treatment of Legionnaires' disease, levofloxacin appears to be associated with better clinical outcomes, including a faster resolution of pneumonia symptoms, a more rapid achievement of clinical stability, and a shorter hospital stay. Thus, monotherapy with levofloxacin might be regarded as first-line antimicrobial for treatment of Legionnaires' disease. However, it must be underlined that the direct comparison of azithromycin and levofloxacin in the treatment of *Legionella* pneumonia has not been performed. Thus, it can be recommended to use fluoroquinolones or azithromycin, rather than older macrolides, for the treatment of Legionnaires' disease (Pedro-Botet *et al.*, 2006).

Delay in the initiation of appropriate antibiotic therapy for *Legionella* pneumonia significantly increases mortality (Heath *et al.*, 1996; Gacouin *et al.*, 2002; Lettinga *et al.*, 2002). It is therefore recommended that anti-*Legionella* agents be included early in the empiric therapy of severe community-acquired pneumonia. However, there are different view points regarding the first line antimicrobial therapy that is generally recommended for treating pneumonia (File *et al.*, 2004).

Conclusions

Increasingly recognized as a major cause of sporadic and epidemic community-acquired and nosocomial pneumonia, *Legionella* represents an important public health problem.

Urinary antigen testing is a valuable diagnostic tool for the rapid detection of Legionnaires' disease. The early recognition of *Legionella* pneumonia by using urine antigen testing may have contributed to the decrease in mortality observed in the last years. Since this test detects mainly infections caused by *L. pneumophila* sg 1, it is especially valuable in outbreak situations. Detection of *Legionella* DNA in respiratory, urine and serum samples is now established and recognized as a rapid and reliable diagnostic assay; as such, it should be used more often to detect sporadic and epidemic cases.

The frequency of diagnosis of legionellosis by culture has decreased significantly in the last years, impairing epidemiological investigations. To reverse this problematic trend, culture of respiratory materials should be performed whenever possible, especially in cases of severe pneumonia.

The detection of antibodies in patient's serum is still a good laboratory test. Although of little use in the acute phase of the illness, it is especially valuable in epidemiological investigations.

Systematic molecular subtyping of all environmental and clinical isolates by using sequence based typing is an important tool in the investigation of Legionnaires' disease outbreaks and will help to identify the sources of the infection rapidly and reliably. By this approach, the occurrence of further cases can be prevented. Furthermore, such investigations will improve our understanding of the epidemiology of legionellosis, knowledge that is fundamental to any scientifically proven risk assessment.

Newer quinolones and macrolides have greater activity on intracellular legionellae in cell culture models and improved efficacy in animal models as compared with older macrolides. _Monotherapy with a quinolone (i.e. levofloxacin) may be the current treatment of choice for Legionnaires' disease, according to results of recent observational studies. However, direct comparison of azithromycin and quinolones has not been performed in clinical studies. _The addition of rifampicin to the treatment with newer quinolones or macrolides probably provides no benefit, but does increase the rate of side-effects. Although the combination of quinolones with azithromycin is supported by some *in vitro* data, the relevance of this observation for patients with severe *Legionella* pneumonia is not yet clear.

Web resources

http://www.cdc.gov
http://www.ewgli.org
http://www.hpa.org.uk
http://www.rki.de

Acknowledgements

Own work was supported by the Federal Ministry of Education and Research of Germany, Network of Competence in Medicine CAPNETZ. I would like to thank Carolin Dix, Sigrid Gäbler, Jutta Paasche, Kerstin Seeliger, Ines Wolf for excellent technical assistance.

References

Bernander, S., Jacobson, K., Helbig, J.H., Lück, P.C., and Lundholm, M. (2003). A hospital-associated outbreak of Legionnaires' disease caused by *Legionella pneumophila* serogroup 1 is characterized by stable genetic fingerprinting but variable monoclonal antibody patterns. J. Clin. Microbiol. *41*, 2503–2508.

Beyrer, K., Lai, S., Dreesman, J., Lee, J.V., Joseph, C.A., Harrison, T.G., Surman-Lee, S., Lück, P.C., Brodhun, B., Buchholz, U., and Windorfer, A. (2006). Legionnaires' disease outbreak associated with a cruise liner, August 2003: epidemiological and microbiological findings. Epidemiol. Infect. in press.

Blanquer, J., Blanquer, R., Borras, R., Nauffal, D., Morales, P., Menendez, R., Subias, I., Herrero, L., Redon, J., and Pascual, J. (1991). Aetiology of community-acquired pneumonia in Valencia, Spain: a multicentre prospective study. Thorax *46*, 508–511.

Blazquez, R.M., Espinosa, F.J., Martnez-Toldos, C.M., Alemany, L., Garca-Orenes, M.C., and Segovia, M. (2005). Sensitivity of urinary antigen test in relation to clinical severity in a large outbreak of *Legionella* pneumonia in Spain. Eur. J. Clin. Microbiol. Infect. Dis. *24*, 488–491

Blázquez-Garrido, R.M., Espinosa-Parra, F.J., Alemany, F.L., Ramos Guevara, R.M., Sanchez-Nieto, J.M., Segovia, H.M., Serrano-Martnez, J.A., and Huerta, F.H. (2005). Antimicrobial chemotherapy for legionnaires disease: Levofloxacin versus Macrolides. Clin. Infect. Dis. *40*, 800–806.

Borchardt, J., Helbig, J.H., Lück P.C. (2006). Epidemiological subtyping of *Legionella pneumophila* strains isolated from patients in Germany. Abstract RFP03, Annual Meeting of the German Society for Hygiene and Microbiology, Würzburg.

Boshuizen, H.C., Den Boer, J.W., De Melker, H., Schellekens, J.F., Peeters, M.F., van Vliet, J.A., and Conyn-Van Spaendonck, M.A. (2003). Reference values for the SERION classic ELISA for detecting *Legionella pneumophila* antibodies. Eur. J. Clin. Microbiol. Infect. Dis. *22*, 706–708.

Buchbinder, S., Leitritz, L., Trebesius, K., Banas, B., and Heesemann, J. (2004). Mixed lung infection by *Legionella pneumophila* and *Legionella gormanii* detected by fluorescent in situ hybridization. Infection *32*, 242–245.

Burnsed, L.J., Hicks, L.A., Smithee, L.M., Fields, B.S., Bradley, K.K., Pascoe, N., Richards, S.M., Mallonee, S., Littrell, L., Benson, R.F., Moore, M.R., and Moore, M.R. (2007). A large, travel-associated outbreak of legionellosis among hotel guests: utility of the urine antigen assay in confirming Pontiac fever. Clin. Infect. Dis. *44*, 222–228.

Cloud, J.L., Carroll, K.C., Pixton, P., Erali, M., and Hillyard, D.R. (2000). Detection of *Legionella* species in respiratory specimens using PCR with sequencing confirmation. J. Clin. Microbiol. *38*, 1709–1712.

Den Boer, J.W. and Yzerman, E.P. (2004). Diagnosis of *Legionella* infection in Legionnaires disease. Eur. J. Clin. Microbiol. Infect. Dis. *23*, 871–878.

Diederen, B.M. and Peeters, M.F. (2006a). Evaluation of Rapid U *Legionella* Plus Test, a new immunochromatographic assay for detection of *Legionella pneumophila* serogroup 1 antigen in urine. Eur. J. Clin. Microbiol. Infect. Dis. *25*, 733–735.

Diederen, B.M., de Jong, C.M., Marmouk, F., Kluytmans, J.A., Peeters, M.F., and Van der Zee.A. (2007). Evaluation of real-time PCR for the early detection of *Legionella pneumophila* DNA in serum samples. J. Med. Microbiol. *56*, 94–101.

Diederen, B.M. and Peeters, M.F. (2006b). Evaluation of two new immunochromatographic assays (Rapid U *Legionella* antigen test and SD Bioline *Legionella* antigen test) for detection of *Legionella pneumophila* serogroup 1 antigen in urine. J. Clin. Microbiol. *44*, 2991–2993.

Doleans, A., Aurell, H., Reyrolle, M., Lina, G., Freney, J., Vandenesch, F., Etienne, J., and Jarraud, S. (2004). Clinical and environmental distributions of *Legionella* strains in France are different. J. Clin. Microbiol. *42*, 458–460.

Dominguez, J.A., Gali, N., Pedroso, P., Fargas, A., Padilla, E., Manterola, J.M., and Matas, L. (1998). Comparison of the Binax Legionella urinary antigen enzyme immunoassay (EIA) with the biotest *Legionella* urine antigen EIA for detection of *Legionella* antigen in both concentrated and nonconcentrated urine samples. J. Clin. Microbiol. *36*, 2718–2722.

Dowling, J.N., McDevitt, D.A., and Pasculle, A.W. (1985). Isolation and preliminary characterization of erythromycin-resistant variants of *Legionella micdadei* and *Legionella pneumophila*. Antimicrob. Agents. Chemother. *27*, 272–274.

Edelstein, P.H. (1995). Review of azithromycin activity against *Legionella* spp. Pathol. Biol. (Paris) *43*, 569–572.

Edelstein, P.H. (2000). Detection of selected fastidious bacteria. Clin. Infect. Dis. *31*, 846.

Edelstein, P.H., Weiss, W.J., and Edelstein, M.A.C. (2003). Activities of tigecycline (GAR-936) against *Legionella pneumophila in vitro* and in guinea pigs with *L. pneumophila* pneumonia. Antimicrob. Agents Chemother. *47*, 533–540.

Euzèby J.P. (2006). List of prokaryotic names with standing in nomenclature. http://www.bacterio.citr.fr

Feddersen, A., Meyer, H.G., Matthes, P., Bhakdi, S., and Husmann, M. (2000). GyrA sequence-based typing of *Legionella*. Med. Microbiol. Immunol. (Berl) *189*, 7–11.

Fields, B.S., Benson, R.F., and Besser, R.E. (2002). *Legionella* and Legionnaires' disease: 25 years of investigation. Clin. Microbiol. Rev. *15*, 506–526.

File, T.M., Garau, J., Blasi, F., Chidiac, C., Klugman, K., Lode, H., Lonks, J.R., Mandell, L., Ramirez, J., and Yu, V. (2004). Guidelines for empiric antimicrobial prescribing in community-acquired pneumonia. Chest *125*, 1888–1901.

Fitzgeorge, R.B., Lever, S., and Baskerville, A. (1993). A comparison of the efficacy of azithromycin and clarithromycin in oral therapy of experimental air-borne Legionnaires' disease. J. Antimicrob. Chemother. *31 Suppl E*, 171–176.

Fry, N.K., Alexiou-Daniel, S., Bangsborg, J.M., Bernander, S., Castellani-Pastoris, M., Etienne, J., Forsblom, B., Gaia, V., Helbig, J.H., Lindsay, D., Lück, P.C., Pelaz, C., Uldum, S.A., and Harrison, T.G. (1999). A multicenter evaluation of genotyping methods for the epidemiologic typing of *Legionella pneumophila* serogroup 1: Results from a pan-European study. Clin. Microbiol. Infect. *5*, 462–477.

Fry, N.K., Bangsborg, J.M., Bergmans, A., Bernander, S., Etienne, J., Franzin, L., Gaia, V., Hasenberger, P., Baladron, J., Jonas, D., Lindsay, D., Mentula, S., Papoutsi, A., Struelens, M., Uldum, S.A., Visca, P., Wannet, W., and Harrison, T.G. (1997). European Working Group on *Legionella* Infection (EWGLI) Amplified Fragment Length Polymorphism types of *Legionella pneumophila* serogroup 1 and results of Intercentre Proficiency Testing using a standard protocol. Microbiol. Infect. Dis. *21*, 722–728.

Gacouin, A., Le, T., Lavoue, S., Camus, C., Hoff, J., Bassen, R., Arvieux, C., Heurtin, C., and Thomas, R. (2002). Severe pneumonia due to *Legionella pneumophila*: prognostic factors, impact of delayed appropriate antimicrobial therapy. Intensive Care Med. *28*, 686–691.

Gaia, V., Fry, N.K., Afshar, B., Lück, P.C., Meugnier, H., Etienne, J., Peduzzi, R., and Harrison, T.G. (2005). Consensus sequence-based scheme for epidemiological typing of clinical and environmental isolates of *Legionella pneumophila*. J. Clin. Microbiol. *43*, 2047–2052.

Gläser, S., Weitzel, T., Schiller, R., Suttorp, N., and Lück, P.C. (2005). Persistent culture-positive *Legionella* infection in an immunocompetent adult. Clin. Infect. Dis. *41*, 765–766.

Guerrero, C., Toldos, C.M., Yague, G., Ramirez, C., Rodriguez, T., and Segovia, M. (2004). Comparison of diagnostic sensitivities of three assays (Bartel's enzyme immunoassay [EIA], Biotest EIA, and Binax NOW Immunochromatographic test) for detection of *Legionella pneumophila* serogroup 1 antigen in urine. J. Clin. Microbiol. *42*, 467–468.

Hammerschlag, M.R., Roblin, P.M., and Bbar, C.M. (2001). Activity of telithromycin, a new ketolide antibacterial, against atypical and intracellular respiratory tract pathogens. J. Antimicrob. Chemother. *48 Suppl T1*, 25–31.

Harrison, T.G., Doshi, N., Fry, N.K., and Joseph, C.A. (2007). Comparison of clinical and environmental isolates of *Legionella pneumophila* obtained in the UK over 19 years. Clin. Microbiol. Infect. *13*, 78–85.

Hayden, R.T., Uhl, J.R., Qian, X., Hopkins, M.K., Aubry, M.C., Limper, A.H., Lloyd, R.V., and Cockerill, F.R. (2001). Direct detection of *Legionella* species from bronchoalveolar lavage and open lung biopsy specimens: comparison of lightcycler pcr, in situ hybridization, direct fluorescence antigen detection, and culture. J. Clin. Microbiol. *39*, 2618–2626.

Heath, C.H., Grove, D.I., and Looke, D.F. (1996). Delay in appropriate therapy of *Legionella* pneumonia associated with increased mortality. Eur. J. Clin. Microbiol. Infect. Dis. *15*, 286–290.

Helbig, J.H., Benson, R.F., Pelaz, C., Jacobs, E., and Lück, P.C. (2007). Identification and serotyping of atypical *Legionella pneumophila* strains isolated from human and environmental sources. J. Appl. Microbiol. *102*, 100–105.

Helbig, J.H., Bernander, S., Castellani, P., Etienne, J., Gaia, V., Lauwers, S., Lindsay, D., Lück, P.C., Marques, T., Mentula, S., Peeters, M.F., Pelaz, C., Struelens, M., Uldum, S.A., Wewalka, G., and Harrison, T.G. (2002). Pan-European study on culture-proven Legionnaires' Disease: Distribution of *Legionella pneumophila* serogroups and monoclonal subgroups. Eur. J. Clin. Microbiol. Infect. Dis. *21*, 710–716.

Helbig, J.H., Engelstädter, T., Maiwald, M., Uldum, S.A., Witzleb, W., and Lück, P.C. (1999). Diagnostic relevance of the detection of *Legionella* DNA in urine samples by the polymerase chain reaction. Eur. J. Clin. Microbiol. Infect. Dis. *18*, 716–722.

Helbig, J.H., Lück, P.C., Kunz, B., and Bubert, A. (2006). Evaluation of the Duopath *Legionella* Lateral Flow Assay for identification of *Legionella pneumophila* and *Legionella* species culture isolates. Appl. Environ. Microbiol. *72*, 4489–4491.

Helbig, J.H., Uldum, S.A., Bernander, S., Lück, P.C., Wewalka, G., Abraham, B., Gaia, V., and Harrison, T.G. (2003). Clinical utility of urinary antigen detection for diagnosis of community-acquired, travel-associated, and nosocomial legionnaires' disease. J. Clin. Microbiol. *41*, 838–840.

Helbig, J.H., Uldum, S.A., Lück, P.C., and Harrison, T.G. (2001). Detection of *Legionella pneumophila* antigen in urine samples by the BinaxNOW immunochromatographic assay and comparison with both Binax *Legionella* Urinary Enzyme Immunoassay (EIA) and Biotest *Legionella* Urin Antigen EIA. J. Med. Microbiol. *50*, 509–516.

Horbach, I., Naumann, D., and Fehrenbach, F.J. (1988). Simultaneous infections with different serogroups of *Legionella pneumophila* investigated by routine methods and Fourier transform infrared spectroscopy. J. Clin. Microbiol. *26*, 1106–1110.

Horn, J. (2001). Binax and Biotest *Legionella* urinary antigen kits. J. Clin. Microbiol. *39*, 1682.

Joly, J.R., McKinney, R.M., Tobin, J.O., Bibb, W.F., Watkins, I.D., and Ramsay, D. (1986). Development of a standardized subgrouping scheme for *Legionella pneumophila* serogroup 1 using monoclonal antibodies. J. Clin. Microbiol. *23*, 768–771.

Jonas, D., Engels, I., Daschner, F.D., and Frank, U. (2000). The effect of azithromycin on intracellular *Legionella pneumophila* in the Mono Mac 6 cell line at serum concentrations attainable *in vivo*. J. Antimicrob. Chemother. *46*, 385–390.

Jonas, D., Engels, I., Friedhoff, C., Spitzmuüller, B., Daschner, F.D., and Frank, U. (2001). Efficacy of moxifloxacin, trovafloxacin, clinafloxacin and levofloxacin against intracellular *Legionella pneumophila*. J. Antimicrob. Chemother. *47*, 147–152.

Jonas, D., Engels, I., Hartung, D., Beyersmann, J., Frank, U., and Daschner, F.D. (2003). Development and mechanism of fluoroquinolone resistance in *Legionella pneumophila*. J. Antimicrob. Chemother. *51*, 275–280.

Jonas, D., Meyer, H.G., Matthes, P., Hartung, D., Jahn, B., Daschner, F.D., and Jansen, B. (2000). Comparative evaluation of three different genotyping methods for investigation of nosocomial outbreaks of Legionnaires' disease in hospitals. J. Clin. Microbiol. *38*, 2284–2291.

Ko, K.S., Lee, H.K., Park, M.Y., Lee, K.H., Yun, Y.J., Woo, S.Y., Miyamoto, H., and Kook, Y.H. (2002). Application of RNA Polymerase {beta}-Subunit Gene (rpoB) Sequences for the Molecular Differentiation of *Legionella* Species. J. Clin. Microbiol. *40*, 2653–2658.

Koide, M., Higa, F., Tateyama, M., Sakugawa, H., and Saito, A. (2004). Comparison of polymerase chain reaction and two urinary antigen detection kits for detecting *Legionella* in clinical samples. Eur. J. Clin. Microbiol. Infect. Dis. 23, 221–223.

Koide, M., Saito, A., Kusano, N., Tateyama, M., Inadome, J., Kyan, Y., Kishaba, T., and Miyagi, S. (1996). Relation between the polymerase chain reaction and the indirect fluorescent antibody method in the diagnosis of *Legionella* infection. Clin. Infect. Dis. *23*, 656–657.

Kurz, R.W., Graninger, W., Egger, T.P., Pichler, H., and Tragl, K.H. (1988). Failure of treatment of *Legionella* pneumonia with ciprofloxacin. J. Antimicrob. Chemother. 22, 389–391.

Lettinga, K.D., Verbon, A., Weverling, G.J., Schellekens, J.F., Den,Boer, Y. W. Yzerman, E.P., Prins, J., Boersma, W.G., van Ketel, R., Prins, J.M., and Speelman, P. (2002). Legionnaires' disease at a Dutch flower show: prognostic factors and impact of therapy. Emerg. Infect. Dis. *8*, 1448–1454.

Lightfoot, N.F., Richardson, I.R., Shrimanker, J., and Farrell, D.J. (1991). Post-mortem isolation of *Legionella* due to contamination. Lancet *337*, 376.

Lindsay, D.S., Abraham, W.H., and Fallon, R.J. (1994). Detection of mip gene by PCR for diagnosis of Legionnaires' disease. J. Clin. Microbiol. *32*, 3068–3069.

Lindsay, D.S., Abraham, W.H., Findlay, W., Christie, P., Johnston, F., and Edwards, G.F.S. (2004). Laboratory diagnosis of Legionnaires ' disease due to *Legionella pneumophila* serogroup 1: comparison of phenotypic and genotypic methods. J. Med. Microbiol. *53*, 183–187.

Lück, P.C., Freier, T., Steudel, C., Knirel, Y.A., Lüneberg, E., Zähringer, U., and Helbig, J.H. (2001). A point mutation in the active site of *Legionella pneumophila* O-acetyltransferase results in modified lipopolysaccharide but does not influence virulence. Int. J. Med. Microbiol. *291*, 345–352.

Lück, P.C., Igel, L., Helbig, J.H., Kuhlisch, E., and Jatzwauk, L. (2004). Comparison of commercially available media for the recovery of *Legionella* species. Int. J. Hyg. Environm. Health *207*, 589–593.

Lück, P.C., Rasch, R., Schmidt, C., Helbig, J.H., Ehret, W., Stockmann, B., and Witzleb, W. (1992). [A minor epidemic due to *Legionella pneumophila* serogroup 1]. Dtsch. Med. Wochenschr. *117*, 460–464.

Lück, P.C., von Baum, H., Helbig, J.H., and Marre, R. (2006). The usefulness of microbiological diagnostic methods for the detection of *Legionella* infections. In *Legionella* – State of the art 30 years of its recognition. N.P. Cianciotto, Y. AbuKwaik, P.H. Edelstein, B.S. Fields, D.F. Geary, T.G. Harrison, C.A. Joseph, R.M. Ratcliff, J.E. Stout, M.S. Swanson, eds (Washington DC:.ASM Press) pp. 15–21.

Lück, P.C., Wenchel, H.M., and Helbig, J.H. (1998). Nosocomial pneumonia caused by three genetically different strains of *Legionella pneumophila* and detection of these strains in the hospital water supply. J. Clin. Microbiol. *36*, 1160–1163.

Lüttichau, H.R., Vinther, C., Uldum, S.A., Moller, J., Faber, M., and Jensen, J.S. (1998). An outbreak of Pontiac fever among children following use of a whirlpool. Clin. Infect. Dis. *26*, 1374–1378.

Maiwald, M., Schill, M., Stockinger, C., Helbig, J.H., Lück, P.C., Witzleb, W., and Sonntag, H.G. (1995). Detection of *Legionella* DNA in human and guinea pig urine samples by the polymerase chain reaction. Eur. J. Clin. Microbiol. Infect. Dis. *14*, 25–33.

Malan, A.K., Martins, T.B., Jaskowski, T.D., Hill, H.R., and Litwin, C.M. (2003). Comparison of two commercial enzyme-linked immunosorbent Assays with an immunofluorescence assay for detection of *Legionella pneumophila* types 1 to 6. J. Clin. Microbiol. *41*, 3060–3063.

Matsiota-Bernard, B., Waser, S., and Vrioni, G. (2000). Detection of *Legionella pneumophila* DNA in urine and serum samples from patients with pneumonia. Clin. Microbiol. Infect. *6*, 223–225.

McDade, J.E., Shepard, C.C., Fraser, D.W., Tsai, T.R., Redus, M.A., and Dowdle, W.R. (1977). Legionnaires' disease: isolation of a bacterium and demonstration of its role in other respiratory disease. N. Engl. J. Med. *297*, 1197–1203.

McNally, C., Hackman, B., Fields, B.S., and Plouffe, J.F. (2000). Potential importance of *Legionella* species as etiologies in community-acquired pneumonia (CAP). Diagn. Microbiol. Infect. Dis. *38*, 79–82.

Murdoch, D.R., Walford, E.J., Jennings,.C., Light, G.J., Schousboe, M.I., Chereshsky, A.Y., Chambers,.T., and Town, G.I. (1996). Use of the polymerase chain reaction to detect *Legionella* DNA in urine and serum samples from patients with pneumonia. Clin. Infect. Dis. *23*, 475–480.

Mykietiuk, A., Carratal, J., Fernndez-Sab, N., Dorca, J., Verdaguer, R., Manresa, F., and Gudiol, F. (2005). Clinical outcomes for hospitalized patients with *Legionella* pneumonia in the antigenuria era: the influence of levofloxacin therapy. Clin. Infect. Dis. *40*, 794–799.

Nielsen, K., Bangsborg, J.M., and Hoiby, N. (2000a). Susceptibility of *Legionella* species to five antibiotics and development of resistance by exposure to erythromycin, ciprofloxacin, and rifampicin. Diagn. Microbiol. Infect. Dis. *36*, 43–48.

Nielsen, K., Hindersson, P., Hoiby, N., and Bangsborg, J.M. (2000b). Sequencing of the *rpo*B Gene in *Legionella pneumophila* and Characterization of Mutations Associated with Rifampin Resistance in the *Legionellaceae*. Antimicrob. Agents Chemother. *44*, 2679–2683.

Pedro-Botet, L. and Yu, V.L. (2006). *Legionella*: macrolides or quinolones? Clin. Microbiol. Infect. *12 Suppl 3*, 25–30.

Plouffe, J.F., Breiman, R.F., Fields, B.S., Herbert, M., Inverso, J., Knirsch, C., Kolokathis, A., Marrie,.J., Nicolle, L., and Schwartz, D.B. (2003). Azithromycin in the treatment of *Legionella* pneumonia requiring hospitalization. Clin. Infect. Dis. *37*, 1475–1480.

Rantakokko-Jalava K. and Jalava, J. (2001). Development of conventional and real-time PCR assays for detection of *Legionella* DNA in respiratory specimens. J. Clin. Microbiol. *39*, 2904–2910.

Ratcliff, R.M., Lanser, J.A., Manning, P.A., and Heuzenroeder, M.W. (1998). Sequence-based classification scheme for the genus *Legionella* targeting the mip gene. J. Clin. Microbiol. *36*, 1560–1567.

Reischl, U., Linde, H.J., Lehn, N., Landt, O., Barratt, K., and Wellinghausen, N. (2002). Direct Detection and Differentiation of *Legionella* spp. and *Legionella pneumophila* in Clinical Specimens by Dual-Color Real-Time PCR and Melting Curve Analysis. J. Clin. Microbiol. *40*, 3814–3817.

Rojas, A., Navarro, M.D., Forns, F.E., Serra, E., Simarro, E., Rojas, J., and Ruiz, J. (2005). Value of serological testing for diagnosis of legionellosis in outbreak patients. J. Clin. Microbiol. *43*, 4022–4025.

Rubin, C.J., Thollesson, M., Kirsebom, L.A., and Herrmann, B. (2005). Phylogenetic relationships and species differentiation of 39 *Legionella* species by sequence determination of the RNaseP RNA gene rnpB. Int. J. Syst. Evol. Microbiol. 55, 2039–2049.

Rudin, J.E., Evans, T.L., and Wing, E.J. (1984). Failure of erythromycin in treatment of *Legionella micdadei* pneumonia. Am. J. Med. *76*, 318–320.

Sabria, M., Pedro-Botet, M.L., Gomez, J., Roig, J., Vilaseca, B., Sopena, N., Banos, V., and for the Legionnaires Disease Therapy Group (2005). Fluoroquinolones vs macrolides in the treatment of Legionnaires Disease. Chest *128*, 1401–1405.

Sopena, N., Sabri, M., and Sabri, M. (2005). Multicenter study of hospital-acquired pneumonia in non-ICU patients. Chest *127*, 213–219.

Sopena, N., Sabri, M., Pedro-Botet, M.L., Reynaga, E., Garca-Nez, M., Domnguez, J., and Matas, L. (2002). Factors related to persistence of *Legionella* urinary antigen excretion in patients with legionnaires disease. Eur. J. Clin. Microbiol. Infect. Dis. *21*, 845–848.

Stout, J.E., Arnold, B., and Yu, V.L. (1998). Activity of azithromycin, clarithromycin, roxithromycin, dirithromycin, quinupristin/dalfopristin and erythromycin against *Legionella* species by intracellular susceptibility testing in HL-60 cells. J. Antimicrob. Chemother. *41*, 289–291.

Stout, J.E., Rihs, J.D., Yu, V.L. (2003). *Legionella*. In Manual of Clinical Microbiology 8th edition. eds. P.R, Murray, E.J. Baron, M.A. Pfaller, J.H. Jorgensen, R.H. Yolken, (Washington DC: ASM Press) pp. 809–823.

Stout, J.E., Sens, K., Mietzner, S., Obman, A., and Yu, V.L. (2005). Comparative activity of quinolones, macrolides and ketolides against *Legionella* species using *in vitro* broth dilution and intracellular susceptibility testing. Int. J. Antimicrob. Agents *25*, 302–307.

Tan, J.S., File, DiPersio, J.R., DiPersio, L.P., Hamor, R., Saravolatz, L.D., and Stout, J.E. (2001). Persistently positive culture results in a patient with community-acquired pneumonia due to *Legionella pneumophila*. Clin. Infect. Dis. *32*, 1562–1566.

Visca, P., Goldoni, P., Lück, P.C., Helbig, J.H., Cattani, L., Giltri, G., Bramati, S., and Castellani, P. (1999). Multiple types of *Legionella pneumophila* serogroup 6 in a hospital heated-water system associated with sporadic infections. J. Clin. Microbiol. *37*, 2189–2196.

Yu, V.L., Greenberg, R.N., Zadeikis, N., Stout, J.E., Khashab, M.M., Olson, W.H., and Tennenberg, A.M. (2004). Levofloxacin efficacy in the treatment of community-acquired legionellosis. Chest *125*, 2135–2139.

Yu, V.L., Plouffe, J.F., Pastoris, M.C., Stout, J.E., Schousboe, M., Widmer, A., Summersgill, J., File, T., Heath, C.M., Paterson, D.L., and Chereshsky, A. (2002). Distribution of *Legionella* species and serogroups isolated by culture in patients with sporadic community-acquired legionellosis: an international collaborative survey. J. Infect. Dis. *186*, 127–128.

Yzerman, E.P., DenBoer, J.W., Lettinga, K.D., Schellekens, J., Dankert, J., and Peeters, M. (2002). Sensitivity of three urinary antigen tests associated with clinical severity in a large outbreak of Legionnaires' disease in The Netherlands. J. Clin. Microbiol. *40*, 3232–3236.

Zou, C.H., Knirel, Y.A., Helbig, J.H., Zähringer, U., and Mintz, C.S. (1999). Molecular cloning and characterization of a locus responsible for O acetylation of the O polysaccharide of *Legionella pneumophila* serogroup 1 lipopolysaccharide. J. Bacteriol. *181*, 4137–4141.

The Epidemiology of Legionnaires' Disease

3

Carol Joseph* and Katherine Ricketts

Abstract

Much has been learnt about the epidemiology of Legionnaires' disease since the organism was first identified in 1976. National surveillance systems and research studies were established early, and in recent years improved ascertainment and changes in clinical methods of diagnosis have contributed to an upsurge in reported cases in many countries. Environmental studies continue to identify novel sources of infection, leading to regular revisions of guidelines and regulations. Between 1995 and 2005 over 32,000 cases of Legionnaires' disease and more than 600 outbreaks were reported to the European Working Group for Legionella Infections (EWGLI). In the future, there may be an increase in cases as the population becomes more elderly and climate change perhaps leads to an increased risk of exposure to infection in many countries. This would put ever more demands on public health resources for the management of this disease. This chapter on epidemiological aspects of Legionnaires' disease will discuss how surveillance, detection, treatment, control and prevention strategies, development and use of guidelines and legislation, and national and international public health initiatives all aim to improve our knowledge about the disease and contribute to reducing its public health burden on the population.

Introduction

History

In 1976 a convention for military veteran legionnaires held at a downtown hotel in Philadelphia, USA, was associated with an outbreak of severe respiratory disease involving 189 cases and 29 deaths. The investigations into the outbreak led to the discovery of the bacteria responsible which was named *Legionella pneumophila*, as a tribute to these men. (Sharrar *et al.*, 1976; Anon, 1997a, Chapter 1). Subsequent reanalysis of sera from patients associated with an outbreak of pneumonia of unknown aetiology in 1947, at a meat packing plant in 1957 and at the same Philadelphia hotel in 1974 showed the organism to be the same as that implicated in the 1976 Philadelphia outbreak (Terranova *et al.*, 1978; Thacker *et al.*, 1978; McDade *et al.*, 1979; Osterholm *et al.*, 1983). In 1973, an outbreak of pneumonia of unknown aetiology occurred among Scottish tourists returning from a holiday in Spain. Three persons died and seven others were affected. All had stayed at the same hotel in the ten days before onset of illness (Reid, 1978). This outbreak was also later shown to be due to *Legionella* infection and was the first in Europe to be retrospectively recognized. Another international outbreak at the same hotel in 1979 confirmed the hotel's water system as the environmental source of infection for the cases (Bartlett *et al.*, 1984).

It soon became apparent that the epidemiology of the disease was important to countries in southern Europe with a large tourist industry. Surveillance programmes began to be established to learn more about its clinical features, and associated programmes focused on discovering more about the microbiology, pathogenicity and virulence of the organism, its ecology and en-

*Corresponding author.

vironmental sources of infection and modes of transmission. Over the last 30 years a wealth of knowledge on *Legionella* infections has emerged from these various research disciplines (Breiman *et al.*, 1998; Fields *et al.*, 2002; Sabria *et al.*, 2003; Edelstein *et al.*, 2005).

Legionella infections – the spectrum of illness

Legionellosis is the term given to describe the spectrum of illness caused by the *Legionella* bacteria. These range from the severe pneumonic form – Legionnaires' disease, which cannot be differentiated at the clinical level from other types of pneumonia, to the mild febrile form known as Pontiac fever (Chapters 1 and 2). Pontiac fever derives its name from an explosive outbreak of flu-like illness that occurred in a County Health Department building in Pontiac, Michigan, USA, in 1968. The organism responsible was found in 1978 to be *L. pneumophila* (Glick *et al.*, 1978).

The incubation period ranges from 2 to 10 days (rarely up to 20 days) for Legionnaires' disease and from 12 to 48 hours for Pontiac fever. The former is characterized by a low attack rate (0.1–5% of the exposed population) compared with the latter where up to 95% of those exposed may develop Pontiac fever. Within the spectrum of legionellosis it is now recognized that a non-pneumonic milder form of Legionnaires' disease (distinct from Pontiac fever), can also occur. It has been increasingly recognized as a consequence of wider testing of suspected cases, particularly during outbreak investigations (Garcia-Fulgueiras *et al.*, 2003; Greig *et al.*, 2004).

Methods of diagnosis and main *Legionella* types to cause infection

Around 50 species of *Legionella* have been described to date, but the number of newly recognized species continue to increase (Harrison, 2005).

Nineteen of these species have been associated with human infection. The most common type to cause disease is *Legionella pneumophila*. This species has 16 serogroups, but serogroup 1 accounts for the majority of human infections and is also widely found in the environment. It can be further subdivided into subtypes using monoclonal antibody subtyping and molecular methods. Those strains reacting to the group of monoclonal antibodies collectively known as the MAb2 'Pontiac' strains are the most common cause of Legionnaires' disease (Harrison *et al.*, 2006; Fields *et al.*, 2002, Chapter 2).

Other *L. pneumophila* serogroups or other *Legionella* species are mainly associated with infection among immunocompromised people, particularly in health care settings (Joseph *et al.*, 1994; Kool *et al.*, 1998). Along with *L. pneumophila*, other *Legionella* species have also given rise to outbreaks of Pontiac fever, e.g. *L. anisa* (Fenstersheib *et al.*, 1990; Jones *et al.*, 2003), *L. micdadei*, (Fields *et al.*, 2001; Huhn *et al.*, 2005), and *L. feeleii* (Herwaldt *et al.*, 1984).

In a patient who meets the clinical criteria for investigation of the disease, *Legionella* infection can be diagnosed by several different methods (Chapter 2), including by culture of the organism from a respiratory specimen, by detection of serum antibodies or urinary antigen, or through detection of bacterial DNA using polymerase chain reaction (PCR). Some *Legionella* species, including *L. pneumophila* can also be cultured from environmental specimens. Positive results from these specimens can be compared with cultures obtained from patients at about the same time to investigate whether the sets of strains have similar or indistinguishable molecular profiles using genotypic and phenotypic methods.

Culture of the organism from a respiratory specimen remains the 'gold standard' method because of the role it can play in microbiological, epidemiological and environmental investigations. However, the proportion of cases detected by this method is small, less than 20% in most European countries (Ricketts *et al.*, 2005). Increasingly, the urinary antigen detection method has become the method of choice for diagnosis of *Legionella* infections; as a consequence, the assay has changed the epidemiological picture of Legionnaires' disease in many countries. The method's ease of use, low cost, high sensitivity and specificity and rapid result has increased patient testing and the yield of cases with confirmed *Legionella* infection, and decreased mortality rates by permitting earlier treatment with appropriate antibiotics (Chapter 2). Notwithstanding the obvious surveillance benefit and improved patient outcome, the urinary antigen method of diagnosis is limited by its high specificity for *L. pneumophila* serogroup 1 infections. Thus, its predominant use at the expense of other methods has probably led to a decline in the diagnosis of Legionnaires' disease caused by other *Legionella* species and other serogroups. Consequently, there is likely to be a bias in the relative proportion of community and travel associated cases detected in many countries compared with nosocomial (hospital acquired) cases, which are more frequently associated with infection from other *Legionella* species or other *L. pneumophila* serogroups.

Sources of infection

Environmental sources of infection are water systems that produce aerosols containing *Legionella* bacteria that are dispersed into the atmosphere. The water in the aerosols evaporates, leaving particles containing *Legionella* bacteria in the air that can be inhaled. For infection to occur in a susceptible host, these particles must be sufficiently small to penetrate the lung alveoli but large enough to contain at least one bacterial cell. Environmental sources of infection are mainly related to industrial sites with water cooling devices, large buildings with wet cooling systems used for air conditioning, hot and cold water systems used in hospitals, hotels, tourist facilities, cruise ships, office buildings, whirlpool spas, gardening facilities, fountains and domestic premises. Food display humidifiers and respiratory therapy equipment also cause *Legionella* infections. (Arnow *et al.*, 1982; Mastro *et al.*, 1991; Maloney *et al.*, 1992; Hahné *et al.*, 2002). Aspiration can occur, and has been observed in hospitalized patients. (Yu, 2000; Blatt *et al.*, 1993). No case to case transmission of Legionnaires' disease has ever been documented.

The management and prevention of *Legionella* infections is through a multidisciplinary approach. To obtain the information necessary to initiate and direct an environmental investigation, cases must be investigated microbiologically and epidemiologically. Although all aspects of the investigations are interdependent, the microbiology is pivotal, as it underpins and connects the epidemiological and environmental data necessary for control and prevention of the disease.

Surveillance

Surveillance has been described as 'the ongoing systematic collection, analysis, and interpretation of health data, essential to the planning, implementation, and evaluation of public health practice, closely integrated with the timely dissemination of these data to those who need to know. The final link in the surveillance chain is the application of these data to prevention and control. A disease surveillance system includes a functional capacity for data collection, analysis, and dissemination linked to public health programmes' (Last, 2000).

Development of surveillance in England and Wales

Surveillance of Legionnaires' disease in England and Wales began in 1979 with comprehensive annual data sets available from 1980 (Bartlett *et al.*, 1983). Initially, the emphasis centred on obtaining information about clinical risk factors and individual susceptibility as well as validation of microbiological diagnostic methods. As risk factors for individuals became better established, the emphasis in the surveillance scheme shifted toward collecting information about possible exposures to *Legionella* bacteria. Currently, upon report of a new case of Legionnaires' disease, local health care staff are advised to collect a two week history from the patient prior to their date of onset of illness, making note of their daily travel routes for work, business or pleasure, their exposure to possible sources of infection and their use of potable water in buildings such as hospitals, hotels, the work place and leisure centres. More recently, domestic water systems in patient homes have been added to the list of potential sites for further investigation. Using the clinical and exposure history as a basis, local staff can focus their environmental investigations. They should also collect sufficient information to report to the national surveillance scheme and to assist in categorizing the case as community, hospital or travel associated. These detailed summaries inform epidemiological analysis at the national level.

Case definitions

No surveillance scheme can develop without agreement on how a case should be defined for the purposes of data collection. This was decided nationally in England and Wales in 1979 but modified in later years to take account of the introduction of new *Legionella* detection methods. A combined clinical and microbiological case definition is used for Legionnaires' disease to differentiate its diagnosis from other types of pneumonic illness with similar clinical manifestations.

The current case definitions of Legionnaires' disease used by the National Surveillance Scheme for residents of England and Wales are as follows: (Joseph *et al.*, 1999).

(I) *Confirmed case*
Clinical or radiological evidence of pneumonia and one or more of the following:
(a) isolation (culture) of *Legionella* species from clinical specimens;
(b) seroconversion (a four-fold rise or greater) by the indirect immunofluorescent antibody test (IFAT) using *L. pneumophila* serogroup 1 antigen;
(c) positive urine ELISA using validated reagents.

(II) *Presumptive case*
Clinical or radiological evidence of pneumonia and one or more of the following:
(a) a single titre of 128 or more using IFAT as above (64 or more in an outbreak);
(b) positive direct fluorescence (DFA) on a clinical specimen using validated monoclonal antibodies;
(c) seroconversion (a fourfold rise or greater) by the indirect immunofluorescent antibody test (IFAT) to *L. pneumophila* other serogroups or other *Legionella* species.

Information on exposure history collected from a 2-week follow-up questionnaire assists with further epidemiological classification of a case. Using these data, cases can be grouped into those suspected or known to be nosocomial, travel associated, community acquired (including domestically acquired) or with unknown exposure. Definitions used in the UK for these groupings are as follows:

- A: Nosocomial Legionnaires' disease
    - (I) Definitely nosocomial: patients who spent all of the 10 days in hospital before onset of symptoms.
    - (II) Probably nosocomial: patients who spent between 1 and 9 of the 10 days in hospital before onset of symptoms and either (a) became ill in a hospital associated with one or more cases of Legionnaires' disease or (b) yielded an isolate that by monoclonal antibody (mAb) subgrouping, or by molecular typing methods, was indistinguishable from isolates obtained from the hospital water system at about the same time.
    - (III) Possibly nosocomial: patients who spent between one and nine of the ten days in hospital before onset of symptoms, in a hospital not known to be previously associated with any other cases of Legionnaires' disease and where no microbiological link has been established between the infection and the hospital.
- B: Travel-associated Legionnaires' disease
    - (I) Single cases: cases who spent one or more nights away from home in the 2–10 days before onset of symptoms, either in their country of residence or abroad. (Overnight stays include accommodation in hotels, camp sites, ships, rented holiday apartments or other tourist facilities.)
    - (II) Clusters: two or more cases who stayed at the same accommodation site in the 2–10 days before onset of symptoms and with onset of illness within the same 2-year period.
- C: Community clusters/outbreaks
Clusters comprise two or more cases linked by area of residence or work, or places visited, if the onsets of illness are sufficiently close to warrant further investigation. (This is a working definition: the decision to follow up cases is usually made locally.) The term 'outbreak' is applied to community clusters where strong epidemiological evidence of a common source of infection

with or without microbiological evidence has been demonstrated and where control measures have been applied to suspected sources of infection.

Development of surveillance in Europe

The World Health Organization's Regional Office for Europe organized the first of a series of European meetings on Legionnaires' disease in 1981. (WHO, 1982). A group comprising epidemiologists and microbiologists and other public health officials from 12 countries strongly recommended the need for European training in laboratory diagnosis of *Legionella* infections, the need for national surveillance programmes and the need for a European collaborating centre for monitoring the international aspects of the disease. The European Working Group for Legionella Infections (EWGLI) was created when this group met in 1986. The European Surveillance Scheme for Travel Associated Legionnaires' Disease, now known as EWGLINET, was established the following year, along with objectives for the surveillance programme, internationally agreed case definitions and reporting procedures (WHO, 1990). EWGLI is now the strongest group world wide for collaborative *Legionella* surveillance, public health research and develop ment initiatives, and for monitoring of control and prevention programmes. It currently comprises 35 European countries covering a population of 550 million (Ricketts *et al.*, 2005). Its collaborations extend beyond Europe when dealing with travel associated outbreaks of Legionnaires' disease. The organization is instrumental in recommending standards for surveillance at the national level and international level and in allowing the pooling of European data for studies of the effectiveness of surveillance and control and prevention programmes within and between countries (Joseph, 2004; Ricketts *et al.*, 2005).

Statutory notification of cases of Legionnaires' disease

Legionnaires' disease is a statutorily notifiable disease in all European countries except Andorra, England and Wales and Romania. Despite this legal requirement to report cases to a national collection point (usually the Ministry of Health or National Institute of Public Health), several countries have used capture–recapture studies to estimate the proportion of cases reported through notification and also identified cases through other means, such as laboratory reports. All have found a discrepancy between the number of notified cases and the total number of cases identified in any one year of surveillance (Infuso *et al.*, 1998; Nardone *et al.*, 2003).

Collection of data about a disease can be through active or passive means, usually depending on whether the disease is rare, of political importance, has a high clinical impact or is an emerging infection requiring a knowledge base to be established. Legionnaires' disease meets all the above criteria to justify active surveillance but with one additional marker, namely that infection is not from person to person but from an environmental source that necessitates identification and investigation to prevent further cases occurring. Thus, whilst active surveillance is the recommended ideal, it is recognized that a vast heterogeneity of national public health surveillance systems exist. The different priorities accorded to specific diseases in the different countries may mean that provision of Legionnaires' disease data will be determined by the country's technical ability to identify cases and produce data and its willingness to prioritize resources to this particular infection. These priorities may also depend on the social, financial and cultural milieu in which the national public health system is historically embedded. (Anon, 1998a; WHO, 1999). There are a wide range of national incidence rates reported within Europe; these may be partly explained by differences in control of environmental sources of infection within countries but more plausibly by the way in which resources are differently apportioned to *Legionella* detection and surveillance (Joseph, 2004).

To summarize, any national surveillance scheme for Legionnaires' disease will be affected by the country's infrastructure, its public health laws, its adopted surveillance principles and standard operating procedures, its notification law, data protection and freedom of information laws, and its policy on patient confidentiality. Schemes that collect microbiological case reports or notifications without epidemiological data on exposure history will remain ineffective blunt tools that simply count cases; lacking any means by which these data can be interpreted or applied to control and prevention strategies.

Estimating incidence

Community-acquired pneumonia studies

There are two main ways of estimating the incidence of Legionnaires' disease worldwide. One is to use studies such as those conducted in Europe, America and Australia which measure the proportion of cases admitted to hospital with a community-acquired pneumonia (CAP) that could be attributed to microbiologically confirmed *Legionella* infection. The other is to use surveillance trend data where a chosen rate per million population is set as the 'gold standard' against which to measure rates from other countries. Both methods are subject to bias. For the CAP studies, the calculations are influenced by the timing, duration or number of cases (*Legionella* has a seasonal component and is a rare infection), or differences in the frequency or type of laboratory detection method applied. Comparisons against a 'gold standard' incidence rate will be affected by differences in ascertainment methods and reporting procedures between national surveillance systems. Nevertheless, as guides to levels of under diagnosis and under reporting within and between countries, both are valuable epidemiological tools.

CAP studies attribute from 0.5–16% of pneumonias to *Legionella* infections; in industrialized countries, the incidence is typically 2–3%. (Anon, 1986b; Bohte *et al.*, 1995; Marston *et al.*, 1997). A 2% estimate indicates that approximately only 10% of the true number of cases is currently being detected and reported in the UK. This difference in the UK, and perhaps elsewhere in Europe, is mainly associated with under diagnosis, as opposed to under reporting. Under diagnosis may be explained by the rapid use of antibiotics to treat hospitalized pneumonia cases. If the antibiotics are effective against *Legionella* (and other bacteria) and the patient recovers, they negate any further need to establish the cause of the infection. Antibiotic treatment for pneumonia in most countries is also less expensive than diagnostic investigation of the pneumonia's cause; hence no microbiological diagnosis is made. This effect is clearly shown in England where unclassified or unspecified pneumonia accounted for 97% of all the primary diagnoses for community-acquired pneumonias admitted to hospital in 2004/05. (http://www.dh.gov.uk/PublicationsAndStatistics)

Other reasons for under diagnosis relate to the predominant use of the urinary antigen detection method (Fields, 2002), the sensitivity and specificity of other methods used to diagnose Legionnaires' disease, resulting in missed cases, and also a lack of recognition of the disease in patients hospitalized with other serious conditions. However, using as proxy markers severity of illness, admission to an intensive care unit, mean length of stay in hospital and mortality rates, studies have shown that these are all greater in patients with Legionnaires' disease compared with other pneumonia cases (Hubbard *et al.*, 1993; Macfarlane *et al.*, 1993; http://www.dh.gov.uk/PublicationsAndStatistics; Ewig *et al.*, 1999).

The 'gold standard' incidence rate

EWGLI has defined a 'gold standard' rate that countries should aspire to of 20 cases per million population, based on historical trends in reporting within Europe. This target is based on data from Denmark where annual reported incidence rates have consistently been around this level (Ricketts *et al.*, 2005). Detection and reporting in Denmark are considered to be very representative of its population since it is a small country that has a long tradition of national surveillance for Legionnaires' disease that includes a high level of testing for *Legionella* in patients with pneumonia. The country also has a centralized *Legionella* Reference Laboratory for diagnosing and reporting cases. Rates at the gold standard level are now beginning to emerge elsewhere in Europe, mainly because of changes in ascertainment and improved surveillance in countries such as the Netherlands, France and Spain. On the other hand, rates of <2 per million continue to be reported from former eastern European countries, such as Estonia, Hungary, Poland, Romania, and Slovakia, highlighting the need for improved diagnosis and perhaps also increased resources and expertise in microbiological reference laboratory capability.

When the 'gold standard' method is applied to the UK, approximately one third of its true number of cases are being detected and reported (i.e., 400 against the expected number of 1200). This compares with 10% of the estimated number of 4000 cases using CAP studies. Which estimate is of greater relevance for epidemiological purposes is unknown. In countries that experience large community outbreaks, the effects of under diagnosis and under reporting rapidly decline, as was observed in the Netherlands and England following major outbreaks in 1999 and 2002, respectively (Den Boer *et al.*, 2002a; Smith *et al.*, 2005; Telford *et al.*, 2006). Recognition of low incidence rates has driven countries to introduce enhanced systems to improve their overall surveillance data (Infuso *et al.*, 1998; Montagna *et al.*, 2005).

Epidemiological trends

Most of the epidemiological trend data discussed next in this chapter has come from the annual data sets submitted by collaborating countries in EWGLINET. Each year these countries are requested to complete a set of standard tables that provide epidemiological and microbiological information for all cases in their annual national dataset. These annual reports collect information on age and sex distribution, deaths, whether the reported cases are hospital, community or travel associated, and details of national outbreaks by source and type. Microbiological data include method of diagnosis for all cases, and the species, serogroup and subgroup of culture confirmed cases. The data are analysed at the EWGLINET coordinating centre in London; this chapter will describe the period 1995–2005. Data from England and Wales from 1980–2005 will also be used to examine in more detail trends within one national surveillance scheme.

In the 11 years between 1995–2005, a total of 32,944 cases of Legionnaires' disease were reported to the European dataset by an annual average of 29 countries (range 24–35 countries; Table 3.1). Over 2600 deaths were recorded. The annual trend in total cases has been upwards, with an almost five fold increase observed over this 11-year period. Most of the rise has occurred since 2001, partly due to more countries contributing data, the detection of large community outbreaks and improved ascertainment in many countries. In 2005, countries with the most cases to report were France (1527 cases), Spain (1229 cases), Italy (869) and Germany (459 cases). These countries, together with England and Wales and the Netherlands, account for most of

Table 3.1 European annual dataset: total reported cases, deaths and rate per million population, 1995 – 2005. (Source: European Working Group for Legionella Infections)

Year	No. of cases	No. of deaths	No. of countries contributing data	Population (millions)	Rate per million population
1995	1255	95	24	339	3.7
1996	1563	77	24	352	4.5
1997	1360	136	24	351	3.9
1998	1442	188	28	333	4.3
1999	2136	193	28	398	5.4
2000	2156	234	28	400	5.4
2001	3470	284	29	455	7.6
2002	4696	284	32	467	10.1
2003	4578	352	34	468	9.8
2004	4588	396	35	557	8.3
2005	5700	377	35	551	10.3
Total	32,944	2616	(mean 29)	(mean 424.6)	(mean 6.7)

the overall rising trend since 2001. At the other extreme, two or three countries per year have no cases at all to report.

Incidence rates

Nine countries in 2005 had an incidence rate of 20 or more cases per million population. These were Andorra, Croatia, the Czech Republic, Denmark, France, Lithuania, Luxembourg, Spain and Switzerland. However, in some of these countries the rate is based on a small geographical regional population rather than the national population. Denmark is the only one of these countries that has more or less maintained this rate since 1995, others have achieved this level very recently: four in 2004, three in 2003, two in 2002 and two in 2001. Reaching the 'gold standard' for surveillance remains an elusive target for the majority of countries in Europe (Fig. 3.1).

In England and Wales, incidence between 1995 and 2001 remained stable at 3.2–3.7 per million population per year (Anon, 1997b; Anon, 1998b; Anon, 1999; Anon, 2000b). In 2002 the rate jumped to 7.2 because of a large outbreak, and it has remained around the 6.5 rate since, due to increased ascertainment subsequent to the outbreak. A similar but more marked picture is evident in the Netherlands, where a new surveillance programme introduced after their outbreak in 1999 has improved ascertainment levels considerably (Bosman *et al.*, 1999; Den Boer *et al.*, 2002a; Health Council of the Netherlands, 2003). Their rate increased from around 3 cases per million before 1999 to 11.2 in 2000, 14.8 in 2004 and 18.7 in 2005. A new surveillance scheme introduced in 1995 in France likewise yielded major changes in ascertainment levels (Hubert *et al.*, 1997).

Cases and rates by age group

Males constitute between 70% and 75% of the total cases reported each year, and the peak age in overall reports for both sexes combined is between 50 and 60 years. However, age standardized rates of infection per 100,000 population show large differences within and between European countries. For all countries combined, age standardized rates rise in line with increasing age, with cases aged 80 years or more having a rate almost double that of those aged 50–59 years (3.8 compared with 2.0 per 100,000 in 2005; Fig. 3.2). Countries with a high absolute number of cases (France, Italy and Spain) are examples of the trend in rising incidence rates among the most elderly. In contrast in countries where overall rates are less than 10 per million population, their age-standardized rates tend to decline in the very elderly compared with younger age groups. The reasons for these differences are unclear but could be related to more generous

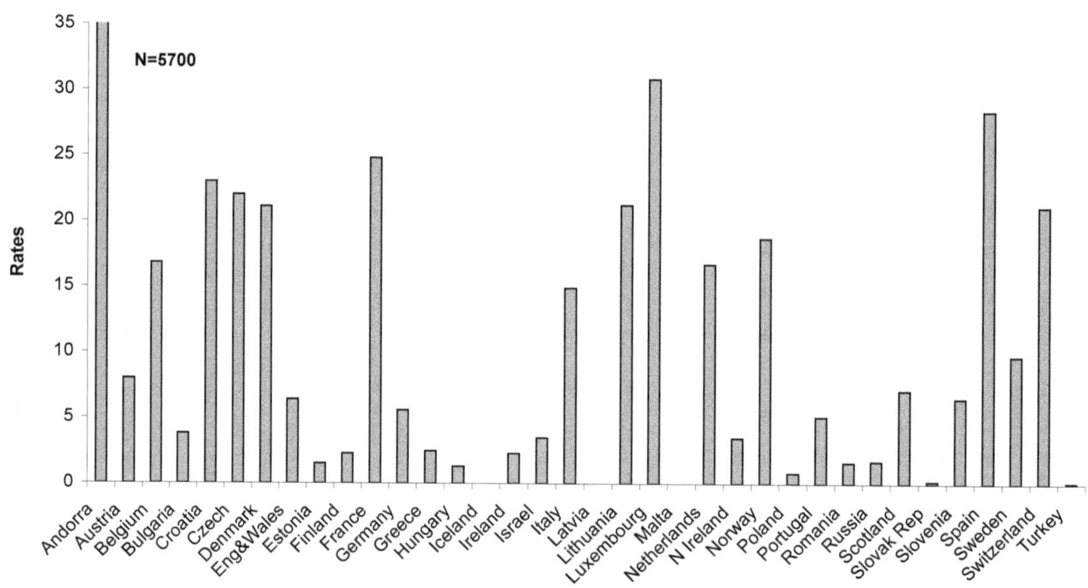

Figure 3.1 European annual dataset: rate per million population by country of report, 2005. (Source: European Working Group for Legionella Infections.)

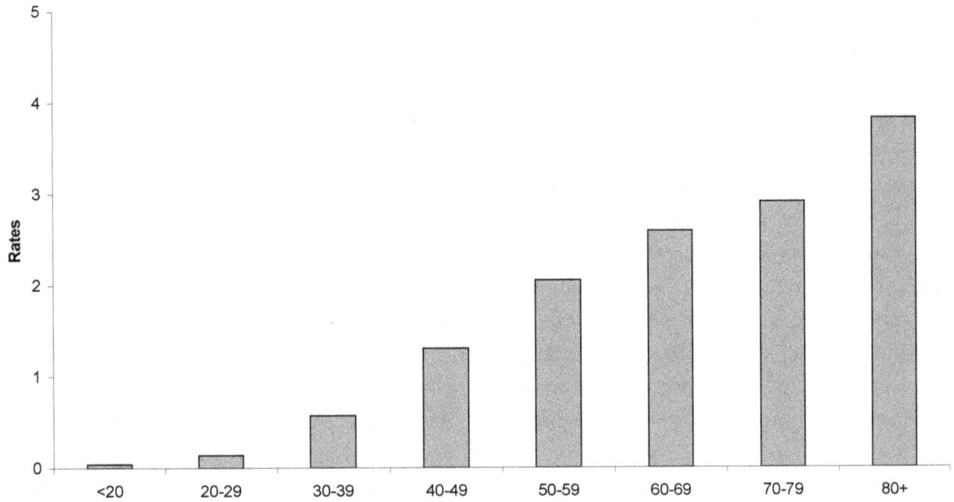

Figure 3.2 European annual dataset: standardized incidence rates per 100,000 population by age, 2005. (Source: European Working Group for Legionella Infections.)

and inclusive testing policies in countries that detect very large numbers of cases.

Mortality rates

The overall case fatality rate of 7.9 (range 5–13%) is regarded as an underestimate in the European dataset because many countries cannot contribute reliable mortality information to it. However, an overall declining trend in deaths has been observed, decreasing from 13% in 1998 to 6.5% in 2005. In recent years it has been noted throughout Europe that, whilst very large outbreaks continue to occur, associated deaths are significantly lower than those from outbreaks in the 1980s and are also proportionately lower than those among current sporadic cases. The reasons for this are clear. Firstly, publicity during an outbreak heightens awareness of the disease, case searching takes place and leads to prompt investigation of potential cases. Secondly, the urinary antigen detection test which is almost universally applied enables a rapid diagnosis to be made and leads to the early treatment of cases with appropriate antibiotic therapy. Thirdly, this diagnostic method has widened the pool of people tested during outbreaks, many of whom include a significant proportion of mildly ill non-pneumonic cases, lowering the case fatality ratio. Urinary antigen testing was not available in the 1980s, a period when diagnostic tests took several days or weeks to confirm the illness. Hence, it is likely that delays in appropriate treatment at that time contributed to a higher mortality among a smaller pool of recognized cases. Sporadic community-acquired cases in England and Wales continue to be associated with mortality rates of 12–15% compared with rates of less than 10% for outbreak cases. In these non-outbreak incidents, cases may have a delayed diagnosis and hence worse outcome, or only the most severely ill ones may be detected.

Seasonality

Cases of Legionnaires' disease occur all year round. In England and Wales, data by month of onset show the typical northern hemisphere pattern of case numbers rising in spring and peaking over the late summer months of August and September. Many of these summer cases are the result of infection acquired abroad, while others are thought to be related to warm weather effects, higher summer humidity levels and exposure to more potential sources of infection such as warm water showers, air conditioning, water features used in gardens, etc. (Hlady *et al.*, 1993; Jones *et al.*, 2003; Fisman *et al.*, 2005). A significant upsurge in cases with onset of illness in August and September 2006 was reported in England and Wales and some other northern European countries (Joseph *et al.*, 2006; Laurance, 2006). This increase in cases is now the subject of a national and international study which aims to identify the possible meteorological and ecological factors responsible.

Main method of diagnosis

In the European dataset, microbiological diagnoses between 1995–2005 show an average of 369 clinical isolates of *Legionella* were obtained each year (range 236–491), representing 12.3% of the total cases. Ninety five per cent of the isolates were reported to be *Legionella pneumophila*, the remainder were reported as *L. anisa, L. bozemanii, L. brunensis, L. cincinnatiensis, L. dumoffii, L. feeleii, L. gormanii, L. jordanis, L. longbeachae, L. micdadei, L. parisiensis and L. saint helensi*. For 110 isolates, the *Legionella* species was not given. The proportion of cases diagnosed primarily by urinary antigen detection has seen a dramatic rise since 1998. In 1995 they represented 14.6% of diagnosed cases; this had risen in 1998 to 33% and in 2005 to 71% of all cases. Serological diagnoses, either by single high titre or by a four fold rise in antibody levels, represent 21% of the 11-year dataset. In recent years, serological diagnoses occurs in only around 9% of cases due to greater use of the urinary antigen detection method. For example, in England and Wales, it is striking how serology has been replaced by urinary antigen detection as the main method of diagnosis (Fig. 3.3).

Cases by category of exposure

The annual European dataset includes data on whether cases are known or assumed to be associated with hospital, community, or travel acquired infection (Fig. 3.4). Community-acquired infection constitutes the majority of cases (42%) and nosocomial the minority of cases (8.3%); the remainder are travel associated (20.6%), or of unknown category (29.1%). The dataset also reports information on national outbreaks associated with each of these categories.

Outbreaks by type

Six hundred and fifty-six outbreaks were reported between 1995 and 2005. One hundred outbreaks were linked to nosocomial infection, 160 to community infection, 390 to travel and six to other causes. In 57 outbreaks, cooling towers were given as the source of infection, 207 were attributed to hot or cold water systems and 33 to spa pools. For the remaining 359, no source was reported.

Hospital-acquired (nosocomial) infections

For all years combined, 8.3% of the cases (2751) were reported as nosocomial; the mean proportion being 11.4% (range 12.5 – 9.1%) between 1995 and 1999, and 7.4% between 2000 and 2005 (range 12.8–5.6%). Although the overall proportion of nosocomial cases has fallen, those associated with clusters or outbreaks have consistently remained at 19% over this period and include a total of 100 outbreaks of hospital acquired Legionnaires' disease reported between 1995–2005. The majority of nosocomial reports came from France, Italy, Spain, Denmark and Germany. No data is collectively available on the range of *Legionella* species and serogroups linked to these cases or outbreaks, but a major review of nosocomial infections in England and Wales between 1980–1992 showed that 27% of the nosocomial cases were diagnosed by culture of the organism compared with 10% and 13% of community and travel associated infections, respectively. When compared to community (83%) and travel cases (81%), a lower proportion of these nosocomial culture-proven cases were due to *L. pneumophila* sg1 infection (45%), supporting the view that hospitalized patients are more susceptible to the non *L. pneumophila* sg1 strains than other groups of patients (Joseph *et al.*, 1994).

Community-acquired infections

Community-acquired infection accounts for the largest proportion of cases in the European dataset and also for the main increase in reported cases since 2001. The annual total of community cases varies each year, depending on the number and magnitude of outbreaks that occur. Between 1995 and 2005 13,694 cases were attributed to community-acquired infection, representing 42% of the total dataset. One hundred and fifty-nine outbreaks involving more than 2000 cases contributed to this total. Many of these outbreaks are memorable for their size or unusual source of infection or because they have generated revised thinking on established beliefs or have led to the development of new or modified guidelines and regulations for control and prevention.

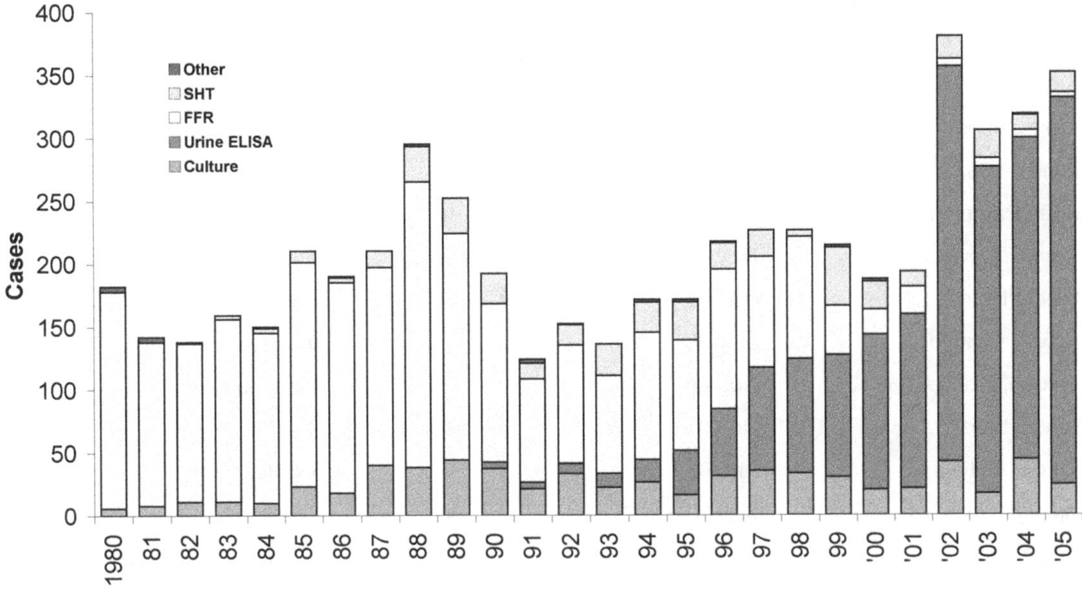

Figure 3.3 Main method of diagnosis: England and Wales, 1980–2005. (Source: National surveillance scheme for Legionnaires' disease in residents of England and Wales.)

Cooling towers are the source of most large community outbreaks because of their ability to disperse *Legionella*-infected aerosols into the atmosphere over long distances. Spa pools (also known as whirlpools or whirlpool spas) can also give rise to large outbreaks if they operate in public buildings (Table 3.2). Other recent community outbreaks of interest in Europe include those linked to a food humidifier (Hahné *et al.*, 2002), spa pools on display at trade fairs (Den Boer *et al.*, 2002b; De Schrijver *et al.*, 2003; Allerberger *et al.*, 2004), and an air scrubber (Nygard *et al.*, 2005).

Some early and notable community outbreaks are described below.

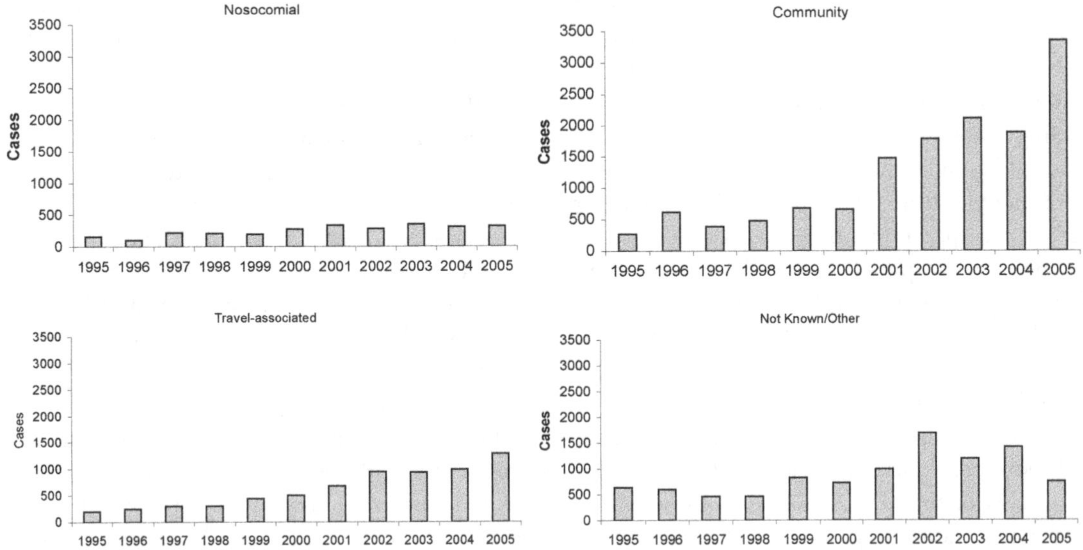

Figure 3.4 European annual dataset: category of cases, 1995–2005. (Source: European Working Group for Legionella Infections).

Table 3.2 European annual dataset: Major recent cooling tower outbreaks in Europe. (Source: European Working Group for Legionella Infections)

Outbreak	Year	Cases	Deaths	Source
France, Paris	Aug 06	29	2	Cooling tower
Netherlands, Amsterdam	July 06	31	3	Cooling tower
Spain, Pamplona	June 06	148	0	Cooling tower
Spain, Torrevieja	Nov 05	34	0	Cooling tower
France, Lyons	May 05	34	0	Cooling tower suspected
Norway, Ostfold	May 05	56	10	Air scrubber
Sweden, Lidköping	Aug 04	32	2	Cooling tower
France, Lens	Dec 03	86	18	Cooling tower
France, Poitiers	Aug 03	24	0	Cooling tower
England, Hereford	Nov 03	27	1	Cooling tower
France, Montpellier	Aug 03	31	4	Cooling tower suspected
Spain, Barcelona	Aug 02	113	2	Cooling tower
England, Barrow-in-Furness	Aug 02	179	7	Cooling tower
France, Meaux	July 02	22	4	Cooling tower
France, Sarlat	July 02	31	6	Cooling tower
Norway, Stavenger	Aug 01	28	7	Cooling tower
Spain, Murcia	July 01	600+*	6	Cooling tower

*Estimated cases, of which 449 were microbiologically confirmed.
Source: European Working Group for Legionella Infections.

Stafford District Hospital, Stafford, England, 1985: magnitude and high mortality. This community outbreak in a hospital setting gave rise to 68 cases and 22 deaths. It led to a Public Enquiry (Committee of Inquiry, 1985) and recommendations for a national register of cooling towers to be established (Health and Safety Commission, 1992). Most of the cases were people from the community attending outpatient appointments who became infected through the hospital's air conditioning system. Follow-up serological studies showed that nearly one-third of the hospital staff had antibodies to *Legionella* compared with 3% of hospital staff in a control group. A case–control study and a study of outpatient department attendees and of hospital staff both concluded that infection had occurred in an area of the hospital supplied by only one of the air conditioning plants (O'Mahony *et al.*, 1990).

BBC Headquarters, London, England, 1988: international profile and demonstration of air-borne spread. A poorly maintained cooling tower at the British Broadcasting Corporation (BBC) headquarters in central London was responsible for this outbreak. Because of the BBC's high profile, the incident received intense media attention in the UK and abroad and generated grave public concern. Seventy cases and three deaths were detected among BBC staff and other people who had walked or driven in the area or who were patients domiciled in a local hospital where windows had been open during the period of exposure. For the first time in the UK, grid maps of the area in conjunction with meteorological data on wind speed, direction and humidity were used in the epidemiological investigations to establish the focal point of exposure. An enquiry into the outbreak's cause and the way the investigations were carried out was held by the House of Commons Select Committee on Employment. The BBC was successfully prosecuted under the Health and Safety at Work Act (Westminster Action Committee, 1988; Employment Committee, 1989).

Bovenkarspel, the Netherlands, 1999: spa pools on display and length of incubation period. A spa pool featured at an annual flower, agriculture and consumer products show attended by over 77,000 visitors was the source of infection. The spa pool water was not changed during the 9 days of the exhibition, and clinical attack rates increased over time, indicating that the infectious dose grew larger each day the pool was in operation. One hundred and eighty-eight confirmed or probable cases, 21 of whom died, occurred among people who spent time near the spa pool. Proven incubation periods ranged from 2–19 days; 16% exceeded 10 days. The outbreak was recognized late, 2 weeks after the first case was diagnosed, partly because cases were dispersed among many different hospitals. Urinary antigen testing was not a common procedure in the Netherlands at the time of the outbreak, but subsequent adoption of this method of diagnosis has improved case ascertainment on an annual basis (Den Boer *et al.*, 2002a).

Murcia, Spain, 2001: magnitude, low mortality and subsequent legislation. The world's largest outbreak occurred in June/July 2001. Over 800 cases were suspected, between 636 and 696 were estimated to be cases, 449 of which were microbiologically confirmed. Only six deaths (1%) were reported. The source of infection came from cooling towers at a northern city hospital. Environmental isolates from these towers were indistinguishable by molecular typing methods from patient isolates. Cooling towers on ten other buildings were also positive for *L. pneumophila* sg1 but were of a different strain to the outbreak strain. Murcia was due to host the European Youth Olympic Games in August 2001. The Olympic authorities requested an international team of experts assess the outbreak's management to ensure that, as a result of the outbreak, the city posed no greater risk to visitors and participants than any other city. A national law for control and prevention of Legionnaires' disease was enacted in Spain immediately following the outbreak (Boletín Oficial del Estado, 2001; De Celestino, 2001; Garcia-Fulgueiras *et al.*, 2003).

Barrow-in-Furness, Cumbria, England, 2002: magnitude and attempted prosecution for corporate manslaughter. The largest outbreak to date in the UK affected almost 200 people, seven of whom died. The contaminated aerosol from a cooling tower owned and operated by Barrow town council was dispersed into a narrow walkway through which most of the town's population passed to reach the shopping area designed for pedestrians. Follow-up studies have been conducted on this geographically isolated population to measure *Legionella* exposure levels and clinical outcomes among cases and controls. A high-profile court case for corporate manslaughter was unsuccessful against one council member of staff with alleged responsibility for the cooling tower maintenance contract, although a prosecution succeeded under the Health and Safety at Work Act, as well as prosecution of the council under Health and Safety Executive legislation (Bennetto, 2006; *The Financial Times*, 2006; *Guardian*, 2006; Telford *et al.*, 2006).

Pas de Calais, France 2003/04: long-distance of aerosol spread and subsequent legislation. Two waves of infection were associated with this outbreak between November and January 2003/04. Cooling towers belonging to a petrochemical plant sited on a large industrial estate in northern France were contaminated with the same strain as that obtained from the patient isolates. After closure and disinfection, a second wave of cases occurred as a consequence of restarting the towers. The high pressure cleaning process created conditions favourable for detaching *Legionella*-contaminated biofilm and air-borne release of aerosols from surface ventilators. In total, 86 cases were detected, 18 of whom died. All patients lived in or visited an area within a 12-km diameter of the industrial plant; many were elderly and did not leave their homes during the period of exposure. Air-borne transmission of *Legionella* infection 6 km from the source is a new public health finding with implications for investigation of other community outbreaks (Nguyen *et al.*, 2006). New national regulations for installation and maintenance of cooling towers as well as risk management guidelines for *Legionella* were issued in France immediately following the outbreak (Ministère de l'Ecologie et du Développement Durable, 2004; Ministère de la Santé et des Solidarités, 2005).

Travel-associated Legionella *infections*

Travel associated Legionnaires' disease accounted for 20.6% of cases in the European annual data set between 1995 and 2005, 13.5% as a result of travel abroad and 7.1% as a result of travel in the same country as the country of residence ('internal travel'). The proportion of cases linked to travel abroad, relative to the total dataset, has remained similar throughout the 11-year period but internal travel associated cases have risen from 1.9% in 1995 to 10.3% in 2005. This rise is due mainly to improved surveillance by countries such as France, Italy, Germany, Spain and the UK that have increased both detection and reporting of these cases. One goal of the national collaborating centre is to detect cases from different countries of residence who may be part of a cluster linked to one accommodation site in a popular holiday resort. For this purpose, each EWGLI participant country is requested to rapidly report their travel cases to the European surveillance scheme for travel associated Legionnaires' disease (EWGLINET). Because of European travel patterns, cases tend to be diagnosed and reported from northern European countries whereas infection is predominantly associated with travel to southern European countries. For example, this pattern is evident when the proportion of cases by category of exposure is compared between England and Wales, the Netherlands, France, Italy and Spain in 2005. In England and Wales and the Netherlands, 42% and 56% of all cases, respectively, were reported as linked to travel abroad, compared with 4.7%, 2% and 1.3% for residents of France, Italy and Spain. In these latter countries people are more likely to take holidays in their own country where they then acquire infection. As a proportion of their total cases, travel associated infection due to travel within the home country was 9.4% for England and Wales, 4% for the Netherlands, 13% for France, 17% for Italy and 13% for Spain.

To detect clusters and implement control measures at suspected sources of infection, cases of travel associated Legionnaires' disease are reported to EWGLINET throughout the year. A change in the EWGLINET cluster case definition in 2001 (Lever *et al.*, 2003) resulted in the detection of 70–100 clusters annually, compared with less than 20 per year from 1995–2000. All clusters are required to be investigated under the European Guidelines for Control and Prevention of Travel Associated Legionnaires' Disease (European Working Group for Legionella Infections, 2002). This task is carried out by the country associated with the infection, irrespective of the nationality of the cases. The guidelines state that, upon receipt of a cluster alert in the country of infection, within two weeks the EWGLINET collaborator must inform the accommodation site, organise a risk assessment to be carried out, and submit a report to the coordinating centre that states that control measures are in progress. Sanctions are applied if this process fails or if the report indicates that no control measures have been initiated. Four weeks later, a second report is requested which must provide information on the range of investigations, control measures carried out, and results of environmental sampling. Again, sanctions are applied if the report is not forthcoming or it states that control measures are inadequate. Sanctions consist of publicizing the cluster and providing the name of the accommodation site on the EWGLI website. This allows members of the public and tour operators to make informed decisions about use of the hotel or other accommodation in the absence of information on its public health safety. The website report is removed when information is provided that control and prevention measures are satisfactory. This 'naming and shaming' approach to *Legionella* cluster alerts is supported by all EU and many non-EU countries. Investigation results each year show that approximately 60% of cluster accommodation sites yield environmental samples that are positive for *Legionella* infection. In the few countries that investigate accommodation sites associated with single cases, 70/114

(61%) were reported to be positive for *Legionella* infection in 2005 (Ricketts *et al.*, 2007).

The EWGLINET data shows that on average around 60 countries are linked to travel associated infection each year, with about 10% of cases visiting more than one country during their incubation period. The majority of clusters occur after travel in France, Italy, Spain and Turkey, but in recent years other clusters have been reported from countries such as Cuba, India, Malaysia, Mexico, Russia, Sri Lanka, Tanzania, United Arab Emirates and Uzbekistan (Ricketts *et al.*, 2006).

Unknown category of case

The remaining 29% of cases in this European dataset were reported as of unknown category of exposure. Although high, this proportion has been falling during this eleven year period from a peak of 50% in 1995 to only 12.5% in 2005, demonstrating a qualitative change in data collection in recent years.

Guidelines for control and prevention of *Legionella* infections

Most countries have developed guidelines or regulations for controlling *Legionella* in water systems, often as a consequence of a large outbreak (Anon, 2000a; Conseil Supérieur d'Hygiène Brussels, 2000a; Conseil Supérieur d'Hygiène Brussels, 2000b; Boletín Oficial del Estado, 2001; Bosman *et al.*, 2002; Ministère de l'Ecologie et du Développement Durable, 2004). Guidelines are advisory and usually include information on risk management programmes whereas regulations are legally enforceable and may outline specific management responsibilities and operator competences for maintenance of water systems. Both guidelines and regulations exist for hospital settings (NHS Estates, 1993), industrial and commercial settings (Control of Substances hazardous to Health Regulations, 2002; Health and Safety Commission, 2000), and the special circumstances linked to short stay accommodation in hotels, camp sites and cruise ships (European Working Group for Legionella Infections, 2002; Centers for Disease Control and Prevention (CDC), 1997). Despite the availability and enforcement of guidelines and regulations, large outbreaks of Legionnaires' disease continue to occur (Decludt, 1999; Garcia-Fulgueiras *et al.*, 2003; Smith *et al.*, 2005; Telford *et al.*, 2006), mainly due to neglect, ignorance or the particular design of cooling towers, spa pools and other water systems (Anon, 1996a; Anon, 2000a; Federal Ministry of Health Germany, 2000; ASHRAE, 2000; Anon, 2001; The Guardian, 2001; The Daily Telegraph, 2005; The Health Protection Agency, 2006).

Conclusions

This review of European data has highlighted the main epidemiological features of Legionnaires' disease and the importance of surveillance programmes to measure and evaluate the success or otherwise of control and prevention policies for this disease. The fact that some European countries have lower incidence rates per million population is not in itself an obvious marker for successful control of *Legionella* bacteria in water systems in these countries compared with others. On the one hand, poor control of environmental sources of infection may be masked by poor ascertainment of clinical cases, while on the other hand high levels of environmental control may exist in countries that also have incident rates above 20 cases per million population and where equilibrium between incidence, detection and reporting rates may have been reached. In between, there probably exists a range of dichotomies between good and bad environmental control and high and low levels of clinical diagnosis. However, if the overall number of detected cases continues to rise across Europe, more clusters at the local level may be identified. Investigation of these clusters may help to monitor and contextualise the long term effects of control and prevention programmes for *Legionella* infection within and between different European countries. As more cases are detected, increases in ascertainment should help to clarify the true incidence rate of Legionnaires' disease in different countries.

With increasing numbers of cases diagnosed and reported each year, major implications for workloads arise. EWGLINET has set standards for surveillance in Europe and, through its guidelines, advocates have applied uniform procedures for following up cases of travel associated

Legionella infections. Individual countries have to prioritize resources for this work and some also have to develop the skills necessary for epidemiological, microbiological and environmental analysis of data. Within the European Union (EU), Decision 2119/98/EC enforces the requirement for epidemiological surveillance and control of a range of communicable diseases in the Community, including Legionnaires' disease. For some new members of the EU, this requirement presents many challenges, especially those where less than twenty cases per year have been reported in their annual return. However, international comparisons of epidemiological data are extremely useful for highlighting where the EC might target extra resources to enable these countries to raise their ascertainment rates in the future.

This review has identified an overall decline in the proportion of cases of hospital acquired infections, but an increase in the number of travel associated infections. An element of bias in detection of travel related cases relative to other cases is probably occurring; nevertheless, there is a growing desire in Europe for all travel cases to be investigated for a possible source of infection. Currently, Italy and the UK are the main countries that meet this expectation. If others were to follow suit, and more countries undertook sampling and control measures based on a single case at a tourist accommodation site, it is possible that a proportion of the clusters identified each year would be prevented. The cost effectiveness of this approach has not been estimated, although a study in Australia has attempted to identify cost savings related to control of *Legionella* in cooling tower systems (Anon, 2001).

The future

In 2003 the European Centre for Disease Prevention and Control (ECDC) was established in Stockholm, Sweden. It has assumed legal responsibility and management of all the European disease specific networks and is currently planning how they will be funded and operated in the future. There is almost certainly more to learn about the epidemiology of Legionnaires' disease at the national and international level. It is hoped that ECDC will continue to facilitate this collaborative approach to new learning and new information in Europe. It is also hoped that collaborations can be extended across continents so that best practice can be utilized for the benefit of a wider population.

New challenges are emerging for the future. Demographic changes in western countries are producing ever increasing proportions of persons in older age groups, and climate change is possibly leading to an increased risk of exposure to *Legionella* infections. Thus, it behoves all nations where possible to ensure that risk management procedures are rigorously applied to all potential environmental sources of infection, that clinical practice optimises its diagnostic capabilities, and that cases among vulnerable population groups are rapidly identified and appropriately treated. Drawing together epidemiological, microbiological and environmental expertise in this work will address these twin challenges of increasing susceptibility and increasing risk of infection.

Web resources

There are many websites offering advice and information on all aspects of *Legionella* infections, as well as data on numbers of cases reported on an annual basis by country. Most national public health institutes offer information on a range of diseases, including Legionnaires' disease. These data can be found at www.hpa.org.uk for England and at www.cdc.gov for the Centers for Disease Control and Prevention, USA. Comprehensive information on control of *Legionella* in water systems is available from www.hse.gov.uk. At the European level, www.ewgli.org provides comprehensive information, as does ECDC (www.ecdc.europa.eu) and the World Health Organization (www.who.int). Weekly bulletins such as those produced by *Eurosurveillance* (www.eurosurveillance.europa.eu) and the MMWR (www.mmwr.cdc.gov) provide timely reports on *Legionella* incidents and outbreaks from Europe and the USA.

Acknowledgements

EWGLINET was funded until December 2006 by the European Commission Health and Consumer Protection Directorate-General and from January 2007 by the European Centre for Disease Prevention and Control. We would like to thank all the collaborators in the scheme for contributing their national annual data sets to this network. We would also like to thank Ms Natalie Zhang for her help with compiling the references used in this chapter.

References

Allerberger, F., Schmid, D., Wewalka, G., and Agentur, Ö. (2004). Legionnaires' disease associated with whirlpools at an exhibition – Austria, March 2004. Eur. Surveill. Wkly 8(15).

Anon. (1986b). Community-acquired pneumonia in adults in British hospitals in 1982–1983: a survey of aetiology, mortality, prognostic factors and outcome. The British Thoracic Society and the Public Health Laboratory Service. Q. J. Med. *62*, 195–220.

Anon. (1996a). Drinking water heating systems and conduits; Technical measures to decrease Legionella growth; rehabilitation and operation. W 552 DVGW Bonn.

Anon. (1997a). Epidemiologic Notes and Reports Respiratory Infection – Pennsylvania. MMWR. *46*, 49–56.

Anon. (1997b). Legionnaires' disease in Europe, 1996. Wkly Epidemiol. Rec. *72*, 253–260.

Anon. (1998a). Inventory of the means of controlling communicable disease in the European Union, Norway and Switzerland. European Commission – Directorate General V, Luxembourg.

Anon. (1998b). Legionnaires' disease in Europe, 1997. Wkly Epidemiol. Rec. *73*, 257–264.

Anon. (1999). Legionnaires' disease in Europe, 1998. Wkly Epidemiol. Rec. *74*, 273–280.

Anon. (2000a). Gazzetta Ufficiale della Repubblica Italiana, serie generale, n. 103. Linee – guida per la prevenzione ed il controlle della legionellosi.

Anon. (2000b). Legionnaires' disease in Europe, 1999. Wkly Epidemiol. Rec. *75*, 347–352.

Anon. (2001). Legionnaires' Disease: managing the health risk associated with cooling tower and warm water systems. Regulatory Impact Statement Health (Legionella) Regulations 2001, Victoria, Australia, Department of Human Sciences.

Arnow, P.M., Chou, T., Weil, D., Shapiro, E.N., and Kretzschmar, C. (1982). Nosocomial Legionnaires' disease caused by aerosolized tap water from respiratory devices. J. Infect. Dis. *146*, 460–467.

ASHRAE (American Society of Heating, Refrigerating and Air-Conditioning Engineers). (2000). Guideline 12–2000 – Minimizing the risk of legionellosis associated with building water systems. Atlanta, Georgia, ASHRAE Inc.

Bartlett, C.L., and Bibby, L.F. (1983). Epidemic legionellosis in England and Wales 1979–1982. Zentralbl. Bakteriol. Mikrobiol. Hyg [A]. *255*, 64–70.

Bartlett, C.L., Swann, R.A., Casal, J, Canada Royo, L., and Taylor, A.G. (1984). Recurrent Legionnaires' Disease from a Hotel Water System. In Legionella. Proceedings for the 2nd International Symposium, C. Thornsberry, ed. (American Society of Microbiology, Washington DC), pp. 237–239.

Bennetto, J. (2006) Council charged with manslaughter after Legionnaire's outbreak. The Independent. 24th November.

Blatt, S.P., Parkinson, M.D., Pace, E., Hoffman, P., Dolan, D., Lauderdale, P., Zajac, R.A., and Melcher, G.P. (1993). Nosocomial Legionnaires' disease: aspiration as a primary mode of disease acquisition. Am. J. Med. *95*, 16–22.

Bohte, R., Van Furth, R., and Van den Broek, P.J. (1995). Aetiology of community-acquired pneumonia: a prospective study among adults requiring admission to hospital. Thorax. *50*, 543–547.

Boletín Oficial del Estado. (2001). Real Decreto del Ministerio de Sanidad y Consumo, por el que se establecen los criterios higiénico-sanitarios para la prevención y control de la legionelosis. Real Decreto 909/2001. July 28.

Bosman, A., Nijhof. M., and Brandsema, P. (2002). Steeds meer legionellose-meldingen. Infectieziekten Bulletin, National Institute of Public Health and the Environment (The Netherlands). *13*, 305–308.

Bosman, A., and Van Vliet, H. (1999) New infectious disease law for the Netherlands.. Eur. Surveill. Wkly 3(23).

Breiman, R.F., and Butler, J.C. (1998). Legionnaires' disease: clinical, epidemiological, and public health perspectives. Semin. Respir. Infect. *13*, 84–89.

Centers for Disease Control and Prevention, CDC. (1997). Final recommendations to minimize transmission of Legionnaires' disease from whirlpool spas on cruise ships. United States Department of health and Human Services, Atlanta, Georgia, Public Health Service, CDC, National Centre for Environmental Health/National Centre for Infectious Disease.

Committee of Inquiry, (1985). First report of the Committee of Inquiry into the outbreak of legionnaires' disease in Stafford in April 1985. (London: HMSO).

Conseil Supérieur d'Hygiène Brussels (2000a). Relatif aux dangers de et aux mesures préventives contre une contamination par Legionella en Belgique (C.S.H.: 4870).

Conseil Supérieur d'Hygiène Brussels (2000b). Recommendations Pour La Prevention Des Infections A Legionella Dans Les Etablissements De Soins No (C.S.H: 7509).

Control of Substances hazardous to health Regulations. (2002). (COSHH)

De Celestino, J.A. (2001). Marques asegura que es la mayor epidemia por legionella del mundo. La Opinion. 13 de Julio.

De Schrijver, K., Dirven, K., Van Bouwel, K., Mortelmans, L., Van Rossom, P., De Beukelaar, T., Vael, C., Fajo, M., Ronveaux, O., Peeters, M.F., Van der Zee, A., Bergmans, A., Ieven, M., and Goossens, H. (2003). An outbreak of Legionnaire's disease among visitors to a fair in Belgium in 1999. Public Health. *117*, 117–124.

Decludt, B., Guillotin, L., Van Gastel, B., Dubrou, S., Jarraud, S., Perrocheau, A., Carlier, D., Reyrolle, M., Capek, I., Ledrans, M., and Etienne, J. (1999). Epidemic cluster of Legionnaires' disease in Paris, June 1998. Eur. Surveill. *4*, 115–118.

Den Boer, J.W., Friesema, I.H., and Hooi, J.D. (2002a). [Reported cases of Legionnaires' Disease in the Netherlands, 1987–2000]. Ned. Tijdschr. Geneeskd. *46*, 315–320.

Den Boer, J.W., Yzerman, E.P., Schellekens, J., Letting. K.D., Boshuizen, H.C., Van, Steenbergen,J.E., Bosman, A., Van den Hof S, Van Vliet, H.A., Peeters, M.F., Van, Ketel R.J., Speelman, P., Kool, J.L., and Conyn-Van Spaendonck, M.A. (2002b). A large outbreak of Legionnaires' disease at a flower show, the Netherlands, 1999. Emerg. Infect. Dis. *8*, 37- 43.

Edelstein, P.H., and Cinaciotto, N.P. (2005). Legionella. In: Mandell, Douglas, and Bennett's principles and practice for infectious disease. G.L. Mandell, J.E. Bennett, R. Dolin, ed. (Philadelphia: Churchill Livingston – Sixth Edition), pp. 2711–2724.

Employment Committee. (1989). Legionnaires' disease at the BBC. (London: HMSO).

European Working Group for Legionella Infections. (2002). European Guidelines for Control and Prevention of Travel Associated Legionnaire's Disease. (London: PHLS).

Ewig. S., and Torres. A., (1999) Severe Community-Acquired Pneumonia. Clin. Chest. Med. *20*, 575–587.

Federal Ministry of Health, Germany. (2000). Protection of Infection Act (IfSG) Act on Prevention and Control of Infectious Diseases in Man.

Fenstersheib, M.D., Miller, M., Diggins, C., Liska, S., Detwiler, L., Werner, S.B., Lindquist, D., Thacker, W.L., and Benson, R.F. (1990). Outbreak of Pontiac fever due to Legionella anisa. Lancet. *336*, 35–37.

Fields, B.S., Benson, R.F., and Besser, R.E. (2002). Legionella and Legionnaires' Disease: 25 Years of Investigation. Clin. Microbiol. Rev. *15*, 506–526.

Fields, B.S., Haupt, T., Davis, J.P., Arduino, M.J., Miller, P.H., and Butler, J.C. (2001). Pontiac fever due to *Legionella micdadei* from a whirlpool spa: possible role of bacterial endotoxin. J. Infect. Dis. *184*, 1289–1292.

Financial Times. (2006). Architect cleared of Legionnaires' deaths. 1st August. Financial Times Group, London.

Fisman, D.N., Lim, S., Wellenius, G.A., Johnson, C., Britz, P., Gaskins, M., Maher, J., Mittleman, M.A., Spain, C.V., Haas, C.N., and Newbern, C. (2005). It's not the heat, it's the humidity: wet weather increases legionellosis risk in the greater Philadelphia metropolitan area. J. Infect. Dis. *192*, 2066–2073.

Garcia-Fulgueiras, A., Navarro, C., Fenoll, D., Garcia, J., Gonzalez-Diego, P., Jimenez-Bunuales, T., Rodriguez, M., Lopez, R., Pacheco, F., Ruiz, J., Segovia, M., Balandron, B., and Pelaz, C. (2003). Legionnaires' Disease Outbreak in Murcia, Spain. Emerg. Infect. Dis. *9*, 915–921.

Glick, T.H., Gregg, M.B, Bernman, B., Mallison, G., Rhodes Jr, W.W., and Kassanoff, I. (1978). Pontiac Fever. An Epidemic of Unknown Etiology in a Health Department: I. Clinical and Epidemiologic Aspects. Am. J. Epidemiol. *107*, 149–160.

Greig, J.E., Carnie, J.A., Tallis, G.F., Ryan, N.J., Tan, A.G., Gordon, I.R., Zwolak, B., Leydon, J.A., Guest, C.S., and Hart, W.G. (2004). An outbreak of Legionnaires' Disease at the Melbourne Aquarium, April 2000: Investigation and case–control studies. Med. J. Aust. *180*, 566–572.

Hahné, S., Watson, P., Bharat, P., Temple, M., Joseph, C., Harrison, T., Lee, J., Ribeiro, D., Smith, R., and Salmon, R. (2002). Outbreak of Legionnaires' Disease linked to a humidifier in a hotel in Wales, United Kingdom. In *Legionella*, R. Marre, Y.A. Kwaik, C. Bartlett, N.P. Cianciotto, B.S. Fields, M. Frosch, J. Hacker, and P.C. Lück, ed. (Washington, DC: ASM Press), pp. 346–348.

Harrison, T.G. (2005). Legionella. In: Topley & Wilson's microbiology and microbial infections, S.P. Borriello, P.M. Murray, G. Funke, ed. (London: Hodder Arnold), pp. 1761–1785.

Harrison, T.G., Doshi, N., Fry, N.K., and Joseph, C.A. (2007). Comparison of clinical and environmental isolates of *Legionella pneumophila* obtained in the UK over 19 years. Clin. Microbiol. Infect. Epub 14th August 2006. J. Clin. Microbiol. Infect. Dis. *13*, 78–85.

Health and Safety Commission (1992). The Notification of Cooling Towers and Evaporative Condensers Regulations 1992. Statutory Instrument, 1992 No 2225, Health and Safety. (London: HMSO).

Health and Safety Commission (2000). The Approved Code of Practice on the prevention and control of legionellosis (including Legionnaires' disease) (L8). (including technical guidance) Health and Safety. (London: HMSO).

Health Council of the Netherlands. (2003). Controlling Legionnaires' Disease. The Hague: Health Care Council of the Netherlands. Publication no. 2003/12.

Health Protection Agency. (2006). Management of Spa Pools: Controlling the Risk of Infection. (London: Health Protection Agency).

Herwaldt, L.A., Gorman, G.W., McGrath, T., Toma, S., Brake, B., Hightower, A.W., Jones, J.S., Reingold, A.L., Boxer, P.A., and Tang, P.W. (1984). A new Legionella species, *Legionella feeleii* species nova, causes Pontiac fever in an automobile plant. Ann. Intern. Med. *100*, 333–338.

Hlady, W.G., Mullen, R.C., Mintz, C.S., Shelton, B.G., Hopkins, R.S., and Daikos, G.L. (1993). Outbreak of Legionnaire's disease linked to a decorative fountain by molecular epidemiology. Am. J. Epidemiol. *138*, 555–562.

http://www.dh.gov.uk/PublicationsAndStatistics (accessed on 01.12.2006)

Hubbard, R.B., Mathur, R.M., and MacFarlane, J.T. (1993). Severe community-acquired *Legionella* pneumonia: treatment, complications and outcome. Q. J. Med. 86, 327–332.

Hubert, H., Infuso, A., and Ledrans, M. (1997). Guide d'investigation d'un ou plusieurs cas de légionellose. Circulaire DG no97/311. BEH nos. 20–22.

Huhn, G.D., Adam, B., Ruden, R., Hilliard, L., Kirkpatrick, P., Todd, J., Crafts, W., Passaro, D., and Dworkin, M.S. (2005). Outbreak of travel-related pontiac fever among hotel guests illustrating the need for better diagnostic tests. J. Travel. Med. *12*, 173–179.

Infuso, A., Hubert, B., and Etienne, J. (1998). Underreporting of legionnaires disease in France: the case for more active surveillance. Eur. Surveill. *3*, 48–50

Jones, T.F., Benson, R.F., Brown, E.W., Rowland, J.R., Crosier, S.C., and Schaffner, W. (2003). Epidemiologic investigation of a restaurant-associated outbreak of Pontiac fever. Clin. Infect. Dis. *37*, 1292–1297

Joseph, C.A. (2004) Legionnaires' disease in Europe 2000–2002. Epidemiol. Infect. *132*, 417–424.

Joseph, C.A., Harrison, T.G., Ilijic-Car, D., and Bartlett, C.L. (1999). Legionnaires' disease in residents of England and Wales: 1998. Commun. Dis. Public. Health. *2*, 280–284.

Joseph, C.A., and Van der Sande, M. (2006). Unexplained summer increase in non-travel related legionellosis in the UK and Netherlands. Eur. Surveill. Wkly 11(10).

Joseph, C.A., Watson, J.M., Harrison, T.G., and Bartlett, C.L.R. (1994). Nosocomial Legionnaires' Disease in England and Wales, 1980–92. Epidemiol. Infect. *112*, 329–345.

Kool, J.L., Fiore, A.E., Kioski, C.M., Brown, E.W., Benson, R.F., Pruckler, J.M., Glasby, C., Butler, J.C., Cage, G.D., Carpenter, J.C., Mandel, R.M., England, B., and Breiman, R.F. (1998). More than 10 years of unrecognized nosocomial transmission of legionnaires' disease among transplant patients. *19*, 898–904.

Last, J.M. (2000). A Dictionary of Epidemiology. (USA: Oxford University Press – Fourth Edition).

Laurance, J., (2006). Climate change blamed for Legionnaires' disease surge. The Independent. 23 November. The Independent News and Media, London.

Lever, F., and Joseph, C.A. (2003). Travel Associated Legionnaires' Disease in Europe in 2000 and 2001. Eur. Surveill. *8*, 65–72.

Macfarlane, J.T, Colville, A., Guion, A., Macfarlane, R.M., and Rose, D.H. (1993). Prospective study of aetiology and outcome of adult lower-respiratory tract infections in the community. Lancet. *341*, 511–514.

Maloney, F., Hoge, C.W., Farley, T.A., Barbaree, J.M., Breiman, R.F., Benson, R.F., and McFarland, L.M. (1992). Communitywide outbreak of Legionnaires' Disease associated with grocery store mist machine. J. Infect. Dis. *165*, 736–739.

Marston, B.J., Plouffe, J.F., File, T.M., Hackman, B.A., Salstrom, S.J., Lipman, H.B., Kolczak, M.S., and Breiman, R.F. (1997). Incidence of community-acquired pneumonia requiring hospitalization: Results of a population-based active surveillance study in Ohio. Arch. Intern. Med. *157*, 1709–1718.

Mastro, T.D., Fields, B.S., Breiman, R.F., Campbell, J., Plikaytis, B.D., and Spika, J.S. (1991). Nosocomial Legionnaires' disease and use of medication nebulizers. J. Infect. Dis. *163*, 667–671.

McDade, J.E., Brenner, D.J., and Bozeman, F.M. (1979). Legionnaires' disease bacterium isolated in 1947. Ann. Intern. Med. *90*, 659–661.

Ministère de l'Ecologie et du Développement Durable. (2004). Arrêté du 13 décembre 2004 relatif aux installations de refroidissement par dispersion d'eau dans un flux d'air soumises à autorisation au titre de la rubrique no. 2921. NOR: DEVP0430480A. Journal Officiel de la République Française 304 du 31 décembre 2004. Available at: http://www.admi.net/jo/20041231/DEVP0430480A.html. Accessed 22 July 2005.

Ministère de la Santé et des Solidarités. (2005). Le risque lié aux légionelles – guide d'investigation et d'aide à la gestion, 1er juillet 2005. Available at: http://www.sante.gouv.fr/htm/pointsur/legionellose/guide_legio_010705.pdf. Accessed 22 July 2005.

Montagna, M.T., Napoli, C., Tatò, D., Spilotros, G., and Barbuti, G. (2005). Results of an active surveillance for legionellosis in southern Italy. 20th Annual Meeting of the European Working Group for Legionella Infections, 16–17 May 2005, Rome, Italy.

Nardone. A., Decludt, B., Jarraud, S., Etienne, J., Hubert, B., Infuso. A., Gallay, A., and Desenclos, J.C. (2003). Repeat capture–recapture studies as part of the evaluation of the surveillance of Legionnaires' disease in France. Epidemiol. Infect. *13*, 647–654.

Nguyen, T.M., Ilef, D., Jarraud, S., Rouil, L., Campese, C., Che, D., Haeghebaert, S., Ganiayre, F., Marcel, F., Etienne, J., and Desenclos, J.C. (2006). A community-wide outbreak of Legionnaires disease linked to industrial cooling towers--how far can contaminated aerosols spread? J. Infect. Dis. *193*, 102–111

NHS Estates. (1993). Health Technical Memorandum 2040: the control of legionellae in healthcare premises – a code of practice. (London: HMSO).

Nygard, K., Outbreak investigation collaborators. (2005). Update: outbreak of legionnaires disease in Norway traced to air scrubber. Eur. Surveill. 10(6).

O'Mahony, M.C., Stanwell-Smith, R.E., Tillett, H.E., Harper, D., Hutchison, J.G., Farrell, I.D., Hutchinson, D.N., Lee, J.V., Dennis, P.J., and Duggal, H.V. *et al.* (1990). The Stafford outbreak of Legionnaires' disease. Epidemiol. Infect. *104*, 361–380.

Osterholm, M.T., Chin, T.D.Y., Osborne, D.O., Dull, H.B., Dean, A.G., Fraswer, D.W., Hayes, P.S., and Hall, W.N. (1983). A 1957 Outbreak of Legionnaires' Disease associated with a Meat Packing Plant. Am. J. Epidemiol. *117*, 60–67.

Reid, D. (1978). Benidorm Episode and Legionnaires' Disease (Editorial). Scott. Med. J. *23*, 118–119.

Ricketts, K.D., and Joseph, C.A. (2005). Legionnaires' disease in Europe 2003–2004. Eur. Surveill. *10*, 256–259.

Ricketts, K.D., McNaught, B., and Joseph, C.A. (2007). Travel-associated Legionnaires' Disease in Europe: 2005. Eur. Surveill. Epub.1(12) January 2007.

Sabria, M., and Campins, M. (2003). Legionnaires' disease: update on epidemiology and management options. Am. J. Respir. Med. 2, 235–243.

Sharrar, R.G. and Parkin, W.E. (1976). Respiratory Infection – Pennsylvania. MMWR. August 6, 244.

Smith, A.F., Wild, C., and Law, J. (2005). The Barrow-in-Furness legionnaires' outbreak: qualitative study of the hospital response and the role of the major incident plan. Emerg. Med. J. *22*, 251–255.

Telford, D., Partridge, S., Cumming, I., Smith, A., and Calvert, N. (2006). The legionnaires' outbreak in Barrow-in-Furness, summer 2002. J. Epidemiol. Community. Health. *60*, 464–466.

Terranova, W., Cohen, M.L., and Fraser, D.W. (1978). 1974 outbreak of Legionnaires' Disease diagnosed in 1977 – Clinical and epidemiological features. Lancet. *2*, 122–124.

Thacker, S.B., Bennett, J.V., Tsai, T.F., Fraser, D.W., McDade, J.E., Shepard, C.C., Williams, K.H. Jr, Stuart, W.H., Dull, H.B., and Eickhoff, T.C. (1978). An outbreak in 1965 of severe respiratory illness caused by the Legionnaires' disease bacterium. J. Infect. Dis. *138*, 512–519.

The Daily Telegraph. (2005). Legionnaire's disease pay-out. 22nd February. Telegraph Group Ltd, London.

The Guardian. (2001). Plastics firm on deaths charge. 11th August. Guardian Media Group, London.

The Guardian. (2006). Council boss fined over legionnaires' outbreak. 1st August. Guardian Media Group, London.

Westminster Action Committee. (1988). Broadcasting House Legionnaires' disease. (London: City of Westminster).

WHO (World Health Organization) (1982). Legionnaires' disease: Report on a WHO Working Group. Copenhagen, WHO Regional Office for Europe, (Euro Reports and Studies, No. 72).

WHO (World Health Organization) (1990). Epidemiology, prevention and control of legionellosis. Memorandum from a WHO meeting. Bulletin of the World Health Organization. *68*, 155–164).

WHO (World Health Organization) (1999). Surveillance Standards, 2nd edition. (WHO/CDS/CSR/ISR/99.2).

Yu, V.L. (2000) Nosocomial legionellosis. Curr. Opin. Infect. Dis. *13*, 385–388.

Developmental Cycle – Differentiation of *Legionella pneumophila*

4

Rafael A. Garduño*, Audrey Chong and Gary Faulkner

Abstract

The discovery that *L. pneumophila* follows a developmental cycle, within which it differentiates into numerous forms with highly altered phenotypes, emphasizes the usefulness of this bacterium as an experimental model for the study of bacterial intracellular differentiation. The developmental cycle of *Chlamydia* species (which are obligate intracellular pathogens) constitutes the current paradigm of bacterial intracellular differentiation. However, unlike *Chlamydia, L. pneumophila* can grow *in vitro* and is amenable to mutagenesis and broad genetic screenings, enhancing our investigative options. Both the extracellular and the intracellular developmental cycles of *L. pneumophila* have been described. The comparative analysis of the differentiation steps associated with these two developmental cycles suggests that the extracellular cycle is an incomplete derivative of the intracellular one. We propose that *L. pneumophila* possesses a single developmental programme controlled by known regulators of bacterial differentiation that respond to changes in the levels of amino acids in the environment, and to other as yet unidentified intracellular factors. However, these regulators seem to play unique roles in *L. pneumophila* that deviate from their known functionality in other bacteria.

Introduction

The natural history of *Legionella pneumophila* indicates that this Gram-negative bacterium has evolved as an intracellular parasite of some amoebal species (Fields 1996). However, *L. pneumophila* owes its rising interest by the scientific community (Steinert *et al.*, 2002) to two other important attributes. On the one hand, *L. pneumophila* has become a significant public health risk as an opportunistic human pathogen that causes outbreaks of Legionnaires' disease associated with high mortality. On the other hand, *L. pneumophila* has attracted the attention of researchers as this bacterium constitutes an excellent model of intracellular bacterial pathogenesis, which is particularly useful in understanding how free-living bacteria may evolve to become intracellular human pathogens.

The utilization of *L. pneumophila* as an experimental research model has been further advanced by the important discovery and subsequent investigative validation that this bacterium differentiates within a developmental cycle. Many investigators are responsible for this discovery, and some details are discussed in Chapter 6, but an exhaustive review of their contributions is not forthcoming in this chapter. Neither is the discussion covering all of the events leading to the establishment of the *L. pneumophila* developmental cycle. Instead, this chapter focuses upon the process of *L. pneumophila* differentiation itself, reviewing the various morphological and functional changes that clearly suggest differentiation, and emphasizing the molecular mechanisms that drive the differentiation of this intriguing intracellular bacterium.

Bacterial differentiation

The term differentiation can be simply defined as the 'development of separate distinguishing

*Corresponding author.

characteristics' (*Webster's Dictionary of the English Language*). When applied to living organisms, differentiation mainly refers to distinct morphological, physiological, or biochemical characteristics expressed by an organism at some particular stage of development. In this context, differentiation could be viewed as the driving mechanism of development. During the development of living organisms, cells displaying specific sets of distinguishing characteristics are produced, which we refer to as 'forms'. Cellular differentiation in multicellular organisms is often 'linear' and irreversible; the forms produced at the end of the process do not give rise to the undifferentiated cells that started the process. In contrast, unicellular organisms usually follow a 'circular' sequence of differentiation steps that may define a developmental cycle. The alternating nature of this developmental process implies that the closing step of any given cycle constitutes the opening step of the next cycle.

Bacteria differentiate, and many follow biphasic developmental cycles in which two forms of the organisms predominate, with one form giving rise to the other. One of the best-known bacterial differentiation processes is sporulation, common to some species of Gram-positive bacteria. A multiplying vegetative cell (the mother cell) differentiates into a resting-stage cell (the spore), which upon germination produces a vegetative cell. Although sporulation is clearly not the focus of this chapter, its study has significantly advanced knowledge on bacterial differentiation. It is thus relevant to emphasize here that the selective expression of a given subset of genes within the bacterial genome, i.e. the synthesis of stage-specific proteins controlled by temporally integrated regulatory networks, can lead to complex forms with markedly different phenotypes (Errington, 2003). Stage-specific proteins may have direct regulatory or structural roles, and (or) be involved in the synthesis of compounds necessary to carry out specific changes that will ultimately determine the differentiation stage of the organism.

Gram-negative bacteria, which are not known to sporulate, still may differentiate into cyst-like, non-replicative forms that effectively survive stress (e.g. desiccation, UV radiation, etc.) and spread into new environmental niches. The development of any specialized form that withstands specific physiological challenges (like starvation), while keeping the ability to quickly resume replication in favourable conditions, is an excellent example of bacterial differentiation. In this context, we discuss the bacterial changes experienced during the stationary phase of growth as a classic example of differentiation. We will focus on Gram-negative bacteria in general, and on *L. pneumophila* in particular.

Stationary phase and differentiation in Gram-negative bacteria

The growth curve of *in vitro*-culturable bacteria is obtained by plotting time versus logarithm of bacterial cell numbers. Two fundamental phases that are always present in a growth curve are: the exponential growth phase, and the post-exponential growth phase or stationary phase.

During exponential growth, bacterial cell numbers increase as a logarithmic function of time. In general, the exponential phase bacterial population is assumed to be relatively homogeneous. Bacteria are actively dividing and all their biosynthetic capacity is primarily dedicated to the production of structural molecules and DNA replication. As the nutrients required to maintain active metabolism are utilized, and their concentration is reduced, the dividing bacteria undergo a metabolic downshift known as the stringent response. This metabolic downshift is initiated by the activation of RelA and (or) other enzymes that synthesize ppGpp, a secondary messenger that activates the expression (among other global regulatory molecules) of the RNA polymerase alternate sigma factor σ^S or RpoS. In *Escherichia coli*, RpoS constitutes a key regulator that controls the expression of 40–50 stationary phase-specific genes (Loewen and Hengge-Aronis, 1994; Lacour and Landini, 2004).

In relation to replicating bacteria in exponential phase, the gene expression pattern of bacteria in stationary phase is radically different (reviewed by Ishihama, 1997, 2000). Importantly, in the stationary phase, active bacterial replication ceases as the bacterial cells prepare for starvation. Some of the morphological, physiological, and biochemical changes that Gram-negative bacteria undergo during stationary phase are quite striking, and remind us that entry into stationary phase is truly a differentiation process. For instance, there are changes in cell volume and shape, cytoplasmic density, and nucleoid structure. Not surprisingly, stationary phase forms accumulate energy- and carbon-storage molecules (e.g. polymers) in preparation for starvation. In addition, stationary phase forms are more resistant to physical and chemical stress than exponential growth phase forms. Enzymes like catalase, or alkyl hydroperoxidase are typically up regulated in stationary phase in a RpoS-dependent manner to protect against oxidative stress. In fact, RpoS was originally discovered as a key protein required for bacterial resistance to stress (Eisenstark *et al.*, 1996), and is currently known to regulate genes involved in resistance to pH changes, UV radiation, radicals, organic solvents, etc.

Finally, a sharp contrast exists between the rather homogeneous exponential growth phase bacteria, and the highly heterogeneous population that emerges once active bacterial replication stops (Finkel and Kolter, 1999). This heterogeneity is in part generated by an increased rate of mutation. The increased rate of mutation in stationary phase bacteria is perhaps a mechanism implemented to ensure the emergence of some beneficial genotypic changes (among numerous unfavourable ones) for improved adaptation to new environmental conditions (i.e. accelerated evolution). One molecular mechanism driving this adaptation-by-hypermutation process seems to be the overproduction of DNA-Pol IV (an error-prone DNA polymerase) during stationary phase (Layton and Foster 2003); but increased DNA recombination is also involved (Hall 1998; Bull *et al.*, 2001; Lombardo *et al.*, 2004). Although RpoS seems to be required for these hypermutation processes to work (including the production of mutagenic double DNA strand breaks), the complex regulatory networks involved have not been as yet unravelled (Lombardo *et al.*, 2004; Ponder *et al.*, 2005).

Stationary phase differentiation in obligate intracellular bacteria

Although obligate intracellular bacterial pathogens cannot grow *in vitro*, their growth inside cells can still be followed by plotting time versus logarithm of bacterial cell numbers. Similar to the curves obtained *in vitro*, the *in vivo* intracellular growth curve shows a standard exponential growth phase and a stationary phase, with the stationary phase forms exhibiting high levels of differentiation. These intracellular stationary phase forms are eventually released from wasted host cells, and their unique characteristics allow them to survive in the environment until they can find a new host cell to infect.

Developmental cycles of intracellular Gram-negative pathogens

The Gram-negative obligate intracellular parasites *Chlamydia* and *Coxiella* display well-defined developmental cycles (McCaul and Williams, 1981; Schachter, 1988). For these pathogens, it is pivotal to assure both the survival of their progeny in the extracellular environment, and the efficient infection of new host cells. Therefore, as intracellular replication ends, due to exhaustion of nutrients (or the accumulation of waste metabolites) in the host cell, these pathogens differentiate into an environmentally resilient and highly infectious form that ensures transmission of the pathogen and its ultimate survival.

Comprehensive reviews have been written describing the drastic changes involved in the bi-phasic developmental cycle of *Chlamydia* (Moulder, 1991; Hatch, 1999), and a recent book has been published on *Chlamydia* (Bavoil and Wyrick, 2006). The reader is referred to these writings to gain a thorough understanding of what is known about the process. Briefly, the extracellular phase of *Chlamydia* (named the elementary body) is resistant to environmental stress, has a very slow metabolism that does not support cell division, and is highly infectious. It is clear, therefore, that the elementary body is the transmissible form of *Chlamydia*. Upon infection of a mammalian host cell, the elementary body is induced to differentiate into the reticulate body. While vulnerable to environmental stress, the reticulate body can undergo active DNA replication and synthesis of molecular precursors required for bacterial cell division. The reticulate body is however unable to initiate infections in new mammalian hosts. The reticulate body is thus the replicative form of *Chlamydia*.

The biphasic developmental cycle of *Chlamydia* constitutes the current paradigm of developmental cycles in intracellular Gram-negative pathogens. *Coxiella* and other obligate intracellular bacterial pathogens follow cycles similar to that of *Chlamydia*. By analogy to this paradigm, the stationary phase form of *Coxiella* corresponds to the so-called small-cell variant (which is resistant to stress, infectious, small, dense, and unable to replicate), whereas the *Coxiella* large-cell variant corresponds to the replicative form (Coleman *et al.*, 2004). However, rigorous investigation of obligate intracellular pathogens has not come without major experimental difficulties. For instance, the inability to grow *Chlamydia* or *Coxiella* on agar plates poses challenges to the application of broad mutagenesis and genetic screenings. Alternate experimental approaches have recently received a boost with the release of the DNA sequences of the entire genomes of various *Chlamydia* species, and the identification of bacterial surrogates for the expression of *Chlamydia* genes (Bavoil and Wyrick, 2006). Nonetheless, the progress achieved in this field continues to be slower in relation to the study of *in vitro*-culturable pathogens.

L. pneumophila presents two differentiated stationary phase forms

L. pneumophila grows extracellularly *in vitro* and possibly (albeit not unequivocally demonstrated) in biofilms, but its best known natural behaviour is as an obligate intracellular pathogen of protozoa. *L. pneumophila* also behaves as an intracellular pathogen of human alveolar macrophages, and multiplies in several mammalian cell lines. Therefore, *L. pneumophila* presents two stationary phase forms: one generated extracellularly when *L. pneumophila* grows in broth cultures, agar plates or at the expense of other microorganisms in the lab (Tison *et al.*, 1980; Pope *et al.*, 1982; Wadosky and Yee, 1983), and a second one when it grows inside amoebae, macrophages or other cells in culture. The distinction between these two differentiated stationary phase forms is necessary because, as it will be discussed below, these forms are essentially different in morphology, physiology and biochemistry. Therefore, in this chapter, the intracellular stationary phase form will be referred to as the mature intracellular form or 'MIF', whereas the extracellular stationary phase form of *L. pneumophila* will be referred to as the 'SP form'.

L. pneumophila differentiation into stationary phase forms

The first evidence of *L. pneumophila* differentiation was noted with the discovery of *L. pneumophila* as the aetiological agent of Legionnaires' disease by McDade *et al.* (1977). Figure 1 of the McDade *et al.* article (showing the first micrograph of *L. pneumophila* ever published) prominently displays the ability of *L. pneumophila* to filament. Filamentation is a morphological change that some bacteria use to cope with stress, or to disseminate. The filamentation of *Proteus* and other Gram-negative bacteria, which is accompanied by profuse flagellation and a multicellular behaviour known as swarming, is perhaps one of the most drastic (and mesmerizing) differentiation processes of Gram-negative pathogens (Rauprich *et al.*, 1996), and has been implicated in the enhanced colonization of the urinary tract, as well as survival (Allison *et al.*, 1994). The differentiation of filamentous, hyperflagellated, non-dividing forms into short vegetative cells that actively multiply, completes a biphasic developmental cycle. The non-dividing, hyperflagellated, filamentous cell technically represents a stationary phase form, and the short, actively dividing cell represents an exponential phase form.

In *L. pneumophila* filamentation is also a stationary phase-related phenotype, because actively dividing legionellae in exponential phase are primarily non-filamentous. Early reports documented the filamentation of *L. pneumophila* in the presence of antibiotics (Smalley *et al.*, 1980; Elliot and Rodgers, 1985), and in a nutrient-limited minimal medium (Warren and Miller 1979). Recently, filamentation was found to significantly increase at 37°C and 42°C when *L. pneumophila* grows on solid surfaces or agar plates (Piao *et al.*, 2006). Although these observations suggest that filamentation may be a survival response of *L. pneumophila* to environmental stress, the molecular signals involved have not been investigated. However, Fettes *et al.* (2001) reported that overexpression of the global regulator CsrA (see below) induces filamentation in *L. pneumophila* late during growth *in vitro*. Based on the fact that filamentation is not always observed in stationary phase cultures of *L. pneumophila*, we assume that signals additional to those that normally trigger entrance into stationary phase are involved. Piao *et al.* (2006) proposed that filamentation is involved in biofilm formation. They also documented a significant difference between *L. pneumophila* and many other *Legionella* species in relation to formation of filamentous biofilms on surfaces, and argued that the increased ability of *L. pneumophila* to filament enhances its survival in temperature-altered waters. In support of this observation, we have also noted that all of our lab strains of *L. pneumophila* often filament when grown to stationary phase on agar plates, but less often when grown as planktonic cells in liquid media (R.A. Garduno, unpublished). Many questions remain unanswered regarding the relevance of filamentation in the biology of *L. pneumophila*, but it is hoped that the need to explain the environmental nature of Legionnaires' disease will promote more studies in this field.

Besides developing filaments, *L. pneumophila* SP forms exhibit many other unique traits. The first comprehensive report that proposed a model for the extracellular differentiation of *L. pneumophila* in stationary phase was that of Byrne and Swanson (1998). The focus of this study was on the cyclical (alternating) adaptation of *L. pneumophila* to conditions that promote bacterial replication and those that do not promote bacterial replication. It was found that *L. pneumophila* responds to low levels of amino acids (in spent culture medium) and enters stationary phase with the concomitant induction of several virulence-related phenotypes, i.e. sensitivity to salt, motility, cytotoxicity, ability to avoid phagosome–lysosome fusion, and consequently, ability to infect cultured macrophages. It seems noteworthy here that *L. pneumophila* heavily relies on amino acids for nutrition (Hoffman 1984), and, therefore, that it would effectively respond to amino acid shortages in the surrounding media. The induction of virulence-related traits in stationary phase led to the proposal that SP forms are the transmissible forms of *L. pneumophila*. Importantly, sensitivity to salt and motility that developed in *L. pneumophila* grown intracellularly (in the late stages of macrophage infection) suggested that the changes observed *in vitro* also occur *in vivo* (refer to next section).

Additional studies have focused on the regulatory networks governing the differentiation of *L. pneumophila* into SP forms. As for other Gram-negative bacteria, *L. pneumophila* appears to undergo a metabolic downshift in response to low concentrations of nutrients. This stringent response is controlled by RpoS (Hales and Shuman, 1999; Bachman and Swanson, 2001) and other previously identified bacterial regulators of stationary phase differentiation. However, the functionality of these regulators may at times appear to be unique in *L. pneumophila*, as discussed below. It is important to note here that the expression of virulence traits mentioned above (e.g. motility, cytotoxicity, stress resistance, sodium sensitivity, increased infectivity, and lysosome evasion) is repressed when nutrients are abundant and *L. pneumophila* is actively multiplying (Molofsky and Swanson, 2003). In fact, the exponential growth phase form of *L. pneumophila* has been confirmed to be non infectious to bone marrow-derived macrophages from A/J mice (Joshi *et al.*, 2001).

The molecular players known to coordinate the induction of virulence-transmission traits and entry into stationary phase in *L. pneumophila* are discussed next.

ppGpp

In the stringent response, *E. coli* inhibits the synthesis of nucleic acids and proteins, which leads to growth arrest (Chatterji and Ojha, 2001). The effector molecule that triggers this response is guanosine 3′,5′-bipyrophosphate or ppGpp, which is synthesized by the ribosome-associated enzyme RelA in response to uncharged transfer RNAs sensed during protein translation. Accumulation of ppGpp induces the expression of RpoS, which in turn activates transcription of stationary phase genes (see below). Analogous to this adaptive response in *E. coli*, expression of virulence traits by *L. pneumophila* is co-ordinated with entry into stationary phase and the accumulation of ppGpp (Hammer and Swanson 1999). Exponential phase *L. pneumophila* forms induced to express plasmid-borne recombinant RelA (from *E. coli*) differentiate into a virulent,

SP-like form (Hammer and Swanson 1999). A *relA L. pneumophila* mutant showed undetectable levels of ppGpp when grown *in vitro*, which were restored by genetic *in trans* complementation, confirming that RelA indeed is responsible for the increased levels of ppGpp during stationary phase in standard lab media. The *relA* mutant was impaired for pigment and flagella production (low transcription of *flaA*), further corroborating a role for RelA in expression of virulence traits by *L. pneumophila* during stationary phase (Zusman *et al.*, 2002). However, *relA* mutants remain salt sensitive, retain expression of pore-forming activity, and are able to infect *A. castellanii* and human macrophages (Zusman *et al.*, 2002; Abu-Zant *et al.*, 2006), suggesting that RelA-independent pathways are also involved in triggering the expression of some transmission traits in *L. pneumophila*.

SpoT is a second ppGpp synthase present in *L. pneumophila* whose function in differentiation has not been possible to address genetically due to the inability to delete its encoding gene from the *L. pneumophila* genome. However, a conditional *relA spoT* double mutant has been recently constructed and preliminary experimental evidence gathered that suggests an important role for SpoT in sensing the nutritional status of *L. pneumophila* in a manner that does not depend on amino acid levels (M. Swanson, 107th General Meeting of the American Society for Microbiology, Colloquium session 008, Toronto, 21–25 May 2007). Although, biochemical evidence is still lacking to unequivocally establish the roles of RelA and SpoT in the process of *L. pneumophila* differentiation, it seems clear that the synthesis and accumulation of ppGpp remains as a central event that triggers such process.

RpoS

The levels of the *L. pneumophila* sigma factor RpoS increase in stationary phase (Hales and Shuman, 1999), and seem to be induced by ppGpp (Bachman and Swanson 2001). However, it has been recently shown that RpoS remains active in a *relA* mutant (Abu Zant *et al.*, 2006), suggesting either that alternate ppGpp synthases, other than RelA, are active (e.g. SpoT), or that *rpoS* is also regulated in a stringent response-independent manner, or both. RpoS is needed for the induction of virulence traits and differentiation into SP forms, but as initially reported by Hales and Shuman (1999), and later confirmed by Bachman and Swanson (2001), RpoS is not required for enhanced resistance to stress. Furthermore, the signalling pathway regulated by RpoS in *L. pneumophila* appears to be functionally different from that regulated by the *E. coli* RpoS (Hales and Shuman 1999; Bachman and Swanson 2004). In addition, RpoS induces only a subset of the virulence traits (sodium sensitivity, motility, lysosome evasion) that characterize the SP form, suggesting that other factors cooperate with RpoS in the full differentiation of SP forms (Hales and Shuman, 1999; Bachman and Swanson, 2001). Therefore, it has been recognized that the differentiation of *L. pneumophila* into SP forms involves RpoS-dependent and RpoS-independent regulatory pathways and factors. Finally, although RpoS is required for intracellular growth in amoebae and mouse macrophages, it is not required for replication in human macrophages (Hales and Shuman, 1999; Bachman and Swanson, 2001; Abu-Zant *et al.*, 2006), indicating that differentiation into SP forms is selectively linked to the development of a subset of virulence traits.

In *E. coli*, RpoS is regulated at the transcriptional, translational and post-translational levels. In fact, the regulatory network that controls the expression and activity of RpoS has been perceived as 'one of the most complex regulatory systems in bacteria' (Venturi, 2003). However, some bacteria, e.g. *Pseudomonas* and *Vibrio* species, show predominance of transcriptional *rpoS* regulation, in which case, two-component regulatory systems homologous to the LetA/S system of *L. pneumophila* (see below), control *rpoS* expression. It is thus not surprising that the regulation of the *L. pneumophila rpoS* gene, and (or) the activity of its gene product RpoS, is emerging as a complex process.

Although RpoS protein levels peak during stationary phase in *L. pneumophila*, one study has shown that maximal levels of *rpoS* transcript seem to be produced during the exponential growth phase *in vitro* (Bachman and Swanson 2004). However, another study based on the use of a whole genome DNA microarray, found that *rpoS* transcript levels do not peak during exponential phase *in vitro*, and are upregulated during stationary phase *in vitro* and late during *A. castellanii* infection (Brüggemann *et al.*, 2006). In spite of this discrepancy in the transcriptional regulation of *rpoS* in *L. pneumophila* (which may be explained by the different strains and experimental methods used in the two studies), a role for *rpoS* in the replicative stage has been suggested. For instance, the presence of multiple *rpoS* copies inhibits expression of transmission traits during exponential phase (Bachman and Swanson, 2004).

As shown for *E. coli* and other Gram-negative bacteria, it seems reasonable to assume that the *L. pneumophila* RpoS controls the expression of many genes, but only a few have been identified so far. Among the *L. pneumophila* genes upregulated by RpoS are the motility-related genes *fliA* (encoding the flagellar sigma factor) and *flaA* (encoding flagellin), as well as *magA*, a gene encoding a putative reductase that may be protective against oxidative stress (Bachman and Swanson 2004; Hiltz *et al.*, 2004). Recently, it was reported that RpoS is also responsible for upregulating the expression of *plaC* (encoding a secreted acyltransferase), and for presumably increasing the levels of phospholipase A and lysophospholipase A (measured as increased enzymatic activity) (Broich *et al.*, 2006). Finally, the expression of the gene lpg0981 of unknown function, and at least 12 horizontally transferred genes encoding putative *L. pneumophila* virulence effectors with distinct eukaryotic motifs, was controlled by RpoS (de Felipe *et al.*, 2005). Interestingly, both RpoS and RelA play a minor role in the regulation of some *dot/icm* genes (Zusman *et al.*, 2002), which encode a type IV secretion system in *L. pneumophila* that is essential for virulence and responsible for the translocation of most virulence effectors.

LetA/S and LetE

The Legionella transmission activator and sensor (LetA/S) is a two-component regulatory system that indirectly induces the expression of virulence-transmission traits in SP forms, by relieving the repression of virulence transmission genes by CsrA (see below) (Hammer *et al.*, 2002; Molofsky and Swanson 2003). The LetA/S system is homologous to two-component regulatory systems previously described in *Pseudomonas aeruginosa*, *Salmonella* and *Vibrio*, but in *L. pneumophila* it is thought to respond to the accumulation of ppGpp under nutrient limiting conditions. However, other factors independent of the stringent response may also be involved in activating LetA/S because *letA* and *relA* mutants do not demonstrate entirely similar phenotypes (Hammer *et al.*, 2002; Zusman *et al.*, 2002).

One accessory factor functionally linked to the activity of RpoS, LetA/S and the differentiation into SP forms is LetE, a putative ~12-kDa activator protein. Although LetE is not essential for differentiation of SP forms, it enhances the expression of the virulence-transmission traits regulated by LetA/S. To this end, it has been proposed that LetE may directly or indirectly interact with the LetA/S system and CsrA (Hammer *et al.*, 2002; Bachman and Swanson 2004a).

L. pneumophila letA mutants grown to stationary phase *in vitro* retain the characteristics of exponential phase bacteria (Hammer *et al.*, 2002; Lynch *et al.*, 2003). Similarly, stationary phase *letE* mutants are less heat resistant, less infectious, less motile and less cytotoxic than its parental strain (Hammer *et al.*, 2002; Bachman and Swanson 2004a). However, both *letA* and *letE* are dispensable for intracellular growth in macrophages (Hammer *et al.*, 2002; Gal-Mor and Segal 2003), but *letA* is required for intracellular growth in *A. castellani* (Gal-Mor and Segal 2003; Lynch *et al.*, 2003). LetA is involved in the regulation of *icmP*, *rpoS*, *flaA*, *plaC* and *ralF* expression (Gal-Mor and Segal 2003; Lynch *et al.*, 2003; Brioch *et al.*, 2006). Reduced secretion of the macrophage infection potentiator (Mip) and of multiple lipolytic and hydrolytic activities of *L. pneumophila*, and increased secretion of the integral membrane protein DotA were also observed in *letA* mutants (Brioch *et al.*, 2006; Shi *et al.*, 2006). The lipolytic and hydrolytic activities regulated by LetA are similarly affected by RpoS (see above), supporting a link between LetA and RpoS in the regulatory cascade that controls differentiation of *L. pneumophila* into SP forms.

CsrA

The carbon storage regulator, CsrA, of *L. pneumophila* is expressed during exponential growth phase and inhibits the expression of transmission traits such as flagellation, heat and osmotic resistance, cytotoxicity, and efficient macrophage infection (Fettes *et al.*, 2001; Molofsky and Swanson 2003; Forsbach-Birk *et al.*, 2004). In addition, the expression of three positive regulators of virulence and transmission traits (*fliA*, *rpoS* and *letE*) is repressed by CsrA during the exponential phase of the growth curve (Forsbach-Birk *et al.*, 2004).

In *E. coli*, CsrA acts as a homodimeric RNA-binding protein that prevents the translation of target mRNAs (Mercante *et al.*, 2006). Untranslated small RNA molecules encoded by *csrB* and *csrC* regulate the activity of CsrA (likely by sequestration) preventing CsrA from binding to target mRNAs. In *Salmonella*, the two component regulatory system SirA/BarA, which is a homologueof the *L. pneumophila* LetA/S system (see above), controls the expression of *csrB* and *csrC* (Fortune *et al.*, 2006). Two CsrA-regulating small RNAs have been identified in *L. pneumophila* through bioinformatics (Kulkarni *et al.*, 2006); thus, it seems reasonable to surmise (by analogy to *Salmonella*) that the LetA/S system relieves CsrA repression via induction of small RNAs expression. However, this hypothesis awaits experimental confirmation.

Genetic manipulation leading to reduced transcription of *csrA* results in the premature expression of transmission phenotypes in replicative *L. pneumophila*, e.g. an increase in the levels of the sigma factor FliA, flagellin, RpoS and LetE (Forsbach-Birk *et al.*, 2004), whereas mutants constitutively overexpressing *csrA* do not express transmission phenotypes (Molofsky and Swanson, 2003). The inability of *letA* mutants to induce virulence-transmission traits (e.g. motility, cytotoxicity, sodium sensitivity and stress resistance) is a phenotypic pattern shared by *L. pneumophila* that constitutively expresses *csrA*. However, this altered phenotype can be overridden by the loss of *csrA* expression (Molofsky and Swanson 2003). Collectively, these observations confirm that CsrA-mediated repression is relieved by the activation of the two component regulatory system LetA/S. CsrA is required for intracellular growth in *A. castellanii* and mouse macrophages, i.e. in the absence of *csrA* expression, *L. pneumophila* (although expressing transmission traits) fails to replicate in mouse macrophages. Therefore, it seems that CsrA repression must be restored during the differentiation of *L. pneumophila* into replicative forms to promote intracellular replication (Molofsky and Swanson, 2003; Forsbach-Birk *et al.*, 2004).

FliA

The flagellar sigma factor FliA of *L. pneumophila* is required not only for the synthesis of flagella and motility (Heuner *et al.*, 2002; Brüggemann *et al.*, 2006), but also for lysosome evasion and cytotoxicity (Hammer *et al.*, 2002; Molofsky and Swanson 2003; Molofsky *et al.*, 2005). Furthermore, *fliA* mutants are inefficient at both replicating within *Dictyostelium discoideum* and *Acanthamoeba castellanii* (Heuner *et al.*, 2002; Brüggemann *et al.*, 2006), and infecting mouse bone marrow-derived macrophages (Hammer *et al.*, 2002). Because motility itself is not required for virulence, it has been suggested that the flagellar sigma factor FliA of *L. pneumophila* may activate promoters of virulence genes in addition to the promoters of the flagellar regulon. For instance, two homologues of the *L. pneumophila* enhanced entry protein, EnhA, were recently identified as FliA-regulated genes in an *in vivo* transcriptional study. However, few other gene expression changes have also been attributed to FliA (Brüggemann *et al.*, 2006). Please refer to Chapter 6 for a more detailed discussion on FliA.

Hfq

The *L. pneumophila hfq* gene (host factor 1, Qβ phage replication) is strongly transcribed during exponential phase, but not during stationary phase (McNealy *et al.*, 2005). Hfq is an RNA chaperone protein that functionally assists one of the major classes of small regulatory RNAs. DsrA is a member of this Hfq-assisted small RNA class, known to positively regulate the translation of *rpoS* transcripts (reviewed by Majdalani *et al.*, 2005). The small regulatory RNAs

CsrB and CsrC belong to a different class of small RNAs that do not require Hfq and directly bind CsrA.

Unlike *E. coli*'s and *Salmonella*'s Hfq, the *L. pneumophila* Hfq homologuedoes not affect the translation efficiency of RpoS (McNealy *et al.*, 2005). Instead, Hfq expression appears to be positively regulated by RpoS in exponential phase, and negatively regulated by LetA in stationary phase. In addition, unlike other systems (Majdalani *et al.*, 2005), the *L. pneumophila* Hfq was found to interact with exponential phase regulators, e.g. *csrA* and *fur*. The *in vitro* growth and pigmentation defects demonstrated by an *hfq* mutant (with reduced levels of *csrA* mRNA during exponential phase growth, but no difference in CsrA protein levels in comparison to its parental strain) were restored by overexpression of *csrA*. The *hfq* mutant, showing reduced *fur* transcripts, grew poorly in medium lacking an iron source in comparison to its parental strain. Thus, Hfq may contribute to the exponential phase regulatory network in *L. pneumophila* by stabilizing exponential phase mRNAs. However, deletion of *hfq* only slightly affected the ability of *L. pneumophila* to invade and replicate in both a mouse alveolar macrophage cell line, and the amoeba *A. castellani*.

L. pneumophila differentiation into MIFs

Rowbotham (1983) first recognized, at the light microscope level, that the intracellular growth of *L. pneumophila* in amoebae produces a differentiated MIF (also reviewed by Rowbotham 1986). It was observed that short, highly motile legionellae emerged from wasted amoebae following the cessation of bacterial replication. The highly motile cells would rapidly infect fresh amoebal hosts to reinitiate bacterial multiplication, in turn followed by another round of differentiation into MIFs. Interestingly, free MIFs did not attach to dying amoebae or already infected ones, suggesting that MIFs possess a mechanism to recognize suitable hosts.

An extensive body of literature has now accumulated that clearly confirms that MIFs are highly differentiated forms of *L. pneumophila*. In particular, the MIFs that emerge from amoebae show: (i) a unique morphology defined by a small size (Rowbotham 1986; Cirillo *et al.*, 1994; Barker *et al.*, 1995), a positive Giménez staining, prominent inclusions, and a thicker envelope with condensed cytoplasm (Cirillo *et al.*, 1994); (ii) a unique physiology defined by resistance to biocides (Barker *et al.*, 1992) and antibiotics (Barker *et al.*, 1995), increased environmental fitness (Abu Kwaik *et al.*, 1997), and increased infectivity to both host cells in culture (Cirillo *et al.*, 1994, 1999) and mice (Brieland *et al.*, 1997; Cirillo *et al.*, 1999); and (iii) a unique biochemistry defined by a specific fatty acid profile, a possibly specific LPS, and a specific protein profile (Barker *et al.*, 1993; Cirillo *et al.*, 1994).

We have described and characterized the MIFs produced by *L. pneumophila* in human HeLa cells. Remarkably, these MIFs display many of the aforementioned characteristics of MIFs produced in amoebae. That is, MIFs from HeLa cells also show: (i) a unique morphology defined by a small size, a positive Giménez staining, prominent inclusions, a thick cell envelope, a dark condensed cytoplasm, and no ultrastructural signs of cell division (Garduno *et al.*, 1998b, 2002b; Faulkner and Garduno 2002); (ii) a unique physiology defined by resistance to detergents and antibiotics, increased environmental fitness, and increased infectivity to host cells in culture (Garduno *et al.*, 2002a and 2002b); and (iii) a unique biochemistry defined by a specific protein profile, a low basal respiration rate, and an inability to increase oxygen consumption in response to utilizable substrates (Garduno, 2002b).

Only a few stage-specific markers of MIF formation have been identified. MagA, a protein regulated by RpoS (see above), is encoded by a gene exclusively carried in Philadelphia-1 strains (Brassinga *et al.*, 2003; Hiltz *et al.*, 2004). MagA is poorly expressed in exponential phase forms of *L. pneumophila*, but highly expressed in MIFs. In fact, analysis of whole cell lysates obtained from MIFs grown in HeLa cells indicates that MagA is the most abundant protein in *L. pneumophila* Philadelphia-1 MIFs (Garduno, 2002b; Hiltz *et al.*, 2004). However, the function of MagA has not been determined. Its obvious absence in non Philadelphia-1 strains (which still differentiate and form infectious MIFs), as well as the lack of an obvious phenotype in a *magA*Δ mutant (Hiltz *et al.*, 2004), indicates that MagA is a dispensable protein. Yet, the question remains as to why MIFs of Philadelphia-1 strains would produce such abundant amounts of a dispensable protein? Other MIF-specific proteins have been identified, but not characterized, e.g. the hypothetical protein Lpg2755 and the σ^{54} modulation protein (Michael Morash PhD thesis, Dalhousie University). Finally, we also know that RpoS and LetA are essential regulators of MIF differentiation. That is, *rpoS*Δ and *letA*Δ mutants neither develop the morphology, nor the physiological traits that characterize MIFs (R.A. Garduno, unpublished). In fact, the *letA*Δ mutant replicates well in HeLa cells, but after replication it remains in ribosome-studded vacuoles within apparently healthy HeLa cells, with no change in morphological features. In addition, this mutant keeps the typical ultrastructure of the Gram-negative envelope (defined by the presence of a cytoplasmic membrane, a clear periplasmic space and an outer membrane) that characterizes the replicative, exponential growth phase form of *L. pneumophila*.

SP forms and MIFs are not equivalent

Having two stationary phase forms in *L. pneumophila* (the extracellular and the intracellular forms) logically invites the question: are these forms similar or different in relation to their morphology, physiology and biochemistry? To answer this question we conducted a direct experimental comparison between two pairs of SP forms and MIFs; one pair derived from a Philadelphia-1 strain, and the other from an Olda strain (not expressing MagA) (Garduno *et al.*, 2002b). We determined that, in relation to MIFs produced in HeLa cells, SP forms expressed a different proteome, showed a remarkably different morphology (at the light and the electron microscope levels), were less infectious to mammalian cells in culture, less resistant to detergents or antibiotics, and had a higher respiration rate. A retrospective comparison of published data also suggested that SP forms were different from MIFs produced in amoebae (e.g. Cirillo *et al.*, 1994; Byrne and Swanson, 1998, versus Barker *et al.*, 1992, 1995). Finally, the levels of MagA in SP forms of the Philadelphia-1 strain were lower than the levels found in MIFs, but higher than the levels found in exponential phase forms (Hiltz *et al.*, 2004). At the time, we proposed that SP forms were immature MIFs. We reasoned that the differentiation process initiated *in vitro* by *L. pneumophila* in response to low concentrations of amino acids (Byrne and Swanson 1998) could not be completed, due to the lack of either a particular nutrient, or additional signals only found in the intracellular environment. However, an alternate explanation was that *L. pneumophila* possessed two developmental programmes, one that controlled the differentiation of exponential growth phase forms into SP forms, and a second one that controlled the differentiation of replicative forms into MIFs (Fig. 4.1). We have now accumulated convincing evidence, both anecdotal and experimental (unpublished data), to support the notion that *L. pneumophila* has a single developmental programme, and consequently, that SP forms are intermediates in the differentiation of actively dividing *L. pneumophila* into MIFs (Fig. 4.1A). A number of morphological traits (e.g. production of flagella, cytoplasmic inclusions, and long invaginations of the cytoplasmic membrane), and physiological traits (e.g. thermal tolerance, resistance to osmotic shock and proteases, and susceptibility to desiccation and sodium) develop in both SP forms and MIFs (Byrne and Swanson, 1998; Garduno *et al.*, 2002b). Moreover, as mentioned before, *rpoS*Δ and *letA*Δ mutants, which are defective in developing SP forms *in vitro*, are also unable to develop MIFs in HeLa cells. Finally, SP forms transferred to distilled-deionized water experience a number of changes that make them more similar to MIFs, including the transition from a dull to a bright red colour after Giménez staining, an increase in the number of large cytoplasmic inclusions, and a more apparent invagination of the cytoplasmic membrane (R.A. Garduno, unpublished). However, the strongest evidence to support the notion that SP forms are intermediates in the development of MIFs comes from recent experiments in the ciliate *Tetrahymena* as explained below.

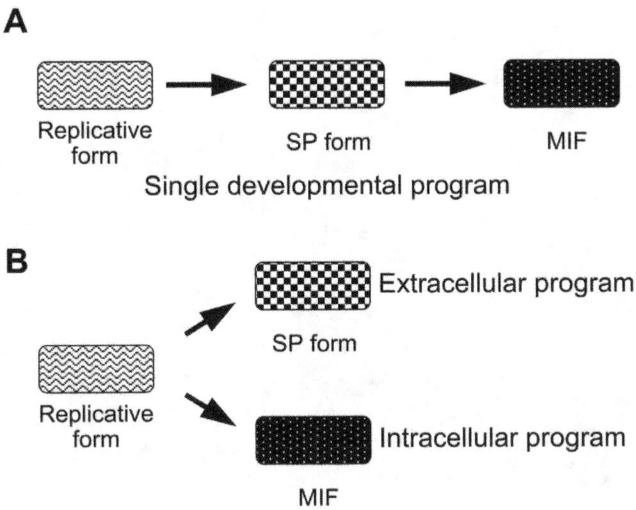

Figure 4.1 Diagram illustrating the two possible relationships between SP forms and MIFs. (A) *L. pneumophila* has a single developmental programme so that SP forms are intermediates in the differentiation of replicative forms into MIFs. (B) SP forms and MIFs independently arise from replicative forms, implying that there are two developmental programmes in *L. pneumophila*: an extracellular programme and an intracellular programme. Experimental evidence indicates that *L. pneumophila* possesses a single developmental programme, as discussed in this chapter.

The ciliate model of *L. pneumophila* differentiation

An as yet unidentified species of the genus *Tetrahymena* that groups in the *T. tropicalis* clade is able to ingest large numbers of *L. pneumophila* cells in suspension and produce numerous free spherical 'pellets' containing up to a hundred live legionellae (McNealy *et al.*, 2002; Garduno *et al.*, 2006). This legionellae-packaging process occurs in the absence of intracellular bacterial multiplication, as determined by viable bacterial cell counts and the lack of ultrastructural evidence of cell division by electron microscopy (Garduno *et al.*, 2006, R.A. Garduno, unpublished). Pellets begin to appear free in the surrounding medium as early as 1 h after feeding the ciliates with a suspension of *L. pneumophila*, suggesting a rapid turnover of the food vacuole contents. The material that holds the pellet together is believed to be composed of both ciliate-derived materials secreted into the food vacuole, and bacterial-derived outer membranes (perhaps the remains of a few digested legionellae).

The most relevant observation to the subject matter of this chapter is that the legionellae contained in the pellets show all the ultrastructural features of MIFs produced in either HeLa cells or amoebae (Fig. 4.2), in addition to the development of physiological traits typical of MIFs that include resistance to detergents and antibiotics and enhanced infectivity to cells in culture (Garduno *et al.*, 2006; R.A. Garduno, unpublished). Feeding *Tetrahymena* sp. with SP forms grown *in vitro* yields numerous pellets of differentiated MIFs, and it should be remembered that this differentiation of SP forms into MIFs happens in the absence of bacterial intracellular replication. That is, it constitutes a direct and rapid process.

What are the signals involved in the differentiation of stationary phase forms into MIFs?

When SP forms are left in their spent culture medium, additional morphological changes are not observed, but these forms rapidly lose viability (or culturability). However, as mentioned above, when SP forms are washed to remove the spent medium and then resuspended in deionized water, they experience a defined morphological change to produce a MIF-like form, and most importantly, they remain viable (or culturable) for weeks. The question thus arises as to why SP forms do not differentiate into MIFs (not

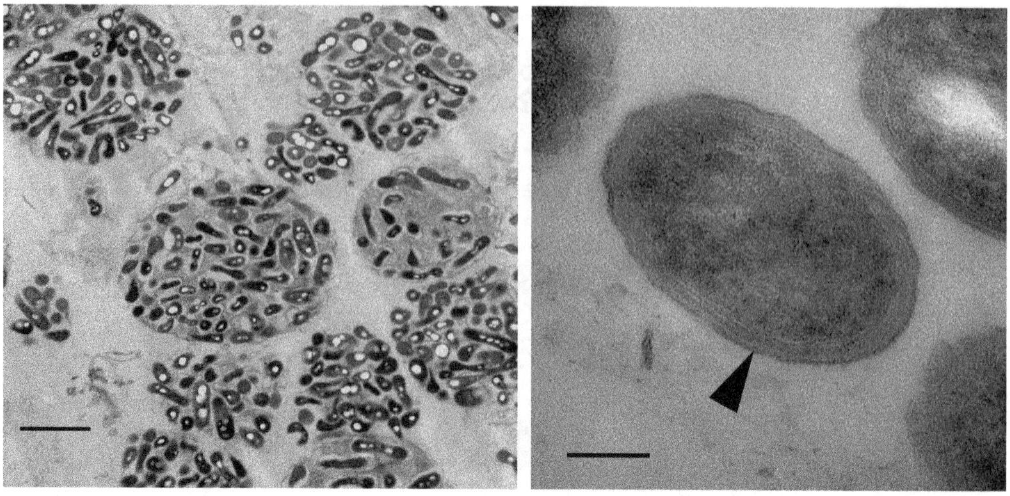

Figure 4.2 Electron micrographs of ultrathin-sectioned pellets expelled by *Tetrahymena* after feeding on *L. pneumophila* strain Lp02. (A) Pellets contain live differentiated *L. pneumophila* showing the morphological characteristics of MIFs. Notice the presence of cytoplasmic inclusions (clear bubble-like zones), dark cytoplasm, and the irregular shape of some bacteria. (B) Detail of a bacterial cell within an expelled pellet showing a thick envelope characterized by a darkened and very straight outer membrane (arrowhead), as well as the stacking of internal membranes under it.

even into MIF-like forms) in spent medium? It seems clear that additional signals or nutrients are required.

The stringent response linked to RelA by genetic evidence, as well as the response to RelA-independent pathways *in vitro* (see above), alone, do not seem to provide the regulation and (or) signals to allow SP forms to fully differentiate. It seems reasonable to propose that a metabolite, either exclusively found in the cytoplasm of eukaryotic cells or produced by *L. pneumophila* in response to a specific change in the host cell environment, may provide a much needed differentiation signal. It is also possible that metabolic signals additional to those provided by low amino acid concentrations *in vitro* are in fact not required, but perhaps a nutrient exclusively found in eukaryotic cells is needed for the synthesis of a structural component required by *L. pneumophila* to proceed to its full differentiation into MIFs.

Of special interest is the fact that sodium-tolerant mutants of *L. pneumophila* (likely, but not necessarily, carrying a defective Dot/Icm virulence system) (Sadosky *et al.*, 1993; Vogel *et al.*, 1996) do not become more similar to MIFs when placed in deionized water, but rather undergo filamentation (R.A. Garduno, unpublished). It is possible that a functional link exists between the Dot/Icm system and differentiation, e.g. the Dot/Icm system may be needed to correctly 'interpret' differentiation signals, and (or) differentiation-associated events may be required for the functionality of the Dot/Icm system or its secreted effectors. Such a link has been experimentally suggested by the following reported observations:

1 A *rpoS*Δ mutant does not develop sensitivity to sodium, and is unable to re-direct the intracellular trafficking of the legionellae-containing vacuole (Bachman and Swanson 2001), two defects associated with a non-functional Dot/Icm system.
2 As previously noted, RpoS seems to play a minor role in the regulation of some *dot/icm* genes (Zusman *et al.*, 2002).
3 A number of substrates of the Dot/Icm system are regulated by RpoS (de Felipe *et al.*, 2005) and, therefore, their expression is linked to *L. pneumophila* differentiation via activation of RpoS.
4 Roy *et al.* (1998) demonstrated, by placing the *dotA* gene under the control of an inducible promoter, that DotA (a cytoplasmic membrane protein) was required for invasion and intracellular establishment, but was dispensable during intracellular

growth. When *dotA* expression was 'turned off' towards the end of the intracellular growth cycle (when differentiation would occur), the resulting progeny (MIFs) were avirulent upon re-infection, elegantly suggesting that differentiated bacteria emerging from a wasted host cell must have a functional Dot/Icm system to assure infection of new hosts.

5 The proteins DotO and DotH appear to transiently form a 'fibrous' structure on the surface of *L. pneumophila* that coincides with the timing for intracellular differentiation into MIFs, or with the conditions leading to extracellular differentiation into SP forms (Watarai *et al.*, 2001).

Further support to the notion that *L. pneumophila* responds to a water-generated signal and that type IV secretion might be involved in differentiation, comes from a recent report by Bandyopadhyay *et al.* (2007), in which *dot/icm* mutants became infectious to amoebae and macrophages in culture after resuspension in distilled-deionized water for 19 h (what they called 'water stress'). Moreover, this response required a functional Lvh system. That is, double *dot/icm*Δ *lvh*Δ mutants were unable to increase their infectivity towards amoebae and macrophages in response to water stress.

Differentiation of *L. pneumophila* in water

Early studies have shown that *L. pneumophila* is capable of surviving for up to 14 months in water with only a modest loss in culturability in the first few months (Skaliy and McEachern, 1979; Schofield, 1985; Lee and West, 1991). This unusual tenacity at surviving under extreme starvation implies that *L. pneumophila* adopts a very robust physiology. In fact, this long-term survival in water may be associated with the formation of viable but non-culturable (VBNC) forms (Paszko-Kolva *et al.*, 1992; Steinert *et al.*, 1997), which constitute a differentiated form of bacteria.

Here we consider the VBNC form of *L. pneumophila* as distinct from the SP form and the MIF, simply because these two latter forms are perfectly capable of forming colonies on agar plates. When MIFs produced in HeLa cells are washed and resuspended in deionized water, they undergo further morphological changes that include a dramatic thinning of the bacterial cytoplasm, and the disappearance of cytoplasmic inclusions (Fig. 4.3). Whether these morphological changes are associated with the formation of VBNC forms is not known at this time, but the ability of the severely starved MIFs to form colonies on agar plates decreases with time (R.A. Garduno, unpublished). Very little is known about the mechanism by which *L. pneumophila* develops the VBNC form. Also no information is available on the signals that trigger entrance into the VBNC state, except that some chemical biocides (e.g. monochloramine) are particularly potent inducers of the formation of VBNC forms (Alleron *et al.*, 2006). Although we could speculate here that the MIF-like form produced by SP forms resuspended in deionized water (see above) may be related to the differentiation process that results in the development of VBNC forms, and that *L. pneumophila* responds to a water-generated signal, experimental proof for this relationship is still lacking.

Differentiation of SP forms, MIFs or VBNC forms into replicative forms

We would like to close this chapter by discussing the differentiation of SP forms, MIFs and VBNC forms into replicative forms.

The SP form resumes growth in fresh media and adapts to this new environment by differentiating into an exponential growth phase form that actively replicates, thus closing the differentiation

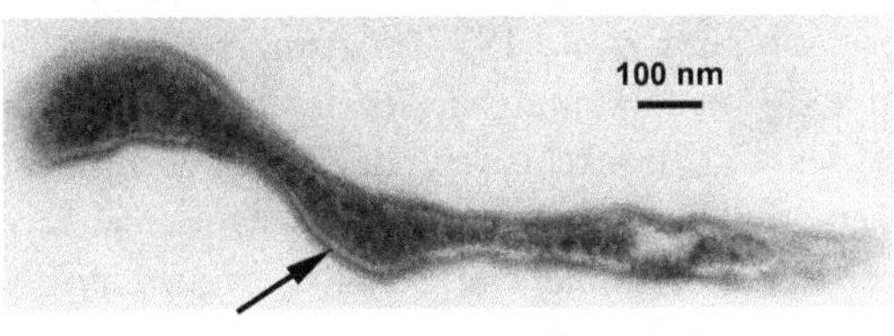

Figure 4.3 Electron micrograph of an ultrathin-sectioned MIF obtained from HeLa cells and placed in deionized water for 42 days. Note how thin the bacterial body is, as well as the lack of cytoplasmic inclusion bodies. The bacterial cell envelope appears intact and displays a characteristic thick layer associated with the inner leaflet of the outer membrane (arrow). The cytoplasm seems to be rich in ribosomes.

cycle *in vitro*. Similarly, when MIFs are placed in a nutrient-rich liquid medium that supports the growth of *L. pneumophila in vitro*, they begin to grow after an extended lag phase. The morphological characteristics of these lag phase forms include the formation of periplasmic vesicles and a change to a Giménez-negative staining (Faulkner and Garduno, 2002; Garduno *et al.*, 2002a). The cell envelope of MIFs is also modified, and the periplasmic space is significantly expanded (Faulkner and Garduno, 2002).

One signal to which MIFs and SP forms respond has been recently elucidated and, not surprisingly, involves the sensing of amino acids. In an elegant study by Sauer *et al.* (2005), it was demonstrated that a phagosomal amino acid transporter encoded by the *phtA* gene is essential for the differentiation of SP forms into the replicative form. In fact, a *phtA* transposon mutant was unable to initiate intracellular growth, in spite of remaining viable inside a vacuole, a defect that could be overcome by either genetic complementation with wild type *phtA* or by supplementing the medium with threonine. Furthermore, physiological characterization indicated that the *phtA* mutant's inability to differentiate into a replicative form could be suppressed by the addition of threonine. Therefore it was concluded that *L. pneumophila* gauges the amount of available amino acids before committing to differentiate into the replicative form (Sauer *et al.*, 2005).

The unsuccessful resuscitation of VBNC forms of *L. pneumophila* in nutrient-rich media suggests that these forms do not respond to amino acid levels alone. The only method known to resuscitate VBNC forms from their dormant state is the mixing of VBNC forms with amoebae (Steinert *et al.*, 1997). Again, it seems that in addition to amino acid levels, *L. pneumophila* senses signals exclusively delivered within the intracellular environment of eukaryotic cells, or perhaps *L. pneumophila* needs to acquire an intracellular eukaryotic metabolite to proceed with its differentiation programme.

Concluding remarks

L. pneumophila has a single developmental programme controlled by RpoS, Let A/S, CsrA and other regulators, previously shown in other bacteria to be involved in stationary phase-related differentiation. However, the global regulators of *L. pneumophila* appear to deviate from their known roles established in other bacteria. For instance, the *L. pneumophila* RpoS is not required to regulate a physiological response to environmental stress.

Although it is clear that *L. pneumophila* differentiates in response to changing levels of amino acids, it is also clear that other environmental signals or factors associated with the intracellular environment of eukaryotic cells are required to allow the complete differentiation of *L. pneumophila*. Since differentiation of *L. pneumophila* into SP forms and MIFs is associated with the expression of virulence determinants, further studies to elucidate both the signals and molecular players involved in this process, and the changes required to develop the full-virulence phenotype, are warranted. Finally, because we could argue that MIFs are the naturally produced transmissive forms that predominate in the water environment (Fig. 4.4), and that MIFs constitute the most virulent form of *L. pneumophila*, understanding the process of MIF development could perhaps lead to the implementation of practical measures to prevent the transmission of Legionnaires' disease.

From a basic research point of view, *L. pneumophila* has the potential to become a novel paradigm of bacterial intracellular differentiation. As mentioned earlier, progress in this field promises to be faster than the progress attained in the study of obligate intracellular pathogens. But to further facilitate such progress, it will be necessary to identify stage-specific proteins that can be used as markers of *L. pneumophila* differentiation. We predict that this will be an area of intense research with the recent implementation of the *L. pneumophila* whole genome DNA microarray. Also, a robust collection of differentiation-defective mutants should be built to facilitate the study of individual genes and their roles in the differentiation process. However, a major limitation for the study of *L. pneumophila* differentiation *in vitro* stems from the fact that this is an incomplete process. Differentiation into MIFs still depends on the intracellular environment, and to date we have not been able to produce *bona fide* MIFs *in vitro*. Finally, the utilization of *L. pneumophila* as an experimental

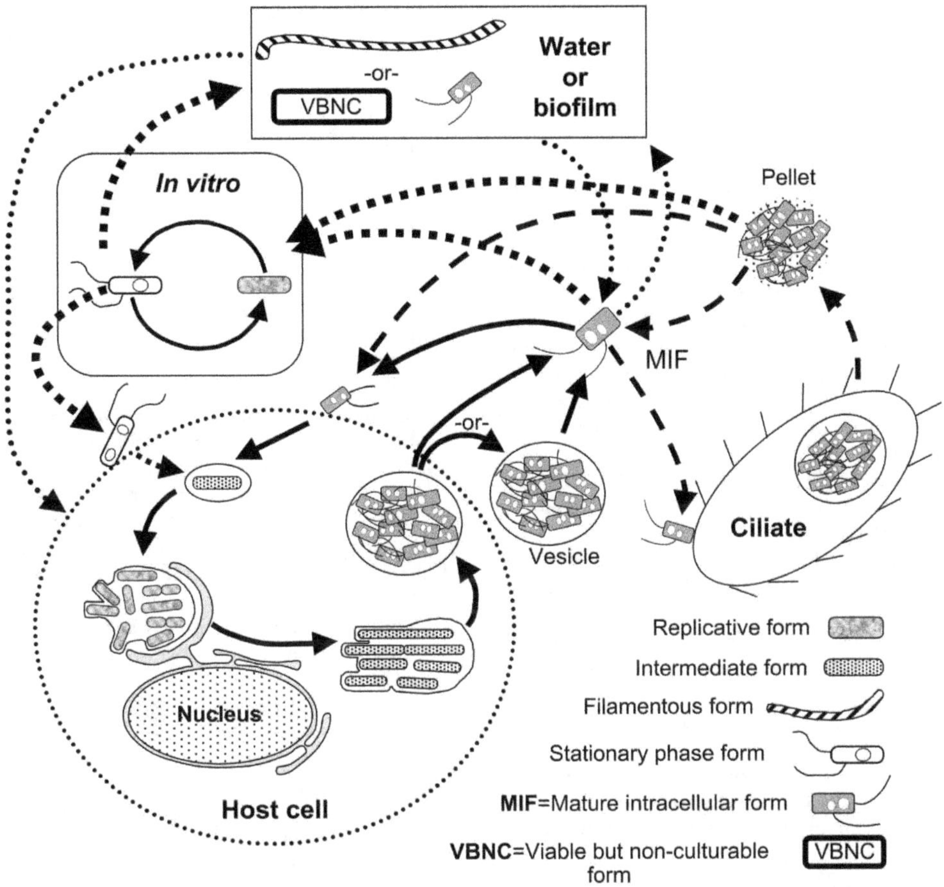

Figure 4.4 Diagrammatic representation of the differentiation network of *L. pneumophila* showing the interconnections between its various morphological forms. The mature intracellular form (MIF), which is the most infectious and environmentally fit form of *L. pneumophila* occupies a central position in this network. Naturally, MIFs are primarily produced as a result of the intracellular growth of *L. pneumophila* in amoebae and are released from amoebae as either vesicles or free MIFs. Free MIFs readily initiate a new round of infection when fresh uninfected amoebae are available, closing the 'host cell' loop indicated with solid line arrows. In this loop of the differentiation network, MIFs alternate with replicative forms passing through one or more morphological intermediates as they differentiate from one form to the other. Free MIFs released into water may become part of a biofilm, thus entering the 'water or biofilm' loop of the differentiation network (indicated by the circular-dot line arrows), or get ingested by ciliates, which act as packagers of *L. pneumophila* into pellets (ciliate loop indicated by the broken line arrows). Ciliate-produced pellets are infectious and able to initiate intracellular infections, as well as grow *in vitro*. When placed in a suitable artificial growth medium, free or pelleted MIFs enter the '*in vitro*' loop (indicated by the squared-dot line arrows) of the differentiation network, where *L. pneumophila* alternates between a replicative form (exponential growth phase form) and the stationary phase form. Stationary phase forms may undergo filamentation or initiate infections, and when placed in water they may form biofilms. In water, either as free bacteria or within biofilms, *L. pneumophila* may differentiate into a filamentous form, or a viable but not culturable form, which may in turn, efficiently persist in the environment or initiate intracellular infections in amoebae (re-entering the host cell loop). Therefore, at least six major differentiated forms (listed in the bottom right corner of the diagram) play important roles in the complex life cycle of *L. pneumophila*.

surrogate for the study of Gram-negative obligate intracellular pathogens like *Coxiella*, remains as a very attractive approach (Vogel, 2004). The fact that *L. pneumophila* grows *in vitro*, and only requires level-2 biohazard containment (in contrast to *Coxiella* which cannot grow *in vitro* and requires level-3 containment), as well as the phylogenetic relatedness between *Coxiella* and *Legionella* (Samuel *et al.*, 2003), constitute significant experimental advantages.

References

Abu Kwaik, Y., Gao, L.-Y., Harb, O.S., and Stone, B.J. (1997). Transcriptional regulation of the macrophage-induced gene (*gspA*) of *Legionella pneumophila* and phenotypic characterization of a null mutant. Mol. Microbiol. *24*, 629–642.

Abu-Zant, A., Asare, R., Graham, J.E., and Abu Kwaik, Y. (2006). Role for RpoS but not RelA of *Legionella pneumophila* in modulation of phagosome biogenesis and adaptation to the phagosomal microenvironment. Infect. Immun. *74*, 3021–3026.

Alleron, L., Frère, J., Merlet, N., and Legube, B. (2006). Monochloramine treatment induces a viable-but-nonculturable state into biofilm and planktonic *Legionella pneumophila* populations. In *Legionella*: State of the art 30 years after its recognition, N.P. Cianciotto, Y. Abu Kwaik, P.H. Edelstein, B.S. Fields, D.F. Geary, T.G. Harrison, C. Joseph, R.M. Ratcliff, J.E. Stout, and M.S. Swanson, eds. (Washington: ASM Press), pp 533–537.

Allison, C., Emödy, L., Coleman, N., and Hughes, C. (1994). The role of swarm cell differentiation and multicellular migration in the uropathogenicity of *Proteus mirabilis*. J. Infect. Dis. *169*, 1155–1158.

Bachman, M.A., and Swanson, M.S. (2001). RpoS co-operates with other factors to induce *Legionella pneumophila* virulence in the stationary phase. Mol. Microbiol. *40*, 1201–1214.

Bachman, M.A. and Swanson M.S. (2004). Genetic evidence that *Legionella pneumophila* RpoS modulates expression of the transmission phenotype in both the exponential phase and the stationary phase. Infect. Immun. 72: 2468–2476. Author's correction published in Infect. Immun. *72*, 6190.

Bachman, M.A. and Swanson, M.S. (2004a). The LetE protein enhances expression of multiple LetA/LetS-dependent transmission traits by *Legionella pneumophila*. Infect. Immun. *72*, 3284–3293.

Bandyopadhyay, P., Liu, S., Gabbai, C.B., Venitelli, Z., and Steinman, H.M. (2007). Environmental mimics and the Lvh type IVA secretion system contribute to virulence-related phenotypes of *Legionella pneumophila*. Infect. Immun. *75*, 723–735.

Barker J., Brown, M.R.W., Collier, P.J., Farrell, I., and Gilbert, P. (1992). Relationship between *Legionella pneumophila* and *Acanthamoeba polyphaga*: physiological status and susceptibility to chemical inactivation. Appl. Environ. Microbiol. *58*, 2420–2425.

Barker, J., Lambert, P.A., and Brown, M.R. (1993). Influence of intra-amoebic and other growth conditions on the surface properties of *Legionella pneumophila*. Infect. Immun. *61*, 3503–3510.

Barker, J., Scaife, H., and Brown, M.R. (1995). Intraphagocytic growth induces an antibiotic-resistant phenotype of *Legionella pneumophila*. Antimicrob. Agents Chemother. *39*, 2684–2688.

Bavoil, P., and Wyrick, P., editors. (2006). *Chlamydia*. Genomics and Pathogenesis. (Norwich, UK: Horizon Bioscience – Horizon Scientific Press).

Brassinga, A.K.C., Hiltz, M.F., Sisson, G.R., Morash, M.G., Hill, N., Garduno, E., Edelstein, P.H., Garduno, R.A., and Hoffman, P.S. (2003). A 65-kilobase pathogenicity island is unique to Philadelphia-1 strains of *Legionella pneumophila*. J. Bacteriol. *185*, 4630–4637.

Brieland, J.K., Fantone, J.C., Remick, D.G., LeGendre, M., McClain, M., and Engleberg, N.C. (1997). The role of *Legionella pneumophila*-infected *Hartmanella vermiformis* as an infectious particle in a murine model of Legionnaires' disease. Infect. Immun. *65*, 5330–5333.

Broich, M., Rydzewski, K., McNealy, T.L., Marre, R., and Flieger, A. (2006). The global regulatory proteins LetA and RpoS control phospholipase A, lysophospholipase A, acyltransferase, and other hydrolytic activities of *Legionella pneumophila* JR32. J. Bacteriol. *188*, 1218–1226.

Brüggemann, H., Hagman, A., Jules, M., Sismeiro, O., Dillies, M.A., Gouyette, C., Kunst, F., Steinert, M., Heuner, K., Coppee, J.Y. and Buchrieser, C. (2006). Virulence strategies for infecting phagocytes deduced from the *in vivo* transcriptional program of *Legionella pneumophila*. Cell. Microbiol., *8*, 1228–1240.

Bull, H.J., Lombardo, M.J., and Rosenberg, S.M. (2001). Stationary-phase mutation in the bacterial chromosome: recombination protein and DNA polymerase IV dependence. Proc. Natl. Acad. Sci. USA *98*, 8334–8341.

Byrne, B., and Swanson, M.S. (1998). Expression of *Legionella pneumophila* virulence traits in response to growth conditions. Infect. Immun. *66*, 3029–3034.

Chatterji, D., and Ojha, A.K. (2001). Revisiting the stringent response, ppGpp and starvation signaling. Curr. Opin. Microbiol. *4*, 160–165.

Cirillo, J.D., Cirillo, S.L.G., Yan, L., Bermudez, L.E., Falkow, S., and Tompkins, L.S. (1999). Intracellular growth in *Acanthamoeba castellani* affects monocyte entry mechanisms and enhances virulence of *Legionella pneumophila*. Infect. Immun. *67*, 4427–4434.

Cirillo, J.D., Falkow, S., and Tompkins, L.S. (1994). Growth of *Legionella pneumophila* in *Acanthamoeba castellani* enhances invasion. Infect. Immun. *62*, 3254–3261.

Coleman, S.A., Fischer, E.R., Howe, D., Mead, D.J., and Heinzen, R.A. (2004). Temporal analysis of *Coxiella burnetii* morphological differentiation. J. Bacteriol. *186*, 7344–7352.

de Felipe, K.S., Pampou, S., Jovanovic, O.S., Pericone, C.D., Ye, S.F., Kalachikov, S., and Shuman, H.A. (2005). Evidence for acquisition of *Legionella* type IV secretion substrates via interdomain horizontal gene transfer. J. Bacteriol. *187*, 7716–7726.

Eisenstark, A., Calcutt, M.J., Becker-Hapak, M., and Ivanova, A. (1996). Role of *Escherichia coli rpoS* and associated genes in defense against oxidative damage. Free Rad. Biol. Med. *21*, 975–993.

Elliot, T.S., and Rodgers, F.G. (1985). Morphological response and growth characteristics of *Legionella pneumophila* exposed to ampicillin and erythromycin. J. Med. Microbiol. *19*, 383–390.

Errington, J. (2003). Regulation of endospore formation in *Bacillus subtilis*. Nat. Rev. Microbiol. *1*, 117–126.

Faulkner, G., and Garduño, R.A. (2002). Ultrastructural analysis of differentiation in *Legionella pneumophila*. J. Bacteriol. *184*, 7025–7041.

Fettes, P.S., Forsbach-Birk, V., Lynch, D., and Marre, R. (2001). Overexpresssion of a *Legionella pneumophila* homologue of the *E. coli* regulator *csrA* affects cell size, flagellation, and pigmentation. Int. J. Med. Microbiol. *291*, 353–360.

Fields, B.S. (1996). Molecular ecology of legionellae. Trends Microbiol. *4*, 286–290.

Finkel, S.E., and Kolter, R. (1999). Evolution of microbial diversity during prolonged starvation. Proc. Natl. Acad. Sci. USA *96*, 4023–4027.

Forsbach-Birk, V., McNealy, T., Shi, C., Lynch, D., and Marre, R. (2004). Reduced expression of the global regulator protein CsrA in *Legionella pneumophila* affects virulence-associated regulators and growth in *Acanthamoeba castellanii*. Int. J. Med. Microbiol. *294*, 15–25.

Fortune, D.R., Suyemoto, M., and Altier, C. (2006). Identification of CsrC and characterization of its role in epithelial cell invasion in *Salmonella enterica* serovar Typhimurium. Infect. Immun. *74*, 331–339.

Gal-Mor, O., and Segal, G. (2003). The *Legionella pneumophila* GacA homologue(LetA) is involved in the regulation of *icm* virulence genes and is required for intracellular replication in *Acanthamoeba castellanii*. Microb. Pathog. *34*, 187–194.

Garduño, E., Faulkner, G., Ortiz-Jiménez, M.A., Berk, S.G., and Garduño, R.A. (2006). Interaction with the ciliate *Tetrahymena* may predispose *Legionella pneumophila* to infect human cells. In *Legionella*: State of the art 30 years after its recognition, N.P. Cianciotto, Y. Abu Kwaik, P.H. Edelstein, B.S. Fields, D.F. Geary, T.G. Harrison, C. Joseph, R.M. Ratcliff, J.E. Stout, and M.S. Swanson, eds. (Washington: ASM Press), pp. 297–300.

Garduño, R.A., Garduno, E., and Hoffman, P.S. (1998a). Surface-associated Hsp60 chaperonin of *Legionella pneumophila* mediates invasion in a HeLa cell model. Infect. Immun. *66*, 4602–4610.

Garduño, R.A., Quinn, F.D., and Hoffman, P.S. (1998b). HeLa cells as a model to study the invasiveness and biology of *Legionella pneumophila*. Can. J. Microbiol. *44*, 430–440.

Garduño, R.A., Garduño, E., Hiltz, M., Allan, D., and Hoffman, P.S. (2002a). Morphological and physiological evidence for a developmental cycle in *Legionella pneumophila*. In *Legionella*, R. Marre, Y. Abu Kwaik, C. Bartlett, N.P. Cianciotto, B.S. Fields, M. Frosch, J. Hacker, and P.C. Lück, eds. (Washington: ASM Press), pp. 82–85.

Garduño, R.A., Garduño, E., Hiltz, M., and Hoffman, P.S. (2002b). Intracellular growth of *Legionella pneumophila* gives rise to a differentiated form dissimilar to stationary phase forms. Infect. Immun. *70*, 6273–6283.

Hales, L.M., and Shuman, H.A. (1999). The *Legionella pneumophila rpoS* gene is required for growth within *Acanthamoeba castellanii*. J. Bacteriol. *181*, 4879–4889.

Hall, B.G. 1998. Activation of the *bgl* operon by adaptive mutation. Mol. Biol. Evol. *15*, 1–5.

Hammer, B.K., and Swanson, M.S. (1999). Co-ordination of *Legionella pneumophila* virulence with entry into stationary phase by ppGpp. Mol. Microbiol. *33*, 721–731.

Hammer, B.K., Tateda, E.S., and Swanson, M.S. (2002). A two-component regulator induces the transmission phenotype of stationary-phase *Legionella pneumophila*. Mol. Microbiol. *44*, 107–118.

Hatch, T.P. (1999). Developmental Biology. In *Chlamydia*: Intracellular biology, pathogenesis and immunity, R.S. Stephens, ed. (Washington: ASM Press), pp. 29–67.

Heuner, K., Dietrich, C., Skriwan, C., Steinert, M., and Hacker, J. (2002). Influence of the alternative σ28 factor on virulence and flagellum expression of *Legionella pneumophila*. Infect. Immun. *70*, 1604–1608.

Hiltz, M.F., Sisson, G.R., Brassinga, A.K.C., Garduno, E., Garduno, R.A., and Hoffman, P.S. (2004). Expression of *magA* in *Legionella pneumophila* Philadelphia-1 is developmentally regulated and a marker of formation of mature intracellular forms. J. Bacteriol. *186*, 3038–3045.

Hoffman, P. (1984). Bacterial physiology. In *Legionella*: Proceedings of the 2nd International Symposium, C. Thornsberry, A. Balows, J.C. Feeley, and W. Jakubowski, eds. (Washington: American Society for Microbiology Press), pp. 61–67.

Ishihama, A. (1997). Adaptation of gene expression in stationary phase bacteria. Curr. Opin. Genet. Dev. *7*, 582–588.

Ishihama, A. (2000). Functional modulation of *Escherichia coli* RNA polymerase. Annu. Rev. Microbiol. *54*, 499–518.

Joshi, A.D., Sturgill-Koszycki, S., and Swanson, M.S. (2001). Evidence that Dot-dependent and –independent factors isolate the *Legionella pneumophila* phagosomes from the endocytic network in mouse macrophages. Cell. Microbiol. *3*, 99–114.

Kulkarni, P.R., Cui, X., Williams, J.W., Stevens, A.M., and Kulkarni, R.V. (2006). Prediction of CsrA-regulating small RNAs in bacteria and their experimental verification in *Vibrio fischeri*. Nucleic Acids Res. *34*, 3361–3369.

Lacour, S., and Landini, P. (2004). σS-dependent gene expression at the onset of stationary phase in *Escherichia coli*: function of σS-dependent genes and identification of their promoter sequences. J. Bacteriol. *186*, 7186–7195.

Layton, J.C., and Foster, P.L. (2003). Error-prone DNA polymerase IV is controlled by the stress-response sigma factor, RpoS, in *Escherichia coli*. Mol. Microbiol. *50*, 549–561.

Lee, J.V., and West A.A. (1991). Survival and growth of *Legionella* species in the environment. J. Appl. Bacteriol. (Symp. Suppl.) *70*, 121S–129S.

Loewen, P.C., and Hengge-Aronis, R. (1994). The role of the sigma factor σS (KatF) in bacterial global regulation. Annu. Rev. Microbiol. *48*, 53–80.

Lombardo, M.-J., Aponyi, I., and Rosenberg, S.M. (2004). General stress response regulator RpoS in adaptive mutation and amplification in *Escherichia coli*. Genetics *166*, 669–680.

Lynch, D., Fieser, N., Glöggler, K., Forsbach-Birk, V., and Marre, M. (2003). The response regulator LetA regulates the stationary phase stress response in *Legionella pneumophila* and is required for efficient infection of *Acanthamoeba castellanii*. FEMS Microbiol. Lett. *219*, 241–248.

McCaul, T. F., and Williams, J.C. (1981). Developmental cycle of *Coxiella burnetii*: structure and morphogenesis of vegetative and sporogenic differentiations. J. Bacteriol. *147*, 1063–1076.

McDade, J.E., Shepard, C.C., Fraser, D.W., Tsai, T.R., Redus, M.A., Dowdle, W.R., and the Laboratory Investigation Team. (1977). Legionnaires' disease. Isolation of a bacterium and demonstration of its role in other respiratory disease. N. Engl. J. Med. *297*, 1197–1203.

McNealy, T., Forsbach-Birk, V., Shi, C., and Marre, R. (2005). The Hfq homologuein *Legionella pneumophila* demonstrates regulation by LetA and RpoS and interacts with the global regulator CsrA. J. Bacteriol. *187*, 1527–1532.

McNealy, T., Newsome, A. L., Johnson, R. A., and Berk, S. G. (2002). Impact of amoebae, bacteria and *Tetrahymena* on *Legionella pneumophila* multiplication and distribution in an aquatic environment. In *Legionella*, R. Marre, Y. Abu Kwaik, C. Bartlett, N.P. Cianciotto, B.S. Fields, M. Frosch, J. Hacker, and P.C. Lück, eds. (Washington: ASM Press), p. 170–175.

Mercante, J., Suzuki, K., Cheng, X., Babitzke, P., and Romeo, T. (2006). Comprehensive alanine-scanning mutagenesis of *Escherichia coli* CsrA defines two subdomains of critical functional importance. J. Biol. Chem. *281*, 31832–31842.

Molofsky, A.B., Shetron-Rama, L.M., and Swanson, M.S. (2005). Components of the *Legionella pneumophila* flagellar regulon contribute to multiple virulence traits, including lysosome avoidance and macrophage death. Infect. Immun. *73*, 5720–5734.

Molofsky, A.B., and Swanson, M.S. (2003). *Legionella pneumophila* CsrA is a pivotal repressor of transmission traits and activator of replication. Mol. Microbiol. *50*, 445–461.

Moulder, J.W. (1991). Interaction of chlamydiae and host cells *in vitro*. Microbiol. Rev. *55*, 143–190.

Paszko-Kolva, C.M. Shahamat, and Colwell, R.R. (1992). Long term survival of *Legionella pneumophila* serogroup 1 under low-nutrient conditions and associated morphological changes. FEMS Microbiol. Ecol. *102*, 45–55.

Ponder, R.G., Fonville, N.C., and Rosenberg, S.M. (2005). A switch from high-fidelity to error-prone DNA double-strand break repair underlies stress-induced mutation. Mol. Cell *19*, 791–804.

Pope, D.H., Soracco, R.J., Gill, H.K., and Fliermans, C.B. (1982). Growth of *Legionella pneumophila* in two-membered cultures with green algae and cyanobacteria. Curr. Microbiol. *7*, 319–321.

Rauprich, O., Matsushita, M., Weijer, C.J., Siegert, F., Esipov, S.E., and Shapiro, J.A. (1996). Periodic phenomena in *Proteus mirabilis* swarm colony development. J. Bacteriol. *178*, 6525–6538.

Rowbotham, T.J. (1983). Isolation of *Legionella pneumophila* from clinical specimens via amoebae, and the interaction of those and other isolates with amoebae. J. Clin. Pathol. *36*, 978–986.

Rowbotham, T.J. (1986). Current views on the relationships between amoebae, legionellae and man. Isr. J. Med. Sci. *22*, 678–689.

Roy, C.R., Berger, K.H., and Isberg, R.R.. (1998). *Legionella pneumophila* DotA protein is required for early phagosome trafficking decisions that occur within minutes of bacterial uptake. Mol. Microbiol. *28*, 663–674.

Sadosky, A.B., Wiater, L.A., and Shuman, H.A. (1993). Identification of *Legionella pneumophila* genes required for growth within and killing of human macrophages. Infect. Immun. *61*, 5361–5373.

Samuel, J.E., Kiss, K., and Varghees, S. (2003). Molecular pathogenesis of *Coxiella burnetii* in a genomics era. Ann. NY Acad. Sci. *990*, 653–663.

Sauer, J.D., Bachman, M.A., and Swanson, M.S. (2005). The phagosomal transporter A couples threonine acquisition to differentiation and replication of *Legionella pneumophila* in macrophages. Proc. Natl. Acad. Sci. USA *102*, 9924–9929.

Schachter, J. (1988). The intracellular life of *Chlamydia*. Curr. Topics Microbiol. Immunol. *138*, 109–139.

Schofield, G.M. (1985). A note on the survival of *Legionella pneumophila* in stagnant tap water. J. Appl. Bacteriol. *59*, 333–335.

Shi, C., Forsbach-Birk, V., Marre, R., and McNealy, T.L. (2006). The *Legionella pneumophila* global regulatory protein LetA affects DotA and Mip. Int. J. Med. Microbiol. *296*, 15–24.

Skaliy, P., and McEachern H.V. (1979). Survival of the Legionnaires' disease bacterium in water. Ann. Intern. Med. *90*, 662–663.

Smalley, D.L., Jaques, P.A., Ourth, D.D., and Layne, J.S. (1980). Antibiotic-induced filament formation of *Legionella pneumophila*. Am. J. Clin. Pathol. *74*, 852.

Steinert, M., Emödy, L., Amann, R., and Hacker, J. (1997). Resuscitation of viable but nonculturable *Legionella pneumophila* Philadelphia JR32 by *Acanthamoeba castellani*. Appl. Environ. Microbiol. *63*, 2047–2053.

Steinert, M., Hentschel, U., and Hacker, J. (2002). *Legionella pneumophila*: an aquatic microbe goes astray. FEMS Microbiol. Rev. *26*, 149–162.

Tison, D.L., Pope, D.H., Cherry, W.B., and Fliermans, C.B. (1980). Growth of *Legionella pneumophila* in association with blue-green algae (cyanobacteria). Appl. Environ. Microbiol. *39*, 456–459.

Venturi, V. (2005). Control of *rpoS* transcription in *Escherichia coli* and *Pseudomonas*: Why so different? Mol. Microbiol. *49*, 1–9.

Vogel, J., Roy, C., and Isberg, R.R. (1996). Use of salt to isolate *Legionella pneumophila* mutants unable to replicate in macrophages. Ann. NY Acad. Sci. *797*, 271–272.

Vogel, J. P. (2004). Turning a tiger into a house cat: Using *Legionella pneumophila* to study *Coxiella burnetii*. Trends Microbiol. *12*, 103–105.

Wadosky, R.M., and Yee, R.B. (1983). Satellite growth of *Legionella pneumophila* with an environmental isolate of *Flavobacterium breve*. Appl. Environ. Microbiol. *46*, 1447–1449.

Warren, W.J., and Miller, R.D. (1979). Growth of Legionnaires' disease bacterium (*Legionella pneumophila*) in chemically defined medium. J. Clin. Microbiol. *10*, 50–55.

Watarai, M., Andrews, H.L., and Isberg, R.R. (2001). Formation of a fibrous structure on the surface of *Legionella pneumophila* associated with exposure of DotH and DotO proteins after intracellular growth. Mol. Microbiol. *39*, 313–329.

Zusman, T., Gal-Mor, O., and Segal, G. (2002). Characterization of a *Legionella pneumophila relA* insertion mutant and roles of RelA and RpoS in virulence gene expression. J. Bacteriol. *184*, 67–75.

The Genetics and Immunology of Host Resistance to *Legionella* Infection

5

Russell Vance* and Thomas R. Hawn

Abstract

Significant progress has been made toward elucidating the critical genetic and immunological components of the host immune response to *Legionella pneumophila*. In this chapter, we review the host immune response to *Legionella*, including the roles of the innate immune response, B cells, T cells, cytokines and chemokines. We examine the use of *in vitro* studies of macrophage infection, *in vivo* animal studies, and more recent human genetic studies that have all contributed to our understanding of the immune response to *Legionella*. We also highlight the distinct biological properties of *Legionella* that have stimulated discoveries of broad relevance to our understanding of host–pathogen interactions. Taken together, numerous genetic and immunological studies have revealed that *Legionella* is a fascinating model for understanding how the host interprets a large number of molecular and cellular cues to provide host defence.

Introduction

Legionella pneumophila is a fastidious, Gram-negative, flagellated pathogen with an atypical LPS structure. *L. pneumophila* evolves in the environment as a parasite of freshwater and soil amoebae, and there is no evidence for animal-to-animal or human-to-human transmission (Katz *et al*., 1982). Nevertheless, adaptation to amoebae has somehow equipped *L. pneumophila* to infect humans, in which it can cause a variety of diseases (Chapter 1). The range of clinical response to *Legionella* infection includes asymptomatic infection, a febrile illness associated with low morbidity (Pontiac fever), and bacterial pneumonia that is associated with significant morbidity and mortality (Legionnaires' disease). The reasons underlying the wide spectrum of host responses to infection are only partially understood but are presumably due to the interaction of bacterial virulence factors, external host factors that cause immunosuppression, and intrinsic genetic factors of the host. Risk factors for human acquisition of Legionnaires' disease include smoking, chronic lung disease, alcohol use, and other chronic medical co-morbidities such as diabetes and kidney failure (Fraser *et al*., 1977; England *et al*., 1981 Straus *et al*., 1996). However, many individuals acquire Legionnaires' disease without any identifiable risk factor, suggesting that there may also be host immunogenetic factors that influence susceptibility.

To explain human susceptibility to Legionnaires' disease, a thorough understanding of the molecular and cellular components of host immunity is required. In this chapter, we will review the role of the innate immune response, B cells, T cells, cytokines and chemokines in host defence to *Legionella*. Table 5.1 provides a convenient summary of many of the key findings from *in vivo* animal infections; in addition, we will also discuss *in vitro* studies and human genetic studies. Beyond reviewing the seminal studies that have shaped our understanding of host resistance to *Legionella* since its discovery in 1976, we will emphasize more recent studies that have included discoveries of how the innate immune response is critical in mediating host protection.

*Corresponding author.

Table 5.1 *L. pneumophila* and *in vivo* infection animal models

Strain (background)	Depletion technique	Susceptibility: Lung CFU and mortality	BAL fluid (B) and histology (H)	Other	Effect[a]	Reference
Strains						
A/J vs. B6, i.v.		↑Mortality (for some doses) and ↑cfu in A/J liver and spleen ~10- to 100-fold			+1/+2	Fujio *et al.* (1992)
A/J vs. B6, i.v.		Minimal effect			0	Tateda *et al.* (2003)
A/J vs. B6, i.n.*		↑cfu ~2-fold at 48 h A/J			0/+1	Zamboni *et al.* (2006)
A/J vs. B6, i.t.		↑cfu ~10-fold at 48 h A/J			+1	Molofsky *et al.* (2006)
A/J vs. B6, i.t.		↑cfu >10-fold at 48 h A/J			+1	Asare *et al.* (2007)
Cells and humoral components						
PMN depletion (A/J), i.t.	Ab	↑cfu ~10- to 100-fold ↑mortality		↓IFN-γ and IL-12	+1/+2	Tateda *et al.* (2001)
PMN depletion (B6), i.t.	Ab	↔cfu			0	Tateda *et al.* (2001)
T cell depletion (A/J) (CD4 and CD8), i.t.	Ab	↑cfu ~10- to 100-fold at 1–3 days ↑cfu ~10^4 fold at 7–11 days			+3	Susa *et al.* (1998)
NK cells, (A/J and B6), i.v.	Ab	↑cfu >10^2		↓ IFN-γ and IL-12	+2	Sporri *et al.* (2006)
Antibody (A/J), i.t.		↑cfu ~10-fold for unopsonized bacteria			+1	Brieland *et al.* (1996)
Complement (A/J), i.t.	Ab	↔cfu			0	Brieland *et al.* (1996)

Innate immune receptors						
MyD88-/- (B6), as	KO	↑cfu ~10^4-fold ↑mortality	B: ↓↓ PMN H: ↓ at d3, but↑↑'d at d6	↑cfu in spleen ↓↓Multiple cytokines within 24 h No difference in IFN-γ	+3	Hawn *et al.* (2006)
MyD88–/– (A/J), i.n.	KO	↑cfu 1000-fold ↑mortality		↑ cfu in spleen and liver ↓↓ IFN-γ at 48 h	+3	Archer *et al.* (2006)
MyD88–/– (A/J), i.v.	KO	↑cfu ~10^2-fold at 48 h		↓↓ IFN-γ, IL-12, TNF at 16 h	+2	Sporri *et al.* (2006)
TLR2–/– (B6), as	KO	↑cfu ~10- to 100-fold	B: ↓ PMN H: ↔ up to 6 days	No cfu in spleen Modest ↓cytokines at 4 h	+1/+2	Hawn *et al.* (2006)
TLR2–/– (A/J), i.n.	KO	↑cfu by <10-fold		↔ IFN-γ	+1	Archer *et al.* (2006)
TLR4–/– (A/J), i.n.	KO	↔cfu			0	Archer *et al.* (2006)
TLR4^{mut} (C3H/HeJ), i.n.	Mut	↔cfu	B: ↔ PMN and MN H: ↔ up to 72 h	↔ TNF-α and KC	0	Lettinga *et al.* (2002)
IPAF–/– (B6), i.t.	KO	↑cfu ~10-fold at 96 h			+1	Amer *et al.* (2006)
Caspase-1–/– (B6), i.n.	KO	↑cfu ~2-fold at 48 h			0/+1	Zamboni *et al.* (2006)
Cytokines/chemokines						
IL-12^{dep} (A/J), i.t.	Ab	↑cfu ~10- to 100-fold at 5 days		↓ TNF-α at 24 h in IL-12^{dep}	+2	Brieland *et al.* (1998)
IL-12^{dep} (A/J), i.t.	mAb	↑cfu ~10-fold at 5 days		↓ IFN-γ at 24-72 h in IL-12^{dep}	+2	Brieland *et al.* (2000.
IL-18R^{dep} (A/J), i.t.	Ab	↔cfu		↓ IFN-γ	0	Brieland *et al.* (2000.
IFN-γ-/– (Balb/c), i.n.	KO	↑cfu <10-fold ↑Mortality	B: ↔ 1st 24 h H: ↑ at 72–96 h in KO	Disseminate to blood in KO ↓ TNF-α, IL-1β at <1 day in KO	+2	Shinozawa *et al.* (2002
IFN-γ-/– (Balb/c), i.t.	KO	↑cfu <10-fold ↔Mortality	B: ↔ 1st 24 h H: ↑ in KO	Disseminate to spleen in KO ↓ TNF-α, IL-1β at <1 day in KO	+2	Heath *et al.* (1996)
IFN-γ-/– (A/J and B6), i.v.	mAb	↑cfu 10^1–10^4-fold in liver and spleen				Fujio *et al.* (1992)
IFN-I-R–/– (129), i.v.	KO	↔ cfu			0	Sporri *et al.* (2006)

Table 5.1 *continued*

Strain (background)	Depletion technique	Susceptibility: Lung CFU and mortality	BAL fluid (B) and histology (H)	Other	Effect[a]	Reference
IFN-II-R (129), i.v.	KO	↑cfu ~>10^2 fold at 36 h		↓ IL-12 at 12–16 h	+2	Sporri *et al.* (2006)
TNF^{dep} Rat, as	Ab	↑cfu >100-fold	B: ↔ PMN and MN H: ↑ at 96 h in TNF^{dep}	↑IFN-γ at 48–96 h in TNF^{dep}	+3	Skerrett *et al.* (1997)
TNF^{dep} (A/J), i.t.	Ab	↑cfu <10-fold			+2	Brieland *et al.* (1995)
NO^{dep} (A/J), i.t.	Inh	↑cfu ~2-fold		↓ TNF-α at 24 h in NO^{dep}	+1	Brieland *et al.* (1995)
NO^{dep} (Balb/c), i.t.	Inh	↔ cfu			0	Heath *et al.* (1996)
$CXCR2^{dep}$ (A/J), i.t.	Ab	↑cfu ~10-fold ↑Mortality		No effect from depletion of MIP-2 or KC	+1	Tateda *et al.* (2001)

[a]Scale: +3, large effect; +2, modest; +1, mild; 0=no effect. Cfu, colony-forming units, i.n., intranasal i.p., intraperitoneal, i.v., intravenous, i.t.,intratrachael, as, aerosolized, B6=C57Bl/6. *mice had an A/J or B6 Naip locus on a mixed background.

Depletion technique indicates method of depletion of gene product: Mut, mutation causing loss of function; K, gene knockout; Ab, antibody depletion of cytokine; Inh, chemical inhibitor. Background refers to mouse background for experiment. BAL refers to degree of inflammation in bronchoalveolar lavage fluid within the first 24 hours. N/, not assessed. $IL\text{-}12^{dep}$, IL-12 depletion; TNF^{dep}, TNF-α depletion. IFN-I-R mice are deficient in the receptors for type 1 and 2 interferon (α/β). IFN-II-R mice are deficient for IFN-γ receptor.

Innate immune response: initial detection of *Legionella*

Macrophages and restriction of intracellular growth of *L. pneumophila* by Naip5

Legionella is capable of growing in diverse cell types. However, since *Legionella* is not efficient at inducing its own uptake into cells, the predominant cell types infected during *in vivo* infections appear to be phagocytic cells, particularly alveolar macrophages. *Legionella* mutants incapable of intracellular growth within macrophages *in vitro* are avirulent *in vivo* (Marra *et al.*, 1992; Brand *et al.*, 1994). Thus, intracellular growth within macrophages is likely a key determinant of *Legionella* pathogenicity.

In contrast to human macrophages, which tend to be highly permissive to *L. pneumophila* replication, macrophages from most inbred mouse strains restrict intracellular replication of *L. pneumophila*. One notable exception is the A/J mouse strain (Yamamoto *et al.*, 1988), whose macrophages can support 3–4 logs of bacterial replication over 3–4 days. Macrophages from some mouse strains (e.g. 129S1, FVB/N) are reported to exhibit intermediate permissiveness (Wright *et al.*, 2003), though other reports indicate these strains can be quite restrictive (Diez *et al.*, 2003). Macrophages from the C57BL/6 (B6) mouse strain exhibit the most consistent ability to restrict *L. pneumophila* replication. B6 macrophages tend to exhibit some permissiveness to *L. pneumophila* replication within the first 24 h of infection, but are subsequently almost entirely restrictive (Derre and Isberg, 2004). Where tested, bone marrow-derived and thioglycollate-elicited peritoneal macrophages exhibit similar strain-dependent abilities to restrict *Legionella* replication.

The dramatic 3–4 log difference in permissiveness between B6 and A/J macrophages has facilitated the use of mouse genetics to dissect *L. pneumophila* pathogenesis. Initial mapping studies demonstrated that a single locus on mouse chromosome 13, called *Lgn1*, was responsible for the entire phenotypic difference between B6 and A/J (Beckers *et al.*, 1995; Dietrich *et al.*, 1995). The intermediate permissiveness of macrophages from 129S1 and FVB/N mice also maps to the *Lgn1* locus (Wright *et al.*, 2003). The restrictive phenotype of B6 was found to be fully dominant over the permissive phenotype of A/J, suggesting that the B6 *Lgn1* locus contained a functional gene (or genes) that conferred resistance to *Legionella*, whereas the $Lgn1^{A/J}$ locus lacks this gene or function. Interestingly, *Lgn1* does not appear to control replication of all *Legionella* species, nor even all strains of *L. pneumophila* (Miyamoto *et al.*, 1996; Izu *et al.*, 1999; Asare *et al.*, 2007).

A 10-year fine-mapping project by two groups revealed that a gene within the *Lgn1* locus called *Naip5* (Neuronal apoptosis inhibitory protein 5) was likely responsible for the differential permissiveness of B6 and A/J macrophages to *L. pneumophila* replication (Scharf *et al.*, 1996; Diez *et al.*, 1997; Endrizzi *et al.*, 2000; Growney and Dietrich, 2000; Diez *et al.*, 2003; Wright *et al.*, 2003). *Naip5* is also sometimes called *Birc1e* (baculoviral inhibitor of apoptosis repeat-containing protein 1e). The A/J allele of *Naip5* exhibits 14 amino acid polymorphisms as compared to the B6 allele and is expressed at lower levels (Wright *et al.*, 2003), but the causative polymorphism(s) has yet to be identified. The mechanism by which *Naip5* restricts replication of *L. pneumophila* remains poorly understood (see below).

In mice, *Naip5* is a member of a small multigene family of related *Naip* genes that vary in number and organization among different mouse strains (Growney and Dietrich, 2000). Humans appear to contain a single functional *NAIP* gene, and comparisons among primate and rodent species suggest that the surrounding genomic region is undergoing rapid evolution (Romanish *et al.*, 2007). Expression of *NAIP* genes in humans and rodents appear to be driven, at least in part, by promoter elements within endogenous retroviral long terminal repeats (LTRs). Interestingly, distinct retroviral insertion events appear to have occurred in primate and rodent lineages (Romanish *et al.*, 2007). In humans and in mice, *NAIP* expression appears to occur in many tissue types (Diez *et al.*, 2000; Romanish *et al.*, 2007). There is no evidence as yet that human *NAIP* is involved in resistance to microbial pathogens.

Although the entire phenotypic difference between B6 and A/J macrophages with respect to permissiveness to *L. pneumophila* can be explained by polymorphisms in *Naip5*, it cannot be concluded that the *Naip5* pathway is the only pathway able to restrict *L. pneumophila* replication. Indeed there is evidence that additional parallel pathways are also able to restrict *L. pneumophila* replication in macrophages (e.g. TLRs, IFN-α/β/γ, TNF-α; Coers *et al.*(2007) and see below). Moreover, although *Naip5* has a large effect on intracellular multiplication of *L. pneumophila* within macrophages *in vitro*, the effect of *Naip5* on replication of *L. pneumophila in vivo* is considerably less dramatic, varying from negligible to approximately 10-fold (see Table 5.1) (Fujio *et al.*, 1992; Tateda *et al.*, 2003; Molofsky *et al.*, 2006; Zamboni *et al.*, 2006; Asare *et al.*, 2007). This is likely because of compensation by several antibacterial responses that are stimulated during *in vivo* infections (e.g. IFN-γ) that are not present during *in vitro* infections of macrophages.

As is often the case, the forward genetic analysis left open a key question: how does *Naip5* function? The answer to this question remains uncertain, though considerable progress has recently been made. An important clue has been that *Naip5* exhibits homology to a large family of pathogen-sensing proteins called the NBS-LRR proteins (nucleotide binding site-leucine rich repeat) proteins (Inohara *et al.*, 2005; Kufer *et al.*, 2005; Ting and Davis, 2005; Fritz *et al.*, 2006). These proteins (also called the NOD-like or CATERPILLAR family of proteins) are essential for innate immune recognition of a variety of bacterial ligands/toxins (Table 5.2). The LRR domain within these proteins appears to function as a pathogen-sensing domain, though it must be emphasized that direct binding of LRRs to a pathogen-derived ligand has not yet been demonstrated for any mammalian NBS-LRR protein, and some evidence suggests an indirect mode of recognition (Mariathasan *et al.*, 2006). The NBS (also sometimes called a NOD or nucleotide oligomerization domain) appears to function as a protein-protein oligomerization domain. NBS-NBS interactions are generally believed to be homotypic, though heterotypic interactions have also been described (Damiano *et al.*, 2004). Oligomerization via the NBS appears to be sufficient to drive signalling downstream of an N-terminal signalling domain (Inohara *et al.*, 2000). In most NBS-LRR proteins, the signalling domain, known as CARD (caspase recruitment domain) or PYD (pyrin domain), recruits other CARD-containing proteins leading to caspase activation and/or NF-κB activation (Table 5.2). However, Naip5 is unique among NBS-LRR proteins in lacking a CARD domain and in containing three N-terminal BIRs (baculovirus inhibitor of apoptosis repeats). The BIRs of human NAIP have been shown to bind and inhibit caspases-3, -7, and -9 (Maier *et al.*, 2002; Davoodi *et al.*, 2004), though a biological role for this interaction remains to be established. Instead, several studies have implicated *Naip5* in activation of caspase-1 in response to *Legionella* (Amer *et al.*, 2006; Molofsky *et al.*, 2006; Ren *et al.*, 2006; Zamboni *et al.*, 2006), as discussed in more detail below. A direct interaction between Naip5 and caspase-1 has not been reported, nor is it clear how Naip5 signalling leads to caspase-1 activation. It should also be noted that human macrophages, which are generally permissive to *Legionella* replication, do not appear to trigger caspase-1 in response to *Legionella* (Santic *et al.*, 2007).

Intracellular sensing of *Legionella* flagellin

The homology of Naip5 to other NBS-LRR proteins suggests that Naip5 might function as a pathogen sensor. It was reasoned that *Legionella* mutants lacking the structure sensed by *Naip5* would be able to evade *Naip5*-mediated restriction and replicate freely in the normally restrictive B6 macrophages. Indeed, two groups, using independent methods, found that *Legionella* flagellin mutants were able to grow in B6 macrophages (Molofsky *et al.*, 2006; Ren *et al.*, 2006). Both groups found that it was the specific loss of the gene encoding flagellin (*flaA*) and not more generic mutations, e.g. those affecting the function or assembly of the flagellum (Chapter 6), that permitted evasion of *Naip5*-mediated immunity. Loss of flagellin also reduced the ability of *Legionella* to activate caspase-1, implying that sensing of flagellin is upstream of caspase-1 activation (Molofsky *et al.*, 2006; Ren *et al.*, 2006). Delivery of *Legionella* flagellin to the macrophage cytosol appears to be essential and sufficient to initiate caspase-1 activation (Molofsky *et al.*, 2006), and K. Lightfield and R. Vance, unpublished), whereas the extracellular flagellin-sensor TLR5 appears not to be involved (Amer *et al.*, 2006; Molofsky *et al.*, 2006; Ren *et al.*, 2006). Interestingly, flagellin appears to access the host cell cytosol independently of the flagellar secretion system, since mutation of *fliI* (required for secretion of flagellin via the flagellar secretion apparatus) did not affect *Naip5*-mediated restriction of *Legionella* or caspase-1 activation (Ren *et al.*, 2006). Indeed, evidence suggests that the type IV Dot/Icm secretion system is required for flagellin to access the host cell cytosol (Molofsky *et al.*, 2006; Ren *et al.*, 2006). Since type IV-mediated secretion of flagellin appears to be quite inefficient, as compared to the secretion of other canonical effector proteins, and provides no obvious advantage to the bacterium, it has been suggested that type IV-mediated secretion of flagellin may not represent a virulence 'strategy' of *Legionella*. Instead, flagellin secretion through the type IV secretion apparatus may simply occur inadvertently or accidentally (Molofsky *et al.*, 2006).

There is no evidence that Naip5 binds flagellin, and in fact, flagellin-sensing is not entirely abolished in A/J macrophages (Molofsky *et al.*, 2006; Ren *et al.*, 2006). This implies either that the A/J allele of *Naip5* is partially functional, or alternatively, that *Naip5* is not essential for flagellin-sensing. Interestingly, it has recently been found that *Salmonella*-encoded flagellin proteins (FliC and FljB) are also sensed intracellularly by macrophages, leading to activation of caspase-1 (Franchi *et al.*, 2006; Miao *et al.*, 2006). Thus, study of *Legionella* has led to discovery of an innate immune sensing pathway that appears to have general significance.

Ipaf and signalling downstream of flagellin recognition

The NBS-LRR protein most closely homologous to Naip5 is Ipaf (ICE-protease activating factor; also called CLAN (CARD LRR and NACHT-containing protein), CARD12, or Nlrc4 (NLR (Nod-like receptor) family, CARD domain containing 4)). Unlike Naip5, Ipaf contains an N-terminal CARD domain and lacks BIRs. *Ipaf*$^{-/-}$ macrophages are as permissive to *Legionella* replication as are A/J macrophages (Amer *et al.*, 2006; Zamboni *et al.*, 2006), and Ipaf and Naip5 have been reported to heterooligomerize when overexpressed (Damiano *et al.*, 2004; Zamboni *et al.*, 2006). Ipaf is required for intracellular sensing of *Salmonella* and *Legionella* flagellin leading to activation of caspase-1 (Amer *et al.*, 2006; Franchi *et al.*, 2006; Miao *et al.*, 2006). Thus, it has been proposed that Ipaf and Naip5 function together to sense flagellin and activate caspase-1. The precise roles of Ipaf and Naip5 have yet to be resolved. Indeed, there is no evidence that sensing of *Salmonella* flagellin requires Naip5. It is possible, for example, that Ipaf is the 'true' flagellin sensor, whereas Naip5 is a regulator of flagellin-sensing.

Despite considerable progress, it remains unclear how Naip5/Ipaf restricts *Legionella* replication. Based on the above evidence, it is tempting to speculate that flagellin triggers Ipaf/Naip5-mediated caspase-1 activation and that caspase-1 then acts to restrict *Legionella* growth. Indeed, caspase-1-deficient macrophages exhibit increased permissiveness to *Legionella* (Zamboni *et al.*, 2006). Caspase-1 is best known as a protease required for cleavage and secretion of interleukins 1, 18 and 33. While these cytokines certainly contribute to inflammation *in vivo*, they do not, however, appear to play a role in restricting *Legionella* replication *in vitro* (Amer *et al.*, 2006; Coers *et al.*, 2007). Caspase-1 is also required for a non-apoptotic necrotic-like cell death, sometimes termed 'pyroptosis', in response to several pathogens, including *Legionella* and *Salmonella* (Fink and Cookson, 2005). Pyroptosis of *Legionella*-infected macrophages occurs rapidly (within 2–4h) and could therefore pre-empt bacterial replication. However, direct evidence that premature cell death is sufficient to restrict *Legionella* replication is lacking. An alternative hypothesis is that caspase-1 may cleave targets in the cell that lead to the induction of autophagy, which may restrict bacterial replication (Swanson and Molofsky, 2005).

Caspase-1 may not be the sole downstream effector of Naip5/Ipaf signalling. Indeed, permissiveness of *Caspase-1*$^{-/-}$ macrophages appears to vary, and is often appreciably less than the permissiveness of A/J macrophages (Fortier *et al.*, 2006). An explanation for the incomplete

Table 5.2 Innate immune detectors and *Legionella*

Detector	Aliases	Domain structure[a] (signalling adaptors)	Senses (directly or indirectly)	Role in sensing *Legionella*?	References[b]
Tlr at cell surface					
Tlr		LRR-TM-TIR (MyD88, TIRAP/MAL)	Lipopeptide +other (*Legionella* LPS?)	Yes	Akamine *et al.* (2005), Girard *et al.* (2003)
Tlr4		LRR-TM-TIR (MyD88,TIRAP/MAL, Trif, TRAM)	LPS	Yes?	Kikuchi *et al.* (2004)
Tlr5		LRR-TM-TIR (MyD88)	Flagellin	Yes	Hawn *et al.* (2003)
Tlr in intracellular compartments					
Tlr3		LRR-TM-TIR (Trif)	Double stranded RNA	Unlikely, but unknown	
Tlr7/8		LRR-TM-TIR (MyD88)	Single stranded RNA	Unlikely, but unknown	
Tlr9		LRR-TM-TIR (MyD88)	Unmethylated CpG DNA	Yes?	Newton *et al.* (2007)
Intracellular NBS-LRR proteins					
Naip5	Birc1e	BIRx3-NBS-LRR	Flagellin?	Yes	Ren *et al.* (2006), Molofsky *et al.* (2006)
Ipaf	Clan, Card12, Nlrc4	CARD-NBS-LRR	Flagellin?	Yes	Zamboni *et al.* (2006)
Nod1	Card4, Nlrc1	CARD-NBS-LRR (RIP2)	iE-DAP	Probably but unknown	

Nod2	Card15, Nlrc2	CARDx2-NBS-LRR (RIP2)	MDP	Probably but unknown	
Nalp1	Card7	PYD-NBD-LRR-CARD[d]	Anthrax lethal toxin	Unknown	
Nalp3	Cias1, cyropyrin, Pypaf, Nlrp3	PYD-NBD-LRR	K^+ efflux, uric acid, toxins, bacterial RNA, MDP?	Unknown	
Intracellular nucleic acid sensors					
RIG-I		CARD-RNA helicase (MAVS[e])	5′ triphosphate RNA	Possible, unknown	
Mda5		CARD-RNA helicase (MAVS)	Poly I:C, picornavirus	Possible, unknown	
'DNA sensor'	DAI, others?	DNA binding domains	Intracellular DNA	Possible, unknown	Stetson *et al.* (2006) Takaoka *et al.* (2007)

[a]L, leucine-rich repeats; TM, transmembrane domain; TIR, Toll/IL-1R domain; BIR, baculovirus inhibitor of apoptosis repeat; NBS, nucleotide binding site (this domain is also sometimes called a NOD or NACHT domain); CARD, caspase recruitment domain; PYD, pyrin domain.

[b]Not all relevant references are cited; the focus is on references with particular relevance for *Legionella*

[c]TLR2 heterodimerizes with TLR1 or TLR6

[d]Mouse Nalp1 does not have a pyrin domain.

[e]MAVS is a signalling adaptor required for RIG-I and Mda5 signalling; it is also called IPS-1, VISA or CARDIF.

restrictiveness of *Caspase-1*$^{-/-}$ macrophages may be that other related caspases compensate for loss of caspase-1. In the mouse, the two caspases most closely related to caspase-1 are caspase-11 and caspase-12. Together, these caspases comprise a sub-family of 'inflammatory' caspases. However, caspase-11 is not expressed in caspase-1 knockouts (Kang *et al.*, 2000), and is therefore unlikely to account for residual resistance of caspase-1 knockouts. Caspase-12, on the other hand, is inducibly expressed, e.g. by LPS. Moreover, given that *Legionella* replicates in an ER-like compartment, it is interesting that caspase-12 has been proposed to be responsive to ER-stress (Nakagawa *et al.*, 2000). Human caspase-12 is not a functional protease, and instead of caspase-11, humans encode two related caspases, caspase-4 and -5. These human caspases have not been reported to play a role in responses to *Legionella*. Moreover, evidence suggests human caspase-1 is not activated in response to *Legionella* (Santic *et al.*, 2007). Instead, a non-inflammatory conventional apoptotic caspase, caspase-3, is activated by *Legionella* (Gao and Abu Kwaik, 1999). However, caspase-3 activation does not seem to restrict *Legionella* replication (Molmeret *et al.*, 2004; Zamboni *et al.*, 2006). Further studies of putative caspase-1-independent mechanisms downstream of Naip5/Ipaf are therefore merited.

Several studies have suggested that Naip5/Ipaf might function to alter intracellular trafficking of *Legionella*-containing phagosomes (Watarai *et al.*, 2001; Amer and Swanson, 2005; Amer *et al.*, 2006; Fortier *et al.*, 2006;). In one study, macrophages from A/J mice were rendered resistant to *Legionella* by expression of a *Naip5*^{B6} transgene (Fortier *et al.*, 2006). At 2h post-infection, *Legionella* colocalized with lysosomal markers (Lamp1 and Cathepsin D) to a greater extent in *Naip5*^{B6}-transgenic than in non-transgenic macrophages. In a different study, *Legionella* trafficked to a Lamp1^{+} compartment in B6 macrophages by 2 h post infection, but avoided this fate in *Ipaf*$^{-/-}$ macrophages (Amer *et al.*, 2006). One other study suggested that Naip5 might induce the rapid acquisition of autophagic markers by *Legionella*-containing phagosomes (Amer and Swanson, 2005). These trafficking differences may explain the reduced capacity of *Legionella* to replicate in Naip5^{+}Ipaf^{+} macrophages. However, since most of the studies are correlative, direct experimental proof for this idea has yet to be provided. Indeed, it is difficult to reconcile the above trafficking studies with others in which *Naip5* does not appear to orchestrate as rapid or dramatic alterations in the maturation of *Legionella*-containing phagosomes (Derre and Isberg, 2004; Molofsky *et al.*, 2006). One suggestion has been that *Naip5*-mediated restriction of *Legionella* replication is 'multifactorial,' and may involve alterations in trafficking and induction of rapid cell death (Derre and Isberg, 2004; Swanson and Molofsky, 2005; Molofsky *et al.*, 2006). These issues will undoubtedly be resolved in future studies.

Toll-like receptors and *in vitro* recognition of *L. pneumophila*

Toll-like receptors (TLRs) constitute a family of transmembrane proteins that differentially recognize pathogen-associated molecular patterns (PAMPs) and initiate inflammatory signalling pathways (Aderem and Ulevitch, 2000; Janeway and Medzhitov, 2002; Akira and Takeda, 2004; Beutler, 2004; Akira, 2006). The bacterial PAMPs that are recognized by TLRs include lipopeptides by TLR2 (as a heterodimer with TLR1 or TLR6), lipopolysaccharide by TLR4, flagellin by TLR5, and bacterial CpG DNA by TLR9 (Table 5.2). Several recent *in vitro* and *in vivo* pneumonia studies suggest that TLRs are critical regulators of the pulmonary immune response to *L. pneumophila*. As a flagellated Gram-negative bacteria, *Legionella* has several features that distinguish it from other pulmonary pathogens including intracellular replication in macrophages, a flagellum, an unusual LPS structure, and fastidious growth requirements (Swanson and Hammer, 2000; Girard *et al.*, 2003). These features suggest that the immune response to *Legionella* will differ in comparison to other lung pathogens. *In vitro* studies indicate that *Legionella* is recognized by several TLRs, including TLR2, TLR5, and probably also TLR4 (Yoshida *et al.*, 1991; Girard *et al.*, 2003; Hawn *et al.*, 2003; Kikuchi *et al.*, 2004; Akamine *et al.*, 2005). Although *Legionella* is a Gram-negative bacterium with LPS, studies suggest that TLR2 is more important than TLR4 in mediating innate immune responses to this pathogen *in vitro* (Hawn *et al.*, 2003; Akamine *et al.*, 2005). It is not currently known if TLR2 cooperates with TLR1 or TLR6 to recognize *Legionella*. One study suggests that *Legionella* LPS may be recognized by TLR2 rather than TLR4 (Girard *et al.*, 2003), although this remains to be confirmed with highly purified LPS preparations used at low doses. There is also some evidence that TLR4 recognizes *Legionella* as Kikuchi *et al* found that TLR4 mediates IL-12 production *in vitro* in murine dendritic cells stimulated with heat-killed (but not formalin-killed or live) *L. pneumophila* (Kikuchi *et al.*, 2004). It is unusual for a Gram-negative bacterium to be predominantly recognized by TLR2 rather than TLR4. Finally, like all bacterial pathogens, *Legionella* has DNA with a stimulatory CpG motif and recent evidence indicates that TLR9 might recognize *Legionella* (Newton *et al.*, 2007). Although deficiency in individual TLRs does not generally have great effects on responses to *Legionella*, deficiency in MyD88, an adaptor required for signalling by several TLRs, appears to result in dramatically decreased IL-6, TNF-α and IL-12 production (Archer and Roy, 2006; Hawn *et al.*, 2006). MyD88-deficient macrophages exhibit either increased permissiveness (Ren *et al.*, 2006; Coers *et al.*, 2007) or no difference in permissiveness in comparison to wild type macrophages on a C57BL/6 (Molofksy *et al.*, 2006) or A/J background (Archer *et al.*, 2006). The experimental details that might account for these different results have not been fully elucidated. Together, these *in vitro* studies suggest that *Legionella* is recognized by multiple TLRs.

TLRs and murine *in vivo* response

Although TLRs and the innate immune system are essential for defending the host against microbes, the degree of redundancy and specificity manifested *in vivo* among different TLR family members is only partially understood. *In vivo* studies indicate a critical role for TLRs in mediating recognition of *Legionella*. MyD88 is an adaptor protein in the TLR pathway that mediates signals from all TLRs, except for TLR3. After aerosol or intranasal infection, in comparison to wild type C57BL/6 mice, *Myd88*$^{-/-}$ strains had dramatically higher bacterial counts in the lungs with decreased neutrophils in the bronchoalveolar lavage fluid as well as absent cytokine and chemokine production at early time points (Hawn *et al.*, 2006). Similar results were obtained when these experiments were done on an A/J background with the permissive Naip5 locus (Archer and Roy, 2006). In a model using an intravenous infection route, there were higher bacterial counts in the spleen of *MyD88*$^{-/-}$ as well as blunted cytokine responses (Sporri *et al.*, 2006). By day 6 of infection in the aerosol model, the *Myd88*$^{-/-}$ mice developed organizing pneumonia with dissemination of *L. pneumophila* to the lymph nodes and spleen (Hawn *et al.*, 2006). *Myd88*$^{-/-}$ mice had both a quantitative decrease in PMN recruitment as well as a localization defect. The neutrophils in the *Myd88*$^{-/-}$ mice were predominantly found in the interstitial and peri-bronchial spaces, but not in the alveoli. The absence of intra-alveolar neutrophils was associated with low intrapulmonary levels of the CXC chemokines MIP-2 and KC.

The defective recognition and inflammatory responses observed in *Myd88*$^{-/-}$ mice could potentially be mediated by several TLRs. Elucidating the distinct roles of each TLR in the innate immune response is critical for understanding how specific innate immune system defects influence the ability of the host organism to successfully control infections. Published studies indicate that *Tlr4*-deficient mice were not significantly more susceptible to *L. pneumophila* and efficiently recruited leukocytes to the lungs (Lettinga *et al.*, 2002a; Archer and Roy, 2006). In contrast, *Tlr2*$^{-/-}$ mice were more susceptible to *L. pneumophila* and had higher bacterial counts in the lung (Archer and Roy, 2006; Hawn *et al.*, 2006). In comparison to *Myd88*$^{-/-}$, the *Tlr2*$^{-/-}$ mice had less severely blunted inflammatory responses and were able to contain the infection, albeit with delayed bacterial clearance. The *Tlr2*$^{-/-}$ mice had only partially reduced numbers of neutrophils in BAL fluid, no defect of PMN migration to the alveoli, no defect in accumulation of mononuclear cells, and clearance of the infection without progression to organizing pneumonia and disseminated disease. These results suggest that TLR2 mediates partial resistance to *L. pneumophila*, indicating that additional MyD88-dependent, TLR2-independent pathways are essential for full protection. A study using an intravenous route of infection found that *Tlr2*$^{-/-}$, *Tlr4*$^{-/-}$, *Tlr2/4*$^{-/-}$, or *Tlr9*$^{-/-}$ mice were still able to control *Legionella* replication (Sporri *et al.*, 2006), implying again that there is redundancy among the various TLRs. The differences in the Sporri *et al.* study for TLR2 (in comparison to the aerosol and intranasal models) may be due to the use of an intravenous route of infection or the examination of only one time point (12–16 h).

Although Toll-like receptor 5 (TLR5), the receptor for bacterial flagellin, recognizes *Legionella*, its precise role during *in vivo* infection remains poorly understood (Hawn *et al.*, 2003). In light of these multiple innate recognition mechanisms, we recently examined whether TLR5 mediates a non-redundant role during the murine host response to infection with *Legionella* (T.R. Hawn and S.J. Skerrett, unpublished). In contrast to bone marrow-derived macrophages, alveolar macrophages recognize flagellin and a majority of the TNF-α response is TLR5-mediated. Furthermore, after an aerosolized infection, *Tlr5*$^{-/-}$ mice had lower numbers of PMNs in their bronchoalveolar lavage fluid in comparison to wild-type mice after 4 h. At 24 and 72 h, the PMN recruitment was similar. Interestingly, when comparing the *Tlr5*$^{-/-}$ and wild-type mice, there was no consistent difference in bacterial counts at any time point. However, at 6 days after infection, the *Tlr5*$^{-/-}$ mice had increased leucocytic infiltrates in the alveolar and peribronchial interstitial spaces. Together, these experiments suggest that TLR5 mediates recognition of *Legionella* in alveolar macrophages and is critical for the *in vivo* pulmonary immune response through regulation of early PMN recruitment and subsequent later development of pneumonia. These data imply that TLR5's role in host defence to *Legionella* is not redundant or readily supplanted by other innate immunity receptors that recognize flagellin.

The molecular and cellular mechanisms underlying the extreme susceptibility of *Myd88*$^{-/-}$ mice to *L. pneumophila* are only partially understood. Previous studies suggest that IFN-γ is critical for an effective immune response to *L. pneumophila* (Skerrett and Martin, 1992; Brieland *et al.*, 1994; Heath *et al.*, 1996; Shinozawa *et al.*, 2002). In two studies, IFN-γ was decreased in *Myd88*$^{-/-}$ mice (Archer and Roy, 2006; Sporri *et al.*, 2006). However, a different study found no differences in IFN-γ levels at early time points (Hawn *et al.*, 2006). Together, these results suggest that the increased susceptibility of *Myd88*$^{-/-}$ mice may be partially related to IFN-γ production. The significant difference in the neutrophil recruitment defect observed between these TLR-deficient mice and the associated difference in outcome suggests that PMN recruitment is another possible mechanism. These data are consistent with previous studies that suggest a critical role for neutrophils during *L. pneumophila* infection (Tateda *et al.*, 2001a,b). Finally, TNF-α- mediated effects may also be critical as levels were significantly depleted in three different studies of *Myd88*$^{-/-}$ mice (Archer and Roy, 2006; Hawn *et al.*, 2006; Sporri *et al.*, 2006).

Innate immune response: cytokines, chemokines, and effector molecules

TNF-α and nitric oxide (NO)

After recognition of *Legionella* by the innate immune system, alveolar macrophages and other host cells secrete a number of cytokines, chemokines, and effector molecules (Park and Skerrett, 2000; Neild and Roy, 2004). Several of these molecules have been identified for their pivotal role in mediating a protective host immune response. Among the initial cytokines and chemokines produced by macrophages during *in vitro* stimulation are molecules such as TNF-α, IL-1β, IL-6, IL-10, MIP-2, KC, and IL-12. Depletion of TNF-α during *in vitro* infections leads to decreased killing of *Legionella* in macrophages (Blanchard *et al.*, 1987; Funai *et al.*, 1997; Matsiota-Bernard *et al.*, 1993; Skerrett and Martin, 1996) and B6-background TNF receptor-deficient macrophages exhibit increased permissiveness to *Legionella* (Coers *et al.*, 2007). The importance of TNF-α for *in vivo* protective immunity has been demonstrated in TNF-α -deficient mice and rats (Brieland *et al.*, 1995; Skerrett and Martin, 1996; Skerrett *et al.*, 1997). In one study, TNF-α was depleted in A/J mice and the mice were infected intratracheally (Brieland *et al.*, 1995). When compared to wild-type mice, the TNF-α depleted mice had similar lung CFUs at 24 h. However, at 72 and 120 h post infection, the TNF-α-deficient mice had significantly increased bacterial counts with around 5-fold higher levels at 72 h. A second study in rats found that TNF-α -deficient animals were more susceptible to *Legionella* after an aerosolized infection with approximately 100-fold more bacteria at 96 h (Skerrett *et al.*, 1997). In addition, histological sections showed increased inflammation in the lungs after 4 days in the TNF-α-deficient animals. Interestingly, the TNF-α-deficient animals had elevated levels of IFN-γ, which suggested that other pathways mediate the protective effect of TNF-α.

One possible mechanism of TNF-α's effect is through nitric oxide (NO). In an *in vitro* study in rat alveolar macrophages, it was found that LPS and IFN-γ cooperate to control *Legionella* replication and that TNF-α depletion partially reverses this effect (Skerrett and Martin, 1996). The effect of TNF could be partially reversed by treatment with *N*-monomethyl-L-arginine (NMMA), an inhibitor of NO synthetase and NO production. Another study also showed the importance of NO for control of *Legionella* replication during *in vitro* infection (Summersgill *et al.*, 1992). In contrast, several other *in vitro* studies have not shown a role for NO in restriction of *Legionella* growth in murine peritoneal macrophages or cell lines (Gebran *et al.*, 1994; Kura *et al.*, 1994; Yamamoto *et al.*, 1996). The importance of NO in controlling *Legionella* infections has also been tested *in vivo* with administration of NMMA. In one study on a BALB/c background, there was no difference in *Legionella* growth in lungs after an intratracheal infection (Heath *et al.*, 1996). In a second study on an A/J background, there was no difference in CFUs at 24 h and a ~2-fold increase of CFUs at 72 h in the NMMA treated mice (Brieland *et al.*, 1995). Together, these *in vitro* and *in vivo* experiments suggest a partial role for NO in control of *Legionella* replication. In addition, these studies suggest that TNF-α may exert its effect, at least in part, through nitric oxide mediated restriction of *Legionella* growth in the macrophage.

IL-10

IL-10 is an inhibitory cytokine that can counter the effects of pro-inflammatory cytokines such as TNF-α and IFN-γ. Park and Skerrett found that IL-10 enhanced the growth of *Legionella* in human mononuclear phagocytes (Park and Skerrett, 1996). In addition, IL-10 was able to reverse the ability of IFN-γ to restrict *Legionella* growth and to block TNF-α secretion. In contrast, the inhibitory cytokines IL-4 and TGF-β did not have any affect on *Legionella* replication. These results suggest that IL-10 induction by *Legionella* may be a virulence mechanism that promotes bacterial replication in monocytes and macrophages.

Chemokines: CXCR2 and neutrophils

Neutrophil recruitment to the lung is a hallmark of *Legionella* infection in humans and animals (Winn and Myerowitz, 1981; Brieland *et al.*, 1994). Despite this recruitment, the importance of neutrophils in host defence has historically been questioned due to the relative resistance of *Legionella* to killing by neutrophils (Horwitz and Silverstein, 1981b; Weinbaum *et al.*, 1983). However, more recent studies suggest an important role for neutrophils and chemokines during a protective immune response to *Legionella*. When neutrophils were depleted from mice on an A/J background, the lethal challenge dose of *Legionella* was decreased by around 100-fold suggesting that neutrophils mediate a critical role in host defence to *in vivo Legionella* infection (Tateda *et al.*, 2001a). Interestingly, there was no difference in lethality when this experiment was done in C57BL/6 mice. The reason for this difference is not clear. The neutrophil-depleted A/J mice also had similar bacterial lung burdens at early time points and then a subsequent rise in counts by around 10–100-fold at 72 h after infection. Interestingly, the PMN-deficient animals had decreased levels of type 1 cytokines (IL-12 and IFN-γ) and increased levels of type 2 cytokines (IL-4 and IL-10) in comparison to the wild-type animals. These surprising results suggested that neutrophils may mediate host protection through an immunomodulatory effect, rather than direct killing of *Legionella*. The antibody used to deplete granulocytes in these studies (RB6–8C5) binds to the Ly6G antigen. Although originally described as a surface marker for neutrophils and other granulocytes, later studies described this protein on plasmacytoid dendritic cells (Hestdal *et al.*, 1991; Asselin-Paturel *et al.*, 2001). Thus, the immunomodulatory effects seen in the *Legionella* infection studies may be related to effects of DCs in addition to, or instead of, neutrophils.

The pulmonary recruitment of PMNs during *Legionella* infection is accompanied by upregulation of CXC chemokines such as keratinocyte-derived chemokine (KC) and macrophage inflammatory protein 2 (MIP-2). Blockade of CXCR2, the receptor for KC and MIP-2, resulted in decreased neutrophil recruitment and decreased survival in comparison to wild-type mice (Tateda *et al.*, 2001b). Similar to the results in neutrophil depleted mice, IL-12 levels were decreased in the CXCR2-deficient mice when compared to wild-type mice at 24–48 h, a period when bacterial counts were the same. These results again suggested an immunomodulatory role for chemokines and neutrophils as opposed to a direct killing mechanism.

Induction of interferon-β via TLR-independent pathway

Type I interferons (IFN-α, IFN-β) are extremely important cytokines in orchestrating innate and adaptive immune defences (Stetson and Medzhitov, 2006b). Among their diverse functions, IFN-α/β promotes NK cell and CD8 T cell cytotoxicity, upregulation of MHC class I molecules on antigen presenting cells, and upregulation of a large number of diverse antimicrobial molecules and enzymes. Although type I interferons are typically considered to be 'antiviral' cytokines, they have also been reported to have antibacterial functions, and may also contribute specifically to defence against *Legionella*. For example, defects in IFN-α/β signalling increase permissiveness to *Legionella* replication of macrophages (Coers *et al.*, 2007) or human A549 cells (Opitz *et al.*, 2006).

Type I interferons are induced by several cell types downstream of TLR3/4/7/8/9 signalling. Since *Legionella* expresses ligands for many, if not all, of these TLRs, it seems likely that TLRs contribute to the induction of type I IFN in response to *Legionella*. However, a role for TLR in induction of type I IFN in response to *Legionella* has not been extensively investigated. In fact, an intriguing recent observation is that $MyD88^{-/-}$/$Trif^{-/-}$ macrophages, which lack all TLR signalling, are still capable of inducing type I IFN in response to *Legionella* (Stetson and Medzhitov, 2006a). *Legionella* mutants lacking the type IV secretion apparatus failed to induce type I IFN in $MyD88^{-/-}$/$Trif^{-/-}$ macrophages (Stetson and Medzhitov, 2006a) or A549 cells (Opitz *et al.*, 2006). The requirement for type IV secretion did not appear to be due to a requirement for proper intracellular trafficking and replication, since a *Legionella icmW* mutant that fails to traffic or replicate normally still induced normal amounts of type I IFN (Stetson and Medzhitov, 2006a). It was proposed that Dot/Icm-dependent delivery of bacterial DNA to the macrophage cytosol might be responsible for the observed induction of type I IFN by *Legionella* (Stetson and Medzhitov, 2006a). Direct evidence for the idea that *Legionella* injects DNA into host cells has not been provided, and the possibility that *Legionella* RNA might be inducing type I IFN via the cytosolic ribonucleic acid sensors RIG-I or Mda5 has not been excluded. It is also possible that a type IV secreted effector expressed by *Legionella* might be responsible for the induction of type I IFN. However, transfection of bacterial DNA in to the cytosol of $MyD88^{-/-}$/$Trif^{-/-}$ cells was sufficient to trigger type I IFN production (Stetson and Medzhitov, 2006a). In A549 cells, induction of type I IFN by *Legionella* was reduced by RNAi-mediated knockdown of the adapter protein MAVS (also called CARDIF, VISA or IPS-1), an adapter required for signalling downstream of the RNA-sensor proteins RIG-I and Mda5 (Opitz *et al.*, 2006). While this result hints that *Legionella* RNA may be exported to the host cell cytosol, the mechanism by which *Legionella* induces type I IFN deserves greater investigation.

Adaptive immune response

Links from Innate to adaptive immunity: IL-12, IL-18, and IFN-γ

IL-12 and IL-18 (originally called IFN-γ inducing factor), which are produced by monocytes and macrophages, have several functions including activation of T cells and induction of IFN-γ production. IL-12 and IL-18 have been shown to act synergistically to cause increased production of IFN-γ by T and NK cells. Brieland *et al* investigated the roles of these molecules during *in vivo Legionella* infection by depleting A/J mice with antibodies to IL-12, the IL-18 receptor (IL-18R), or both (Brieland *et al.*, 1998; Brieland *et al.*, 2000). Although inhibition of either IL-12 or IL-18R led to decreased production of IFN-γ at 24 hours, the inhibition was more complete and persistent with IL-12 depletion and even greater with depletion of both molecules. Depletion of IL-18R did not alter the bacterial burden in the lungs. In contrast, IL-12 depletion did cause a significant increase in *Legionella* replication in comparison to wild-type mice. Together, these results suggest that IL-12 is more important than IL-18 in mediating a protective immune response to *Legionella*.

Numerous studies have shown that IFN-γ, which is produced by T cells and NK cells, activates macrophages and restricts growth of *Legionella in vitro*. The mechanism of inhibition is likely multifactorial and includes inhibition of cellular entry (Horwitz and Silverstein, 1981a), inhibition of complement mediated phagocytosis (Horwitz, 1992), promotion of phagolysosomal fusion (Horwitz, 1983a; Horwitz, 1983b; Horwitz, 1984; Santic *et al.*, 2005), sequestration of nutrients (iron and L-tryptophan) (Gebran *et al.*, 1994), activation of oxygen-dependent killing mechanisms (Locksley *et al.*, 1982; Jacobs *et al.*, 1984), and activation of nitric oxide killing mechanisms (Summersgill *et al.*, 1992; Skerrett and Martin, 1996). There appear to be differences between human and mouse macrophages with respect to the mechanism of IFN-γ.

The essential role of IFN-γ during *in vivo* infection has been shown directly with knockout and antibody-depleted mice, which have increased bacterial lung burdens as well as extrapulmonary dissemination (Fujio *et al.*, 1992; Heath *et al.*, 1996; Shinozawa *et al.*, 2002). In the earliest study, which used an i.v. route of infection, there were increased CFUs in the liver and spleen that ranged from 10 to over 10,000 fold (Fujio *et al.*, 1992). The effects of IFN-γ depletion were seen in both A/J and C57BL/6 strains, though the magnitude was higher in the A/J strain. In a study using an intranasal route of infection in IFN-$\gamma^{-/-}$ mice, there was also increased mortality in the deficient animals (Shinozawa *et al.*, 2002). In contrast to studies with MyD88-deficient animals, there was no difference in the early recruitment of cells to the lungs as assessed in the BAL fluid. However, at later time points, the IFN-$\gamma^{-/-}$ mice exhibited increased lung inflammation and persistent pneumonitis with alveolar exudates that were predominantly neutrophils. The increased bacterial lung burdens were present from early times points (24 h) and persisted for at least 10 days. Based on studies of other pathogens, the cellular source of IFN-γ during *Legionella* infection is likely NK/NKT and effector T cells at the early and late time points, respectively. Indeed, evidence suggests that NK cells are important early producers of IFN-γ in response to *Legionella* (Blanchard *et al.*, 1988; Sporri *et al.*, 2006). In addition, NK-depleted mice exhibited increased growth of *Legionella in vivo* (Sporri *et al.*, 2006). Interestingly, it was shown that IFN-γ production by NK cells required MyD88 function in the NK cells themselves, though it was not clear whether MyD88 function was required downstream of TLR signalling, implying NK cells might directly recognize *Legionella*, or downstream of IL-18, which is an IFN-γ inducing cytokine that requires MyD88 for signalling.

Antigen presentation

Stimulation of T cell responses requires the recognition of foreign peptide antigens presented by major histocompatibility complex (MHC) molecules on the surface of professional antigen presenting cells (dendritic cells and macrophages). Peptides presented by MHC class I are typically derived from cytosolic protein antigens and stimulate cytotoxic $CD8^+$ T cells. In contrast, peptides presented by MHC class II are typically derived from proteins degraded in acidic endocytic compartments and stimulate helper $CD4^+$ T cells. Since *Legionella* resides within an ER-like compartment and is not thought to have direct access to cytosolic or class II processing compartments, it is interesting to ponder how *Legionella* antigens are presented by MHC class I or II molecules to T cells. At present, this discussion is speculative, since *Legionella*-derived peptide antigens have yet to be identified.

There are at least two possible explanations for how *Legionella* antigens might be presented by MHC class I. First, *Legionella* is known to secrete >30 effector proteins into the host cell cytosol via its Dot/Icm type IV secretion system. These proteins would have direct access to the MHC class I antigen presentation pathway, which begins with proteasome-mediated degradation of cytosolic proteins. Second, certain antigen presenting cells (such as dendritic cells) exhibit a special capacity to present phagocytosed antigens on MHC class I. This capacity is called cross-presentation, and its mechanism is unknown. One suggestion has been that cross-presentation occurs when ER-resident proteins are retrotranslocated into the cytosol via the ER-associated degradation (ERAD) pathway (Ackerman *et al.*, 2006). It is plausible that *Legionella* antigens access the MHC class I pathway via ERAD-mediated retrotranslocation, though this remains to be established.

Although it is clear that *Legionella* stimulates specific CD4+ T cells (Neild and Roy, 2003; Neild *et al.*, 2005), the mechanism by which *Legionella* antigens are presented on MHC class II remains uncertain. Assembly of peptide–MHC class II complexes is thought to occur exclusively in an acidified endocytic compartment, but upon entry into cells, *Legionella* evades this compartment and instead trafficks to an ER-like compartment. Given this, several possible mechanisms by which *Legionella* antigens could be presented by MHC class II can be envisaged (Neild *et al.*, 2005). One unlikely possibility is that some functional MHC class II complexes are generated in the non-acidic *Legionella*-containing vacuole. Alternatively, proteins secreted by *Legionella* may exit the *Legionella*-containing vacuole and traffic to acidic compartments in which peptide:MHC class II complexes are assembled. A further possibility, consistent with previous observations (Sturgill-Koszycki and Swanson, 2000), is that the ER-like compartment containing *Legionella* may itself acidify and merge with lysosomes during the replication period and become competent to assemble peptide:MHC class II complexes. The observation that neutralization of endocytic compartments with chloroquine prevented presentation of *Legionella*-derived antigens on MHC class II suggests that an acidic compartment is indeed a critical player in class II MHC presentation (Neild *et al.*, 2005) but does not identify the nature of that acidic compartment. Another possibility is that the small number of bacteria-containing phagosomes that do not successfully evade lysosomal fusion may account for most of the antigen-presenting capacity of infected macrophages. Interestingly, however, *Legionella dotA* mutants are not efficient at presenting *Legionella* antigens to T cells primed with wild-type *Legionella* (Neild *et al.*, 2005), implying that antigen access to MHC class II requires Dot/Icm-dependent trafficking of *Legionella*.

Given the critical role that dendritic cells (DCs) play in stimulating T cell responses, another interesting finding is that DCs differ from macrophages in their permissiveness to *Legionella* replication (Neild and Roy, 2003). Whereas macrophages from A/J mice support vigorous intracellular replication of *Legionella*, dendritic cells from A/J mice permit virtually no bacterial replication. *Legionella* avoided lysosomal fusion in dendritic cells, and were metabolically active (Neild and Roy, 2003). It was speculated that DCs might restrict the intracellular growth of (but did not kill) *Legionella* to allow the bacteria to synthesize proteins that could be presented by DCs after they migrate from sites of infection to lymph nodes (Neild and Roy, 2003).

The role of B cells and T cells in *L. pneumophila* infection

The role of the humoral immune response to *Legionella* during murine infection is suggested by the recruitment of B cells to the lungs and the development of anti-*Legionella* antibodies by day 5 of infection (Brieland *et al.*, 1996). More direct evidence of the importance of the humoral immune response has been obtained with opsonization studies. When *Legionella* was opsonized with an anti-*Legionella* antibody prior to infection, the bacterial burden in the lungs was significantly decreased in comparison to mock-opsonized bacteria treated with a control antibody (Brieland *et al.*, 1996). Opsonization with anti-*Legionella* antibody was associated with increased phagocytosis, which may have subsequently led to an increased ability of alveolar macrophages to restrict replication. These effects were independent of complement. However, earlier work in human monocytes indicated that *Legionella* opsonization does not significantly alter bacterial replication (Horwitz and Silverstein, 1981c). The different methodologies used in these studies (human vs. mouse and *in vitro* vs. *in vivo* infections) likely accounts for the disparate results. Together, these results suggest that the humoral response can restrict *Legionella* growth, although the significance of this in humans is less clear.

The role of T cells was examined with depletion studies which showed that reduction of either CD4+ or CD8+ T cells led to increased lung CFUs and lethality in A/J mice (Susa *et al.*, 1998). Depletion of both subsets led to a greater effect in comparison to depletion of either subset alone suggesting a role for both CD4 and CD8 cells in immunity to *Legionella*.

Vaccination and memory responses

There is no human vaccine for *Legionella*. As is evident from the above discussion, *Legionella* clearly stimulates adaptive immune responses, and these responses are important for clearance of the infection. However, it remains uncertain whether vaccination could stimulate appropriate long-lasting responses that would protect against subsequent exposure. Prior infection can (Breiman and Horwitz, 1987; Blander *et al.*, 1989) but does not necessarily induce protective memory responses (Meyer *et al.*, 1980).

Several vaccine candidates have been tested using the guinea pig model of infection. Vaccination with the major secreted protein (Msp) of *Legionella*, or with a *Legionella* membrane preparation, was found to induce protective humoral and cell-mediated immune responses (Blander and Horwitz, 1989; Blander and Horwitz, 1991a,b). Another group found that immunization with a *Legionella* outer membrane protein (OmpS) provoked strong T cell responses that were protective against subsequent lethal challenge (Weeratna *et al.*, 1994). Interestingly, humoral responses in OmpS-immunized animals were weak, further reinforcing the notion that cell-mediated responses are likely to be more critical than antibody responses for clearance of *Legionella*. Immunization of mice with *Legionella* flagellin protein (Ricci *et al.*, 2005), or with *Legionella*-pulsed DCs (Kikuchi *et al.*, 2004), was also found to induce T cell responses and/or protective immunity against a subsequent lethal challenge with virulent *Legionella*.

Human immune response to *Legionella*

Murine studies of *Legionella* have significantly advanced our understanding of the host immune response. Despite these insights, there remain important differences between mice and humans and the relevance of murine infection studies for human disease needs to be interpreted with caution (Casanova and Abel, 2004). Murine studies generally use high doses of *Legionella*, often with strains that are attenuated by laboratory passage, and sometimes with artificial routes of inoculation. The mouse is usually inbred and often only one strain is studied. Furthermore, experimental mice have a standardized age, are rarely vaccinated, and have a short life span. All of these factors influence the interpretation of results and their relevance for human disease. In this section, we will describe findings from studies of human infection, including more recent genetic studies of the role of the TLR pathway in human *Legionella* infection.

Human vs. mouse macrophage phenotype

There are several important differences between the human and murine macrophage response to *Legionella* infection. First, whereas human alveolar macrophages are highly permissive for growth of *Legionella*, only the A/J mouse background has been shown to be highly permissive among murine strains. Second, although IFN-γ activation of human macrophages can inhibit bacterial growth, net bacterial killing is only observed in murine macrophages (Klein *et al.*, 1991; Skerrett and Martin, 1992; Summersgill *et al.*, 1992; Yamamoto *et al.*, 1992; Kura *et al.*, 1994). Third, nitric oxide killing mechanisms have been demonstrated in murine macrophages, but not humans (Summersgill *et al.*, 1992). This latter difference is not specific to *Legionella* as it has been found for other pathogens as well (Schneemann *et al.*, 1993; Schoedon *et al.*, 1995). However, the extent and significance of differences in NO killing is not clear since many human studies have been performed with *in vitro* culture conditions where NO production is low. Furthermore, macrophages obtained from patients *in vivo* express iNOS and suggest that human macrophages contain an NO pathway that is important for pathogen killing (MacMicking *et al.*, 1997).

Legionella and the immune response during human infection

Very few studies have examined the immune response to *Legionella* during infection in humans. Lung pathology studies demonstrate an acute pneumonia that can be patchy or more commonly involve several lobes, often in a lobular or confluent lobular distribution (Winn *et al.*, 1978; Winn and Myerowitz, 1981). Occasional gross features include nodules (12/42 cases), macroscopic abscesses (10/42), and pleural effusions. Microscopically, there is an acute alveolitis and bronchiolitis with equal numbers of neutrophils and macrophages. The predominant inflammatory exudates are in the alveoli and alveolar ducts as well as the interstitium. Abscess cavities are present microscopically in a minority of cases (10/53). In addition, there are often signs of leucocytoclasis of the cellular exudate (33/53) and small vessel vasculitis (16/53). Together, these features are characteristic of, but not pathognomonic for, Legionnaires' disease.

In addition to autopsy studies, there have been two studies of the cytokine response during *in vivo* infection. In one study of 6 individuals with pneumonia, cytokine levels were examined for seven days after study entry (Fernandez-Serrano *et al.*, 2003). Although *Legionella* infection was associated with elevated levels of IL-6 and IL-10, there was no significant difference in these cytokines levels when analysed by outcome or compared to other causes of pneumonia such as pneumococcus. In a separate study of 14 individuals, cytokine levels were measured for up to 40 days after infection (Tateda *et al.*, 1998). Interestingly, IFN-γ and IL-12, which are associated with Th1 type T cell cytokine responses, were readily detectable in many subjects, whereas the Th2 type T cell cytokines IL-4 and IL-10 were not. Sample numbers were not large enough to examine whether cytokine patterns correlated with different outcomes. Larger studies will be needed to examine whether *Legionella* infection is associated with unique cytokine patterns and whether these profiles have prognostic value.

Human genetic studies

Although molecular, cellular, and *in vivo* animal studies are critical for advancing our knowledge of the pathophysiology of disease, one critical question is whether these findings have relevance for human disease. Risk factors for human acquisition of Legionnaire's disease include smoking (odds ratio range of 1.6–4.2), chronic obstructive pulmonary disease (odds ratio ~3.5), alcohol use (OR 1.3–4.8), and other chronic medical co-morbidities such as diabetes and kidney failure (OR ~2.0) (Fraser *et al.*, 1977; England *et al.*, 1981; Straus *et al.*, 1996). However, many individuals have LD without any identifiable risk factors, suggesting that there may be host genetic factors that influence susceptibility. Recent human studies of *Legionella* have assessed whether genetic factors influence susceptibility to disease(Hawn *et al.*, 2003; Hawn *et al.*, 2005). The establishment of the Human Genome Project in 1990 propelled the scientific community into a new era of genomics, laying the groundwork for genetic mapping and the development of large-scale gene identification technologies that facilitate the study of molecular polymorphisms. The most common type of human genetic variation is the single nucleotide polymorphism (SNP), where two alternative bases occur at appreciable frequency (>1%) in the population. Most recently, several studies have utilized SNPs in candidate genes to find associations with susceptibility to different infectious diseases, including pneumonia, HIV, sepsis, malaria, and tuberculosis (Cooke and Hill, 2001; Cook *et al.*, 2004).

Recent studies have investigated whether polymorphisms in TLRs influence susceptibility to Legionnaires' disease (Hawn *et al.*, 2003). In sequencing studies of the coding region of TLR5 from healthy individuals, a variant (C1174T) was detected in 6.0% of subjects that changed an arginine to a stop codon at amino acid 392 ($TLR5^{392STOP}$). $TLR5^{392STOP}$ was unable to mediate flagellin-induced NF-κB activation in transfected CHO and HEK293 cells, while full-length TLR5 conferred flagellin responsiveness. In addition, in peripheral blood mononuclear cells (PBMCs) stimulated with flagellin, heterozygous stop codon individuals (392R*) had significantly decreased IL-6 production in comparison to those with the wild type arginine at amino acid 392. This result indicated that TLR5392STOP exerts a dominant effect and suggests that heterozygous TLR5 stop codon individuals may be more susceptible to infection with flagellated bacteria.

To determine whether TLR5 is associated with susceptibility to Legionnaires' disease, a case–control cohort from the Bovenkarspel epidemic was examined (Boshuizen *et al.*, 2001; Den Boer *et al.*, 2002; Lettinga *et al.*, 2002b). Cases included individuals with radiologically confirmed pneumonia occurring during the epidemic time period around the West Friese Flower Show in the town of Bovenkarspel in the Netherlands in 1999. Controls were drawn from a pool of exhibitioners who were likely to be exposed to the contaminated water product that caused the epidemic. The combined frequency of 1174TC heterozygotes and 1174TT homozygotes was 16.7% in the cases compared to 9.5% in the control group. The odds ratio for the association of genotype 1174T(C,T) with developing Legionnaires' disease was 1.90 (95% CI 1.06–3.42, P = 0.03). A second polymorphism (A1775G, N592S) was also found to associated with Legionnaires' disease in this study, although the functional significance of this variant was not apparent. Together, these results demonstrated that the TLR5 stop codon was associated with increased susceptibility to Legionnaires' disease and was the first example of an association of TLR5-deficiency with any disease.

TLR4 polymorphisms were also examined by comparing the allele frequencies of two SNPs (A896G and C1196T) that had been previously characterized and studied (Arbour *et al.*, 2000; Cook *et al.*, 2004). Allele 896G was associated with Legionnaires' disease susceptibility with a frequency of 6.5% in the control group versus 2.5% in cases (odds ratio of 0.36, 95% confidence interval 0.14–0.91, P = 0.025). An analysis of genotype frequencies (896 AA vs. AG and GG) demonstrated similar protective associations (case vs. control group comparison, OR = 0.35, 95% CI 0.14–0.89, P = 0.02; matched control group comparison, OR = 0.25, 95% CI 0.09–0.71, P = 0.006). Allele 1196T co-segregated with allele 896G and thus had identical associations. Although previous studies suggest that these TLR4 SNPs are associated with an increased risk of infection, this was the first study to demonstrate an association with resistance. This protective association illustrated that an innate immune receptor can mediate either beneficial or deleterious inflammatory responses and that these outcomes vary with different pathogens.

Taken together, these results suggest that TLR4 and TLR5 polymorphisms are associated with Legionnaires' disease. While these odds ratios are modest, the majority of studies which have been published on the association of immunological genes with susceptibility to common infections have reported modest ratios, ranging from 1.5 to 4 with a large number of these clustered around 2. These findings may reflect the multi-factorial nature of common diseases with anticipated polygenic and environmental influences on susceptibility. Evolutionary pressures strongly select against mutations that result in extreme susceptibility reflected by high odds ratios. Thus, it is precisely these apparently modest associations that may tip the balance towards protection or susceptibility.

Conclusions and future trends

Legionella pneumophila has proven to be a valuable model organism for the genetic analysis of innate immunity. Its utility has been illustrated both by human genetic studies and by the recent discovery of two important innate immune pathways involved in the initial sensing of pathogens, namely, intracellular sensing of flagellin (via Naip5/Ipaf) and cytosolic sensing of nucleic acids. It is already clear that principles elucidated from the study of *Legionella* will be applicable to other pathogens. As for many host–pathogen interactions, innate and adaptive immunity to *Legionella* appears to require a complex collaboration among a diverse set of cytokines, immune cell types and effector mechanisms (Fig. 5.1). A future challenge facing the study of immunity to pathogens in general is to assemble what we have

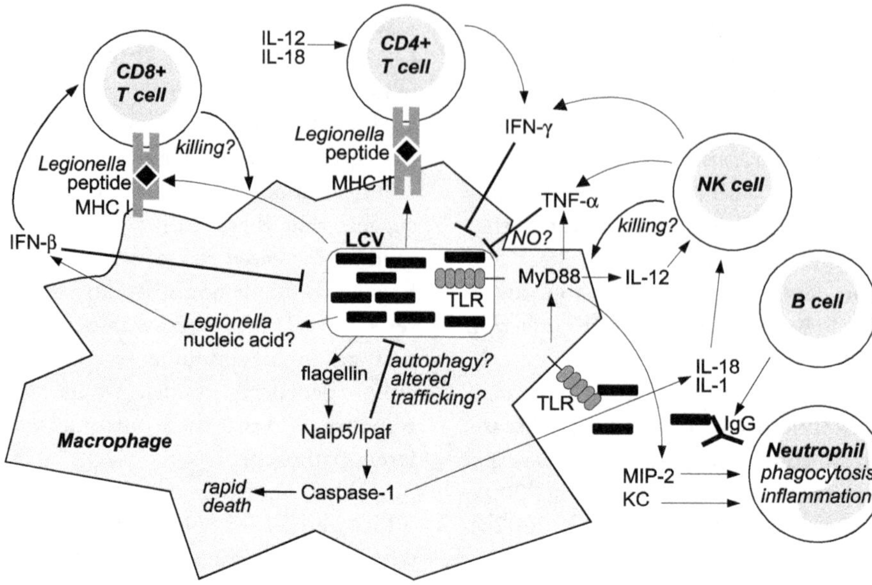

Figure 5.1 Summary of immune responses to *Legionella pneumophila*. Depicted are the molecules and cells that have been identified as important mediators of immunity to *Legionella*. See text for more details. Legend: IgG, immunoglobulin G; Lp, *Legionella pneumophila*; NO, nitric oxide; LCV, *Legionella*-containing vacuole; TLR, Toll-like receptor; MHC I, Major Histocompatibility Complex class I; MHC II, Major Histocompatibility Complex class II. After ingestion by the alveolar macrophage, *Legionella* is sensed by various pathways. For example, *Legionella* expresses several ligands that activate TLRs (Table 5.2), in turn leading to signalling downstream of the MyD88 adaptor protein. It has also been proposed that the *Legionella* Dot/Icm type IV secretion system translocates *Legionella* nucleic acids and flagellin to the macrophage cytosol, leading to the production of IFN-β and activation of caspase-1, respectively. Activated macrophages secrete cytokines and chemokines that either activate macrophages (TNF-α), T cells (IL-12 and IL-18), neutrophils (KC, MIP-2), or natural killer (NK) cells. T cells and NK cells can kill infected cells and can also produce IFN-γ. IFN-γ and TNF-α then activate macrophages to produce NO, which may contribute to clearance of *Legionella*. Neutrophil activation (through the CXCR2 receptor) and B cell production of anti-*Legionella* IgG may also contribute to protective immunity.

learned about these multiple diverse processes into a coherent and unified understanding of how hosts resist infection. It is to be expected that the study of immunity to *Legionella* will continue to shed light on this fundamental and important area of biology.

Acknowledgements

Russell Vance would like to thank Bill Dietrich for his pioneering vision and support, Tao Ren, Dario Zamboni, Jörn Coers, and Craig Roy for being fabulous collaborators, and Karla Lightfield and Jenny Persson for sharing unpublished data. Thomas Hawn thanks Shawn Skerrett, Alan Aderem, Annelies Verbon, Kamilla Lettinga, and Lue Ping Zhao for their outstanding collaborative support.

References

Ackerman, A.L., Giodini, A., and Cresswell, P. (2006). A role for the endoplasmic reticulum protein retrotranslocation machinery during crosspresentation by dendritic cells. Immunity *25*, 607–617.

Aderem, A., and Ulevitch, R.J. (2000). Toll-like receptors in the induction of the innate immune response. Nature *406*, 782–787.

Akamine, M., Higa, F., Arakaki, N., Kawakami, K., Takeda, K., Akira, S., and Saito, A. (2005). Differential roles of Toll-like receptors 2 and 4 in in vitro responses of macrophages to *Legionella pneumophila*. Infect. Immun. *73*, 352–361.

Akira, S. (2006). TLR signaling. Curr. Top. Microbiol. Immunol. *311*, 1–16.

Akira, S., and Takeda, K. (2004). Toll-like receptor signalling. Nat. Rev. Immunol. *4*, 499–511.

Amer, A., Franchi, L., Kanneganti, T.D., Body-Malapel, M., Ozoren, N., Brady, G., Meshinchi, S., Jagirdar, R., Gewirtz, A., Akira, S., and Nunez, G. (2006). Regulation of *Legionella* phagosome maturation and infection through flagellin and host IPAF. J. Biol. Chem. *281*, 35217–35223

Amer, A.O., and Swanson, M.S. (2005). Autophagy is an immediate macrophage response to *Legionella pneumophila*. Cell. Microbiol. *7*, 765–778.

Arbour, N., Lorenz, E., Schutte, B., Zabner, J., Kline, J., Jones, M., Frees, K., Watt, J., and Schwartz, D. (2000). TLR4 mutations are associated with endotoxin hyporesponsiveness in humans. Nat. Gen. *25*, 187–191.

Archer, K.A., and Roy, C.R. (2006). MyD88-dependent responses involving toll-like receptor 2 are important for protection and clearance of *Legionella pneumophila* in a mouse model of Legionnaires' disease. Infect. Immun. *74*, 3325–3333.

Asare, R., Santic, M., Gobin, I., Doric, M., Suttles, J., Graham, J.E., Price, C.D., and Abu Kwaik, Y. (2007). Genetic susceptibility and caspase activation in mouse and human macrophages are distinct for *Legionella longbeachae* and *L. pneumophila*. Infect. Immun. *75*, 1933–1945.

Asselin-Paturel, C., Boonstra, A., Dalod, M., Durand, I., Yessaad, N., Dezutter-Dambuyant, C., Vicari, A., O'Garra, A., Biron, C., Briere, F., and Trinchieri, G. (2001). Mouse type I IFN-producing cells are immature APCs with plasmacytoid morphology. Nat. Immunol. *2*, 1144–1150.

Beckers, M.C., Yoshida, S., Morgan, K., Skamene, E., and Gros, P. (1995). Natural resistance to infection with *Legionella pneumophila*: chromosomal localization of the Lgn1 susceptibility gene. Mamm. Genome. *6*, 540–545.

Beutler, B. (2004). Inferences, questions and possibilities in Toll-like receptor signalling. Nature *430*, 257–263.

Bhardwaj, N., Nash, T.W., and Horwitz, M.A. (1986). Interferon-gamma-activated human monocytes inhibit the intracellular multiplication of *Legionella pneumophila*. J. Immunol. *137*, 2662–2669.

Blanchard, D.K., Djeu, J.Y., Klein, T.W., Friedman, H., and Stewart, W.E., 2nd (1987). Induction of tumor necrosis factor by *Legionella pneumophila*. Infect. Immun. *55*, 433–437.

Blanchard, D.K., Friedman, H., Stewart, W.E., 2nd, Klein, T.W., and Djeu, J.Y. (1988). Role of gamma interferon in induction of natural killer activity by *Legionella pneumophila* in vitro and in an experimental murine infection model. Infect. Immun. *56*, 1187–1193.

Blander, S.J., Breiman, R.F., and Horwitz, M.A. (1989). A live avirulent mutant *Legionella pneumophila* vaccine induces protective immunity against lethal aerosol challenge. J. Clin. Invest. *83*, 810–815.

Blander, S.J., and Horwitz, M.A. (1989). Vaccination with the major secretory protein of *Legionella pneumophila* induces cell-mediated and protective immunity in a guinea pig model of Legionnaires' disease. J. Exp. Med. *169*, 691–705.

Blander, S.J., and Horwitz, M.A. (1991a). Vaccination with *Legionella pneumophila* membranes induces cell-mediated and protective immunity in a guinea pig model of Legionnaires' disease. Protective immunity independent of the major secretory protein of *Legionella pneumophila*. J. Clin. Invest. *87*, 1054–1059.

Blander, S.J., and Horwitz, M.A. (1991b). Vaccination with the major secretory protein of *Legionella* induces humoral and cell-mediated immune responses and protective immunity across different serogroups of *Legionella pneumophila* and different species of *Legionella*. J. Immunol. *147*, 285–291.

Boshuizen, H.C., Neppelenbroek, S.E., van Vliet, H., Schellekens, J.F., den Boer, J.W., Peeters, M.F., and Conyn-van Spaendonck, M.A. (2001). Subclinical *Legionella* infection in workers near the source of a large outbreak of Legionnaires disease. J. Infect. Dis. *184*, 515–518.

Brand, B.C., Sadosky, A.B., and Shuman, H.A. (1994). The *Legionella pneumophila* icm locus: a set of genes required for intracellular multiplication in human macrophages. Mol. Microbiol. *14*, 797–808.

Breiman, R.F., and Horwitz, M.A. (1987). Guinea pigs sublethally infected with aerosolized *Legionella pneumophila* develop humoral and cell-mediated immune responses and are protected against lethal aerosol challenge. A model for studying host defense against lung infections caused by intracellular pathogens. J. Exp. Med. *165*, 799–811.

Brieland, J., Freeman, P., Kunkel, R., Chrisp, C., Hurley, M., Fantone, J., and Engleberg, C. (1994). Replicative *Legionella pneumophila* lung infection in intratracheally inoculated A/J mice. A murine model of human Legionnaires' disease. Am. J. Pathol. *145*, 1537–1546.

Brieland, J.K., Engleberg, N.C., Huffnagle, G.B., Remick, D.G., and Fantone, J.C. (2000). Host pathogen interactions in Legionnaires' disease: lessons learned from a murine animal model. Immunopharmacology *48*, 249–252.

Brieland, J.K., Heath, L.A., Huffnagle, G.B., Remick, D.G., McClain, M.S., Hurley, M.C., Kunkel, R.K., Fantone, J.C., and Engleberg, C. (1996). Humoral immunity and regulation of intrapulmonary growth of *Legionella pneumophila* in the immunocompetent host. J. Immunol. *157*, 5002–5008.

Brieland, J.K., Remick, D.G., Freeman, P.T., Hurley, M.C., Fantone, J.C., and Engleberg, N.C. (1995). In vivo regulation of replicative *Legionella pneumophila* lung infection by endogenous tumor necrosis factor alpha and nitric oxide. Infect. Immun. *63*, 3253–3258.

Brieland, J.K., Remick, D.G., LeGendre, M.L., Engleberg, N.C., and Fantone, J.C. (1998). In vivo regulation of replicative *Legionella pneumophila* lung infection by endogenous interleukin-12. Infect. Immun. *66*, 65–69.

Casanova, J.L., and Abel, L. (2004). The human model: a genetic dissection of immunity to infection in natural conditions. Nat. Rev. Immunol. *4*, 55–66.

Coers, J., Vance, R.E., Fontana, M.F., and Dietrich, W.F. (2007). Restriction of *Legionella pneumophila* growth in macrophages requires the concerted action of cytokine and Naip5/Ipaf signalling pathways. Cell. Microbiol. *9*, 2344–2357.

Cook, D.N., Pisetsky, D.S., and Schwartz, D.A. (2004). Toll-like receptors in the pathogenesis of human disease. Nat. Immunol. *5*, 975–979.

Cooke, G.S., and Hill, A.V. (2001). Genetics of susceptibility to human infectious disease. Nat Rev. Genet. *2*, 967–977.

Damiano, J.S., Oliveira, V., Welsh, K., and Reed, J.C. (2004). Heterotypic interactions among NACHT domains: implications for regulation of innate immune responses. Biochem. J. *381*, 213–219.

Davoodi, J., Lin, L., Kelly, J., Liston, P., and MacKenzie, A.E. (2004). Neuronal apoptosis-inhibitory protein does not interact with Smac and requires ATP to bind caspase-9. J. Biol. Chem. *279*, 40622–40628.

Den Boer, J.W., Yzerman, E.P., Schellekens, J., Lettinga, K.D., Boshuizen, H.C., Van Steenbergen, J.E., Bosman, A., Van den Hof, S., Van Vliet, H.A., Peeters, M. F. Van Ketel, R.J., Speelman, P., Kool, J.L., and Conyn-Van Spaendonck, M.A. (2002). A large outbreak of Legionnaires' disease at a flower show, the Netherlands, 1999. Emerg. Infect. Dis. *8*, 37–43.

Derre, I., and Isberg, R.R. (2004). Macrophages from mice with the restrictive Lgn1 allele exhibit multifactorial resistance to *Legionella pneumophila*. Infect. Immun. *72*, 6221–6229.

Dietrich, W.F., Damron, D.M., Isberg, R.R., Lander, E.S., and Swanson, M.S. (1995). Lgn1, a gene that determines susceptibility to *Legionella pneumophila*, maps to mouse chromosome 13. Genomics *26*, 443–450.

Diez, E., Beckers, M.C., Ernst, E., DiDonato, C.J., Simard, L. R., Morissette, C., Gervais, F., Yoshida, S.I., and Gros, P. (1997). Genetic and physical mapping of the mouse host resistance locus Lgn1. Mamm. Genome. *8*, 682–685.

Diez, E., Lee, S.H., Gauthier, S., Yaraghi, Z., Tremblay, M., Vidal, S., and Gros, P. (2003). Birc1e is the gene within the Lgn1 locus associated with resistance to *Legionella pneumophila*. Nat. Genet. *33*, 55–60.

Diez, E., Yaraghi, Z., MacKenzie, A., and Gros, P. (2000). The neuronal apoptosis inhibitory protein (Naip) is expressed in macrophages and is modulated after phagocytosis and during intracellular infection with *Legionella pneumophila*. J. Immunol. *164*, 1470–1477.

Endrizzi, M.G., Hadinoto, V., Growney, J.D., Miller, W., and Dietrich, W.F. (2000). Genomic sequence analysis of the mouse Naip gene array. Genome. Res. *10*, 1095–1102.

England, A.C., 3rd, Fraser, D. W., Plikaytis, B.D., Tsai, T. F., Storch, G., and Broome, C.V. (1981). Sporadic legionellosis in the United States: the first thousand cases. Ann. Intern. Med. *94*, 164–170.

Fernandez-Serrano, S., Dorca, J., Coromines, M., Carratala, J., Gudiol, F., and Manresa, F. (2003). Molecular inflammatory responses measured in blood of patients with severe community-acquired pneumonia. Clin. Diagn. Lab. Immunol. *10*, 813–820.

Fink, S.L., and Cookson, B.T. (2005). Apoptosis, pyroptosis, and necrosis: mechanistic description of dead and dying eukaryotic cells. Infect. Immun. *73*, 1907–1916.

Fortier, A., de Chastellier, C., Balor, S., and Gros, P. (2006). Birc1e/Naip5 rapidly antagonizes modulation of phagosome maturation by *Legionella pneumophila*. Cell. Microbiol. 9, 910–923.

Franchi, L., Amer, A., Body-Malapel, M., Kanneganti, T.D., Ozoren, N., Jagirdar, R., Inohara, N., Vandenabeele, P., Bertin, J., Coyle, A., Grant, E.P., and Núñez, G. (2006). Cytosolic flagellin requires Ipaf for activation of caspase-1 and interleukin 1beta in salmonella-infected macrophages. Nat. Immunol. *7*, 576–582.

Fraser, D.W., Tsai, T.R., Orenstein, W., Parkin, W.E., Beecham, H.J., Sharrar, R.G., Harris, J., Mallison, G.F., Martin, S.M., McDade, J.E, Shepard, C.C., and Brachman, P.S. (1977). Legionnaires' disease: description of an epidemic of pneumonia. N. Engl. J. Med. *297*, 1189–1197.

Fritz, J.H., Ferrero, R.L., Philpott, D.J., and Girardin, S.E. (2006). Nod-like proteins in immunity, inflammation and disease. Nat. Immunol. *7*, 1250–1257.

Fujio, H., Yoshida, S., Miyamoto, H., Mitsuyama, M., and Mizuguchi, Y. (1992). Investigation of the role of macrophages and endogenous interferon-gamma in natural resistance of mice against *Legionella pneumophila* infection. FEMS Microbiol. Immunol. *4*, 183–191.

Funai, N., Shimamoto, Y., Watanabe, M., Yoshida, S.I., and Kohashi, O. (1997). Marked inhibition of the intracellular multiplication of *Legionella pneumophila* in monocytes isolated from carriers of human T lymphotrophic virus type I. J. Leukoc. Biol. *61*, 133–140.

Gao, L.Y., and Abu Kwaik, Y. (1999). Activation of caspase 3 during *Legionella pneumophila*-induced apoptosis. Infect. Immun. *67*, 4886–4894.

Gebran, S.J., Yamamoto, Y., Newton, C., Klein, T.W., and Friedman, H. (1994). Inhibition of *Legionella pneumophila* growth by gamma interferon in permissive A/J mouse macrophages: role of reactive oxygen species, nitric oxide, tryptophan, and iron(III). Infect. Immun. *62*, 3197–3205.

Girard, R., Pedron, T., Uematsu, S., Balloy, V., Chignard, M., Akira, S., and Chaby, R. (2003). Lipopolysaccharides from *Legionella* and *Rhizobium* stimulate mouse bone marrow granulocytes via Toll-like receptor 2. J. Cell. Sci. *116*, 293–302.

Growney, J.D., and Dietrich, W.F. (2000). High-resolution genetic and physical map of the Lgn1 interval in C57BL/6J implicates Naip2 or Naip5 in *Legionella pneumophila* pathogenesis. Genome Res. *10*, 1158–1171.

Hawn, T.R., Smith, K.D., Aderem, A., and Skerrett, S.J. (2006). Myeloid differentiation primary response gene (88)- and toll-like receptor 2-deficient mice are susceptible to infection with aerosolized *Legionella pneumophila*. J. Infect. Dis. *193*, 1693–1702.

Hawn, T.R., Verbon, A., Janer, M., Zhao, L.P., Beutler, B., and Aderem, A. (2005). Toll-like receptor 4 polymorphisms are associated with resistance to Legionnaires' disease. Proc. Natl. Acad. Sci. USA. *102*, 2487–2489

Hawn, T.R., Verbon, A., Lettinga, K.D., Zhao, L.P., Li, S.S., Laws, R.J., Skerrett, S.J., Beutler, B., Schroeder, L., Nachman, A., Ozinsky, A., Smith, K.D., and Aderem, A. (2003). A common dominant TLR5 stop codon polymorphism abolishes flagellin signaling and is associated with susceptibility to legionnaires' disease. J. Exp. Med. *198*, 1563–1572.

Heath, L., Chrisp, C., Huffnagle, G., LeGendre, M., Osawa, Y., Hurley, M., Engleberg, C., Fantone, J., and Brieland, J. (1996). Effector mechanisms responsible for gamma interferon-mediated host resistance to *Legionella pneumophila* lung infection: the role of endogenous nitric oxide differs in susceptible and resistant murine hosts. Infect. Immun. *64*, 5151–5160.

Hestdal, K., Ruscetti, F.W., Ihle, J.N., Jacobsen, S.E., Dubois, C.M., Kopp, W.C., Longo, D.L., and Keller, J.R. (1991). Characterization and regulation of RB6–8C5 antigen expression on murine bone marrow cells. J. Immunol. *147*, 22–28.

Horwitz, M.A. (1983a). Formation of a novel phagosome by the Legionnaires' disease bacterium (*Legionella pneumophila*) in human monocytes. J. Exp. Med. *158*, 1319–1331.

Horwitz, M.A. (1983b). The Legionnaires' disease bacterium (*Legionella pneumophila*) inhibits phagosome-lysosome fusion in human monocytes. J. Exp. Med. *158*, 2108–2126.

Horwitz, M.A. (1984). Phagocytosis of the Legionnaires' disease bacterium (*Legionella pneumophila*) occurs by a novel mechanism: engulfment within a pseudopod coil. Cell *36*, 27–33.

Horwitz, M.A. (1992). Interactions between macrophages and *Legionella pneumophila*. Curr. Top. Microbiol. Immunol. *181*, 265–282.

Horwitz, M.A., and Silverstein, S.C. (1981a). Activated human monocytes inhibit the intracellular multiplication of Legionnaires' disease bacteria. J. Exp. Med. *154*, 1618–1635.

Horwitz, M.A., and Silverstein, S.C. (1981b). Interaction of the Legionnaires' disease bacterium (*Legionella pneumophila*) with human phagocytes. I. *L. pneumophila* resists killing by polymorphonuclear leukocytes, antibody, and complement. J. Exp. Med. *153*, 386–397.

Horwitz, M.A., and Silverstein, S.C. (1981c). Interaction of the legionnaires' disease bacterium (*Legionella pneumophila*) with human phagocytes. II. Antibody promotes binding of *L. pneumophila* to monocytes but does not inhibit intracellular multiplication. J. Exp. Med. *153*, 398–406.

Inohara, Chamaillard, McDonald, C., and Nunez, G. (2005). NOD-LRR proteins: role in host-microbial interactions and inflammatory disease. Annu. Rev. Biochem. *74*, 355–383.

Inohara, N., Koseki, T., Lin, J., del Peso, L., Lucas, P. C., Chen, F.F., Ogura, Y., and Nunez, G. (2000). An induced proximity model for NF-kappa B activation in the Nod1/RICK and RIP signaling pathways. J. Biol. Chem. *275*, 27823–27831.

Izu, K., Yoshida, S., Miyamoto, H., Chang, B., Ogawa, M., Yamamoto, H., Goto, Y., and Taniguchi, H. (1999). Grouping of 20 reference strains of *Legionella* species by the growth ability within mouse and guinea pig macrophages. FEMS. Immunol. Med. Microbiol. *26*, 61–68.

Jacobs, R.F., Locksley, R.M., Wilson, C.B., Haas, J.E., and Klebanoff, S.J. (1984). Interaction of primate alveolar macrophages and *Legionella pneumophila*. J. Clin. Invest. *73*, 1515–1523.

Janeway, C.A., Jr., and Medzhitov, R. (2002). Innate immune recognition. Annu. Rev. Immunol. *20*, 197–216.

Jensen, W.A., Rose, R.M., Wasserman, A.S., Kalb, T.H., Anton, K., and Remold, H.G. (1987). In vitro activation of the antibacterial activity of human pulmonary macrophages by recombinant gamma interferon. J. Infect. Dis. *155*, 574–577.

Kang, S.J., Wang, S., Hara, H., Peterson, E. P., Namura, S., Amin-Hanjani, S., Huang, Z., Srinivasan, A., Tomaselli, K.J., Thornberry, N.A., Moskowitz, M.A., and Yuan, J. (2000). Dual role of caspase-11 in mediating activation of caspase-1 and caspase-3 under pathological conditions. J. Cell. Biol. *149*, 613–622.

Katz, S.M., Habib, W.A., Hammel, J.M., and Nash, P. (1982). Lack of air-borne spread of infection by *Legionella pneumophila* among guinea pigs. Infect. Immun. *38*, 620–622.

Kikuchi, T., Kobayashi, T., Gomi, K., Suzuki, T., Tokue, Y., Watanabe, A., and Nukiwa, T. (2004). Dendritic cells pulsed with live and dead *Legionella pneumophila* elicit distinct immune responses. J. Immunol. *172*, 1727–1734.

Klein, T.W., Yamamoto, Y., Brown, H.K., and Friedman, H. (1991). Interferon-gamma induced resistance to *Legionella pneumophila* in susceptible A/J mouse macrophages. J. Leukoc. Biol. *49*, 98–103.

Kufer, T.A., Fritz, J.H., and Philpott, D.J. (2005). NACHT-LRR proteins (NLRs) in bacterial infection and immunity. Trends. Microbiol. *13*, 381–388.

Kura, F., Suzuki, K., Watanabe, H., Akamatsu, Y., and Amano, F. (1994). Difference in *Legionella pneumophila* growth permissiveness between J774.1 murine macrophage-like JA-4 cells and lipopolysaccharide (LPS)-resistant mutant cells, LPS1916, after stimulation with LPS. Infect Immun *62*, 5419–5423.

Lettinga, K.D., Florquin, S., Speelman, P., van Ketel, R., van der Poll, T., and Verbon, A. (2002a). Toll-like receptor 4 is not involved in host defense against pulmonary *Legionella pneumophila* infection in a mouse model. J. Infect. Dis. *186*, 570–573.

Lettinga, K.D., Verbon, A., Weverling, G.J., Schellekens, J.F., Den Boer, J.W., Yzerman, E.P., Prins, J., Boersma, W.G., Van Ketel, R. J., Prins, J.M., and Speelman, P. (2002b). Legionnaires' disease at a dutch flower show: prognostic factors and impact of therapy. Emerg. Infect. Dis. *8*, 1448–1454.

Locksley, R.M., Jacobs, R.F., Wilson, C.B., Weaver, W.M., and Klebanoff, S.J. (1982). Susceptibility of *Legionella pneumophila* to oxygen-dependent microbicidal systems. J. Immunol. *129*, 2192–2197.

MacMicking, J., Xie, Q.W., and Nathan, C. (1997). Nitric oxide and macrophage function. Annu. Rev. Immunol. *15*, 323–350.

Maier, J.K., Lahoua, Z., Gendron, N.H., Fetni, R., Johnston, A., Davoodi, J., Rasper, D., Roy, S., Slack, R.S., Nicholson, D.W., and MacKenzie, A.E. (2002). The neuronal apoptosis inhibitory protein is a direct inhibitor of caspases 3 and 7. J. Neurosci. 22, 2035–2043.

Mariathasan, S., Weiss, D.S., Newton, K., McBride, J., O'Rourke, K., Roose-Girma, M., Lee, W.P., Weinrauch, Y., Monack, D.M., and Dixit, V.M. (2006). Cryopyrin activates the inflammasome in response to toxins and ATP. Nature *440*, 228–232.

Marra, A., Blander, S.J., Horwitz, M.A., and Shuman, H.A. (1992). Identification of a *Legionella pneumophila* locus required for intracellular multiplication in human macrophages. Proc. Natl. Acad. Sci. USA *89*, 9607–9611.

Marra, A., Horwitz, M.A., and Shuman, H.A. (1990). The HL-60 model for the interaction of human macrophages with the Legionnaires' disease bacterium. J. Immunol. *144*, 2738–2744.

Matsiota-Bernard, P., Lefebre, C., Sedqui, M., Cornillet, P., and Guenounou, M. (1993). Involvement of tumor necrosis factor alpha in intracellular multiplication of *Legionella pneumophila* in human monocytes. Infect. Immun. *61*, 4980–4983.

Meyer, R.D., Edelstein, P.H., Kirby, B.D., Louie, M.H., Mulligan, M.E., Morgenstein, A.A., and Finegold, S.M. (1980). Legionnaires' disease: unusual clinical and laboratory features. Ann. Intern. Med. *93*, 240–243.

Miao, E.A., Alpuche-Aranda, C.M., Dors, M., Clark, A.E., Bader, M.W., Miller, S.I., and Aderem, A. (2006). Cytoplasmic flagellin activates caspase-1 and secretion of interleukin 1beta via Ipaf. Nat. Immunol. *7*, 569–575.

Miyamoto, H., Maruta, K., Ogawa, M., Beckers, M.C., Gros, P., and Yoshida, S. (1996). Spectrum of *Legionella* species whose intracellular multiplication in murine macrophages is genetically controlled by Lgn1. Infect. Immun. *64*, 1842–1845.

Molmeret, M., Zink, S.D., Han, L., Abu-Zant, A., Asari, R., Bitar, D.M., and Abu Kwaik, Y. (2004). Activation of caspase-3 by the Dot/Icm virulence system is essential for arrested biogenesis of the *Legionella*-containing phagosome. Cell. Microbiol. *6*, 33–48.

Molofsky, A.B., Byrne, B.G., Whitfield, N.N., Madigan, C.A., Fuse, E.T., Tateda, K., and Swanson, M.S. (2006). Cytosolic recognition of flagellin by mouse macrophages restricts *Legionella pneumophila* infection. J. Exp. Med. *203*, 1093–1104.

Nakagawa, T., Zhu, H., Morishima, N., Li, E., Xu, J., Yankner, B.A., and Yuan, J. (2000). Caspase-12 mediates endoplasmic-reticulum-specific apoptosis and cytotoxicity by amyloid-beta. Nature *403*, 98–103.

Nash, T.W., Libby, D.M., and Horwitz, M.A. (1988). IFN-gamma-activated human alveolar macrophages inhibit the intracellular multiplication of *Legionella pneumophila*. J. Immunol. *140*, 3978–3981.

Neild, A., Murata, T., and Roy, C.R. (2005). Processing and major histocompatibility complex class II presentation of *Legionella pneumophila* antigens by infected macrophages. Infect. Immun. *73*, 2336–2343.

Neild, A.L., and Roy, C.R. (2003). *Legionella* reveal dendritic cell functions that facilitate selection of antigens for MHC class II presentation. Immunity *18*, 813–823.

Neild, A.L., and Roy, C.R. (2004). Immunity to vacuolar pathogens: what can we learn from *Legionella*? Cell. Microbiol. *6*, 1011–1018.

Newton, C.A., Perkins, I., Widen, R.H., Friedman, H., and Klein, T.W. (2007). Role of Toll-like receptor 9 in *Legionella pneumophila*-induced interleukin-12 p40 production in bone marrow-derived dendritic cells and macrophages from permissive and nonpermissive mice. Infect. Immun. *75*, 146–151.

Opitz, B., Vinzing, M., van Laak, V., Schmeck, B., Heine, G., Gunther, S., Preissner, R., Slevogt, H., N'Guessan, P.D., Eitel, J., Goldmann, T., Flieger, A., Suttorp, N., and Hippenstiel, S. (2006). *Legionella pneumophila* induced IFNbeta in lung epithelial cells via IPS-1 and IRF3 which also control bacterial replication. J. Biol. Chem. *281*, 36173–36179.

Park, D.R., and Skerrett, S.J. (1996). IL-10 enhances the growth of *Legionella pneumophila* in human mononuclear phagocytes and reverses the protective effect of IFN-gamma: differential responses of blood monocytes and alveolar macrophages. J. Immunol. *157*, 2528–2538.

Park, D.R., and Skerrett, S.J. (2000). Cytokines in *Legionella pneumophila* infections, In Cytokines in pulmonary disease: Infection and inflammation, S. Nelson, and T.R. Martin, eds. (New York: Marcel Dekker, Inc), pp. 155–188.

Ren, T., Zamboni, D.S., Roy, C.R., Dietrich, W.F., and Vance, R.E. (2006). Flagellin-deficient *Legionella* mutants evade caspase-1- and Naip5-mediated macrophage immunity. PLoS Pathog. *2*, e18.

Ricci, M.L., Torosantucci, A., Scaturro, M., Chiani, P., Baldassarri, L., and Pastoris, M.C. (2005). Induction of protective immunity by *Legionella pneumophila* flagellum in an A/J mouse model. Vaccine *23*, 4811–4820.

Romanish, M.T., Lock, W.M., de Lagemaat, L.N., Dunn, C.A., and Mager, D.L. (2007). Repeated Recruitment of LTR Retrotransposons as Promoters by the Anti-Apoptotic Locus NAIP during Mammalian Evolution. PLoS Genet. *3*, e10.

Santic, M., Asare, R., Doric, M., and Abu Kwaik, Y. (2007). Host-Dependent Trigger of Caspases and Apoptosis by *Legionella pneumophila*. Infect. Immun. *75*, 2903–2913.

Santic, M., Molmeret, M., and Abu Kwaik, Y. (2005). Maturation of the *Legionella pneumophila*-containing phagosome into a phagolysosome within gamma interferon-activated macrophages. Infect. Immun. *73*, 3166–3171.

Scharf, J.M., Damron, D., Frisella, A., Bruno, S., Beggs, A.H., Kunkel, L.M., and Dietrich, W.F. (1996). The mouse region syntenic for human spinal muscular atrophy lies within the Lgn1 critical interval and contains multiple copies of Naip exon 5. Genomics *38*, 405–417.

Schneemann, M., Schoedon, G., Hofer, S., Blau, N., Guerrero, L., and Schaffner, A. (1993). Nitric oxide synthase is not a constituent of the antimicrobial armature of human mononuclear phagocytes. J. Infect. Dis. *167*, 1358–1363.

Schoedon, G., Schneemann, M., Walter, R., Blau, N., Hofer, S., and Schaffner, A. (1995). Nitric oxide and infection: another view. Clin. Infect. Dis. *21* Suppl. 2, S152–157.

Shinozawa, Y., Matsumoto, T., Uchida, K., Tsujimoto, S., Iwakura, Y., and Yamaguchi, K. (2002). Role of interferon-gamma in inflammatory responses in murine respiratory infection with *Legionella pneumophila*. J. Med. Microbiol. *51*, 225–230.

Skerrett, S.J., Bagby, G.J., Schmidt, R.A., and Nelson, S. (1997). Antibody-mediated depletion of tumor necrosis factor-alpha impairs pulmonary host defenses to *Legionella pneumophila*. J. Infect. Dis. *176*, 1019–1028.

Skerrett, S.J., and Martin, T.R. (1992). Recombinant murine interferon-gamma reversibly activates rat alveolar macrophages to kill *Legionella pneumophila*. J. Infect. Dis. *166*, 1354–1361.

Skerrett, S.J., and Martin, T.R. (1996). Roles for tumor necrosis factor alpha and nitric oxide in resistance of rat alveolar macrophages to *Legionella pneumophila*. Infect. Immun. *64*, 3236–3243.

Sporri, R., Joller, N., Albers, U., Hilbi, H., and Oxenius, A. (2006). MyD88-dependent IFN-gamma production by NK cells is key for control of *Legionella pneumophila* infection. J. Immunol. *176*, 6162–6171.

Stetson, D.B., and Medzhitov, R. (2006a). Recognition of cytosolic DNA activates an IRF3-dependent innate immune response. Immunity *24*, 93–103.

Stetson, D.B., and Medzhitov, R. (2006b). Type I interferons in host defense. Immunity *25*, 373–381.

Straus, W.L., Plouffe, J.F., File, T.M., Jr., Lipman, H.B., Hackman, B.H., Salstrom, S.J., Benson, R.F., and Breiman, R.F. (1996). Risk factors for domestic acquisition of Legionnaires' disease. Ohio Legionnaires' Disease Group. Arch. Intern. Med. *156*, 1685–1692.

Sturgill-Koszycki, S., and Swanson, M.S. (2000). *Legionella pneumophila* replication vacuoles mature into acidic, endocytic organelles. J. Exp. Med. *192*, 1261–1272.

Summersgill, J.T., Powell, L.A., Buster, B.L., Miller, R.D., and Ramirez, J.A. (1992). Killing of *Legionella pneumophila* by nitric oxide in gamma-interferon-activated macrophages. J. Leukoc. Biol. *52*, 625–629.

Susa, M., Ticac, B., Rukavina, T., Doric, M., and Marre, R. (1998). *Legionella pneumophila* infection in intratracheally inoculated T cell-depleted or -nondepleted A/J mice. J. Immunol. *160*, 316–321.

Swanson, M.S., and Hammer, B.K. (2000). *Legionella pneumophila* pathogesesis: a fateful journey from amoebae to macrophages. Annu. Rev. Microbiol. *54*, 567–613.

Swanson, M.S., and Molofsky, A.B. (2005). Autophagy and inflammatory cell death, partners of innate immunity. Autophagy *1*, 174–176.

Tateda, K., Deng, J.C., Moore, T.A., Newstead, M.W., Paine, R., 3rd, Kobayashi, N., Yamaguchi, K., and Standiford, T.J. (2003). Hyperoxia mediates acute lung injury and increased lethality in murine *Legionella* pneumonia: the role of apoptosis. J. Immunol. *170*, 4209–4216.

Tateda, K., Matsumoto, T., Ishii, Y., Furuya, N., Ohno, A., Miyazaki, S., and Yamaguchi, K. (1998). Serum cytokines in patients with *Legionella* pneumonia: relative predominance of Th1-type cytokines. Clin. Diagn. Lab. Immunol. *5*, 401–403.

Tateda, K., Moore, T.A., Deng, J.C., Newstead, M.W., Zeng, X., Matsukawa, A., Swanson, M.S., Yamaguchi, K., and Standiford, T.J. (2001a). Early recruitment of neutrophils determines subsequent T1/T2 host responses in a murine model of *Legionella pneumophila* pneumonia. J. Immunol. *166*, 3355–3361.

Tateda, K., Moore, T.A., Newstead, M.W., Tsai, W.C., Zeng, X., Deng, J.C., Chen, G., Reddy, R., Yamaguchi, K., and Standiford, T.J. (2001b). Chemokine-dependent neutrophil recruitment in a murine model of *Legionella* pneumonia: potential role of neutrophils as immunoregulatory cells. Infect. Immun. *69*, 2017–2024.

Ting, J.P., and Davis, B.K. (2005). CATERPILLER: a novel gene family important in immunity, cell death, and diseases. Annu. Rev. Immunol. *23*, 387–414.

Watarai, M., Derre, I., Kirby, J., Growney, J.D., Dietrich, W.F., and Isberg, R.R. (2001). *Legionella pneumophila* is internalized by a macropinocytotic uptake pathway controlled by the Dot/Icm system and the mouse Lgn1 locus. J. Exp. Med. *194*, 1081–1096.

Weeratna, R., Stamler, D.A., Edelstein, P.H., Ripley, M., Marrie, T., Hoskin, D., and Hoffman, P.S. (1994). Human and guinea pig immune responses to *Legionella pneumophila* protein antigens OmpS and Hsp60. Infect. Immun. *62*, 3454–3462.

Weinbaum, D.L., Bailey, J., Benner, R.R., Pasculle, A.W., and Dowling, J.N. (1983). The contribution of human neutrophils and serum to host defense against *Legionella micdadei*. J. Infect. Dis. *148*, 510–517.

Winn, W.C., Jr., Glavin, F.L., Perl, D.P., Keller, J.L., Andres, T.L., Brown, T.M., Coffin, C.M., Sensecqua, J.E., Roman, L.N., and Craighead, J.E. (1978). The pathology of Legionnaires' disease. Fourteen fatal cases from the 1977 outbreak in Vermont. Arch. Pathol. Lab. Med. *102*, 344–350.

Winn, W.C., Jr., and Myerowitz, R.L. (1981). The pathology of the *Legionella* pneumonias. A review of 74 cases and the literature. Hum. Pathol. *12*, 401–422.

Wright, E.K., Goodart, S.A., Growney, J.D., Hadinoto, V., Endrizzi, M.G., Long, E.M., Sadigh, K., Abney, A.L., Bernstein-Hanley, I., and Dietrich, W.F. (2003). Naip5 Affects Host Susceptibility to the Intracellular Pathogen *Legionella pneumophila*. Curr. Biol. *13*, 27–36.

Yamamoto, Y., Klein, T.W., and Friedman, H. (1996). Immunoregulatory role of nitric oxide in *Legionella pneumophila*-infected macrophages. Cell. Immunol. *171*, 231–239.

Yamamoto, Y., Klein, T.W., Newton, C., and Friedman, H. (1992). Differing macrophage and lymphocyte roles in resistance to *Legionella pneumophila* infection. J. Immunol. *148*, 584–589.

Yamamoto, Y., Klein, T.W., Newton, C.A., Widen, R., and Friedman, H. (1988). Growth of *Legionella pneumophila* in thioglycolate-elicited peritoneal macrophages from A/J mice. Infect. Immun. *56*, 370–375.

Yoshida, S., Goto, Y., Miyamoto, H., Fujio, H., and Mizuguchi, Y. (1991). Association of Lps gene with natural resistance of mouse macrophages against *Legionella pneumophila*. FEMS Microbiol. Immunol. 4, 51–56.

Zamboni, D.S., Kobayashi, K.S., Kohlsdorf, T., Ogura, Y., Long, E.M., Vance, R.E., Kuida, K., Mariathasan, S., Dixit, V.M., Flavell, R.A., Dietrich, W.F., and Roy, C.R. (2006). The Birc1e cytosolic pattern-recognition receptor contributes to the detection and control of *Legionella pneumophila* infection. Nat. Immunol. 7, 318–325.

The Flagellar Regulon of *Legionella pneumophila* and the Expression of Virulence Traits

6

Klaus Heuner* and Christiane Albert-Weissenberger

Abstract

Legionella is a ubiquitous inhabitant of aquatic habitats, and it is believed that motility is an important feature in the life cycle of *L. pneumophila* in its environment. To survive, *L. pneumophila* must be able to respond to different environmental factors and to modulate gene expression. Furthermore, very early after the first report of Legionnaires' disease, researchers discovered that most *Legionella* strains are flagellated and that regulation of the flagellum and the expression of virulence traits are linked. Expression of the flagellum is modulated by various environmental factors. The flagellar regulon, comprising genes encoding the structural components of the flagellum, is controlled by a cascade of regulators. This chapter will focus on the composition of the flagellar complex and the regulation of flagellar operons, including the virulence traits of *L. pneumophila* that are coordinately expressed with the flagellar genes. The contribution of the flagellar system to the virulence of *L. pneumophila* will also be discussed.

Introduction

It is fascinating to watch amoebae infected to see how fast *L. pneumophila* move circularly in the host cell at the end of the infection cycle, due to their flagella. Since *Legionella* is a ubiquitous inhabitant of aquatic habitats and replicates within different protozoa, it is believed that motility is needed to find a new host cell after its release from the exhausted host. If *Legionella* is successful, a next round of intracellular multiplication starts. If not, *Legionella* is well prepared to survive for a long time in its environment, perhaps within biofilm communities, until the next host approaches. Motility may also play a role in biofilm formation and development, as shown for other bacteria (O'Toole *et al.*, 2000; Kirov *et al.*, 2004).

Shortly after the first described outbreak of Legionnaires' disease and the identification of *L. pneumophila* as the causative agent (for details, see Chapter 1), it was reported that legionellae are flagellated (Fig. 6.1A and B) and that they exhibit pili-like structures on the surface (Chandler *et al.*, 1980a; Rodgers *et al.*, 1980). It was also postulated very early that *Legionella* flagellar expression and virulence are associated. Motility may contribute to the rapid spread of bacteria within the lung, since flagellated bacteria are apparent in the alveolar space of diseased patients (Chandler *et al.*, 1980b). Later it was shown that flagellar expression is linked to the developmental stage of *Legionella.* Rowbotham (1986) described the differentiation of *Legionella* from a non-motile replicating form to a non-dividing, motile form. This differentiation occurs at the end of intracellular replication, before the host cell is lysed and flagellated legionellae are released from the host (Fig. 6.1C and D). As these bacteria were found to be more infectious for amoebae than the non-motile forms, it was speculated that the expression of the virulent form of *Legionella* is linked to the expression of the flagellum (Rowbotham, 1986). For further details of *Legionella* differentiation, see Chapter 4. A regulatory link between the flagellum and the expression of the virulent phenotype of

*Corresponding author.

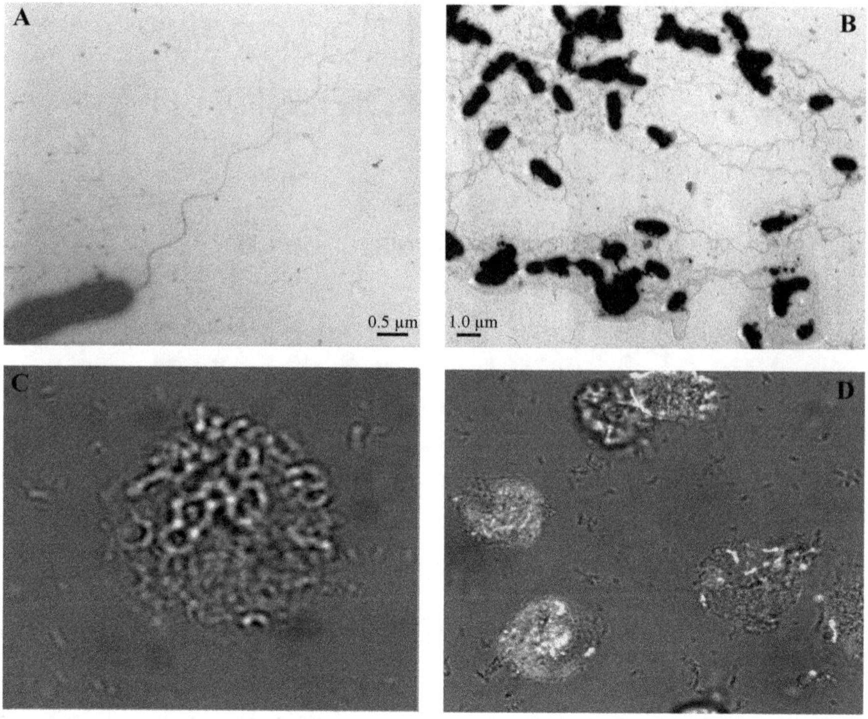

Figure 6.1 Electron microscopy (A, B) and fluorescence microscopy (C, D) of *L. pneumophila*. Micrographs A and B show the flagellation of *L. pneumophila* Corby. Bacteria were grown on BCYE agar plates at 30°C for 4 days, suspended in distilled water, transferred to Pioloform-coated copper grids and shadowed with platinium-palladium. Bacteria were examined with a transmission electron microscope at 60 kV. (C, D) Flagella expression of *L. pneumophila* Paris during infection of *Acanthamoeba castellanii* (C, after 8 h of infection; D, after 15 h of infection), visualized with a polyclonal anti-FlaA antiserum and processed for immunofluorescence (second antibody labelled with Alexa Fluor 488). Bacteria were examined with an LSM 510 confocal laser scanning microscope. A colour version of this figure is located in the plate section at the back of the book.

Legionella was subsequently confirmed by several molecular genetic studies (Pruckler *et al.*, 1995; Bosshardt *et al.*, 1997; Gao *et al.*, 1997; Byrne and Swanson, 1998; Swanson and Hammer, 2000; Hammer *et al.*, 2002; Heuner *et al.*, 2002a; Molofsky and Swanson, 2004; Molofsky *et al.*, 2005; Heuner *et al.*, 2006).

A correlation of motility and virulence has been described for various bacterial pathogens. The flagellum is involved in the adhesion to and invasion of host surfaces, successful colonization of hosts, and biofilm formation (Drake and Montie, 1988; Liu *et al.*, 1988; Allison *et al.*, 1992; Grant *et al.*, 1993; Yao *et al.*, 1994; Mobley *et al.*, 1996; O'Toole *et al.*, 2000; Young *et al.*, 2000; Kirov *et al.*, 2004; Goon *et al.*, 2006). Furthermore, the flagellar secretion system is a type III-like secretion system that can also secrete non-flagellar proteins and virulence factors (Macnab, 1999; Young *et al.*, 1999; Aizawa, 2001; Young and Young, 2002; Konkel *et al.*, 2004).

On the other hand, the flagellin protein is highly antigenic, and its N- and C-terminal regions are well conserved. Mammalian hosts recognize these domains of the flagellin monomer by membrane-bound Toll-like receptors (TLR5) and cytosolic NOD-like receptors (Naip5), which trigger proinflammatory and adaptive immune responses (Gewirtz *et al.*, 2001a,b; Hayashi *et al.*, 2001; Smith and Ozinsky, 2002, Hawn *et al.*, 2003; Smith *et al.*, 2003; Molofsky *et al.*, 2006; Amer *et al.*, 2006; Ren *et al.*, 2006; Akamine *et al.*, 2007). For details see Chapter 5.

Since the flagellar system is composed of more than 50 genes, its expression represents a high metabolic cost for the bacterial cell, making tight regulation of flagellar gene expression necessary (Chilcott and Hughes, 2000). As discussed below, the expression of the flagellum is modulated by different environmental factors, including growth phase, temperature, and the nutrient supply (Heuner *et al.*, 1999). Down-regulation of flagellin expression inside a host cell may also contribute to *L. pneumophila* virulence for mammals by enabling intracellular bacteria to avoid host surveillance mechanisms. Thus, its ability to modulate flagellar gene expression equips *L. pneumophila* to survive not only in its natural environment but also within mammalian hosts.

Since a number of *L. pneumophila* virulence traits are co-regulated with the flagellar system, identifying and characterizing the motility and the flagellar regulon of this bacterium is an important area of research. This chapter will focus on the composition of the flagellar complex, the regulation of the flagellar operons and the link between flagellar expression and virulence of *L. pneumophila*.

The flagellum of legionellae

The flagellum of Gram-negative bacteria

In general, the flagellar system is very complex. In most bacteria, it is composed of more than 50 genes (Chilcott and Hughes, 2000; Aldridge and Hughes, 2002; McCarter, 2006). The flagellum consists of a basal body, the hook structure (FlgE, FlgD) and the filament (FlaA, FliD, FlgK, FlgL; Fig. 6.2A). For details see recent review articles (Bardy *et al.*, 2003; Macnab, 2003; Pallen *et al.*, 2005). Typically, the filament consists of thousands (~20,000) of copies of one single protein, flagellin (FlaA or FliC), and it is about 20 nm in diameter (Minamino and Namba, 2004). The cap protein FliD is needed to assemble the FlaA protein into the filament. The basal structure

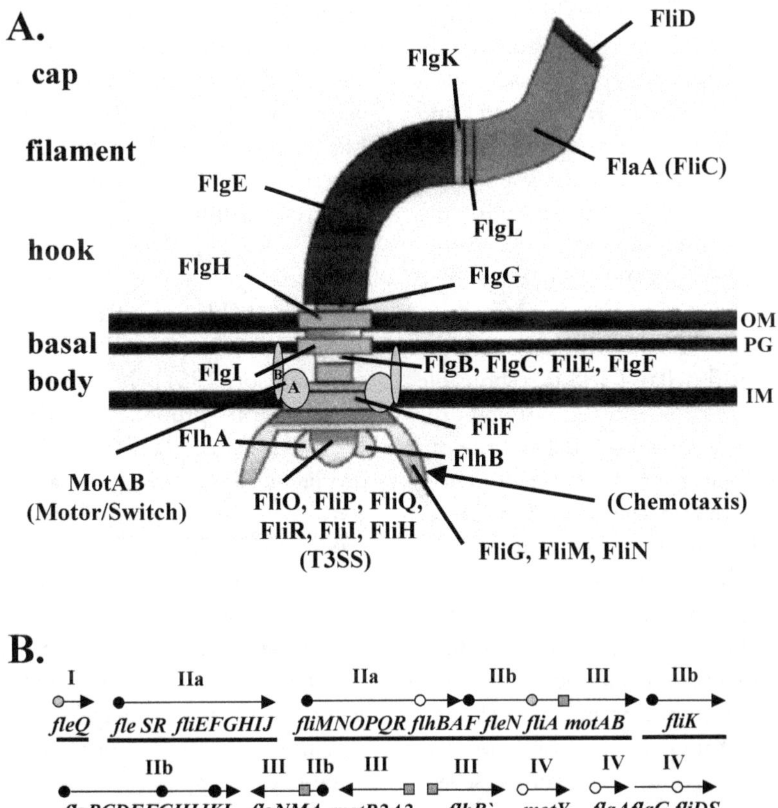

Figure 6.2 (A) Schematic of a flagellum and (B) the flagellar operon structure of *L. pneumophila* (Aizawa, 2001; Heuner and Steinert, 2003; modified). Bars represent the direction of putative operons in *L. pneumophila*. Underlined genes are found together in one chromosomal region. I-IV, classes of the regulatory cascade, see also Fig. 6.4; black dot, σ54 promoter element; white dot, σ28 promoter element; grey dot, σ70 promoter element; grey square, class III promoter, putatively FleR/RpoN-dependent; IM, inner membrane; OM, outer membrane; PG, peptidoglycan layer. (Heuner *et al.*, 2006; modified.)

consists of a rod, various rings (MS, P, L), the flagellum-specific export apparatus, the switch complex and the Mot proteins (Fig. 6.2A). The basal body anchors the flagellum to both membranes (MS ring, cytoplasmic membrane; P ring, peptidoglycan; L ring, outer membrane). The Mot proteins (MotA, MotB) generate the power for rotating the flagellum. For many species of bacteria, the switch proteins (FliG, FliM, FliN) allow microbes to switch rotation in response to attractants or repellents recognized by methyl-accepting chemotaxis proteins (MCPs) through a complex chemotaxis system (for details see Blair, 1995; Manson *et al.*, 1998; Szurmant and Ordal, 2004; Baker *et al.*, 2006). Cells run when the flagellar filaments rotate counterclockwise and tumble when the flagella turn clockwise.

The growth of the filament is unusual, because the flagellin monomers are exported through the filament structure and are assembled at the distal tip of the filament. FliD is needed for this process, since *fliD* mutant strains release flagellin into the medium and do not assemble a flagellum (Shimizu *et al.*, 2003; Minamino and Namba, 2004). The flagellum-specific export apparatus is a type III-like secretion system and has become a model system to study type III secretion (Minamino and Macnab, 1999; Aizawa, 2001; Macnab, 2004). In general, type III secretion systems (T3SSs) secrete proteins (virulence-associated effectors) directly into the target host cell. The flagellar T3SS is responsible for the export of the hook and rod forming proteins, the flagellum forming proteins and the cap protein. But non-flagellar proteins and virulence-associated factors are also secreted by the secretion system of the flagellum. For example, the phospholipase (YplA) of *Yersinia enterocolitica*, which can cause gastrointestinal illnesses in humans, is secreted by the T3SS of the flagellum. YplA is not required for motility but affects survival of *Y. enterocolitica* during infection (Young and Young, 2002; Young *et al.*, 1999). *Bacillus thuringensis*, which sporadically causes severe human infections, secretes virulence-associated factors (haemolysin BL, phospholipase C) by a mechanism that depends on its flagellar secretion system (Ghelardi *et al.*, 2002). *Salmonella typhimurium* and *Campylobacter jejuni* also secrete non-flagellar proteins by the flagellar T3SS.

The signal for the secretion via T3SSs is still unclear. Proteins secreted by T3SSs lack a cleavable signal sequence and do not enter the general secretory pathway (Sec-translocon). Recently it was shown that flagellin secretion is facilitated by the 173-bp untranslated DNA fragment upstream of *fliC*, supporting a model in which the mRNA directs the secretion of newly synthesized proteins via T3SSs. However, there are also signals within the flagellin protein that direct its T3SS-dependent secretion (Majander *et al.*, 2005).

The flagellum of *Legionella*

L. pneumophila exhibits a single monopolar flagellum (Fig. 6.1A). The flagellum is composed of the flagellin subunit FlaA, a protein of approximately 48 kDa (Chandler *et al.*, 1980a; Elliott and Johnson, 1981; Heuner *et al.*, 1995). The flagellin-encoding gene (*flaA*) is found in nearly all *Legionella* strains tested so far (Heuner *et al.*, 1995), and most *Legionella* species produce flagella (Elliott and Johnson, 1982; Bornstein *et al.*, 1991; Ott *et al.*, 1991; Bangsborg *et al.*, 1995; Heuner *et al.*, 1995). Within these studies there are some conflicting reports, perhaps because culture conditions in some studies were not optimal for flagellum expression. For example, *Legionella* grown on supplemented Mueller–Hinton (SMH) agar plates become non-virulent and non-motile (Chandler *et al.*, 1980a; Elliott and Johnson 1981). In one study, using a *flaA* specific DNA probe for Southern blot analysis, *L. oakridgensis*, *L. longbeachae* SG1 and *L. israeliensis* did not exhibit a *flaA*-homologous gene, they also were non-flagellated and did not produce a flagellin protein (Heuner *et al.* 1995). Interestingly, *L. oakridgensis* is also unique among the legionellae in its ability to grow on agar plates without supplementation with L-cysteine (Orrison *et al.*, 1983). As we will see below, the amino acid supply influences flagellum expression.

In a first report about the flagellar regulon, some flagellar genes (*flgAMN*, *motY*, *fliK*, *flhB'*) were not detected (Heuner and Steinert, 2003). Now that the genomes of four *L. pneumophila* species are sequenced (strains Philadelphia, Lens, Paris and Corby; Chien *et al.*, 2004; Cazalet *et al.*, 2004 and Glöckner *et al.*, 2007), more genes predicted to belong to the flagellar system of *L. pneumophila* were identified using *in silico* methods and *in vivo* transcriptome analysis (Brüggemann *et al.*, 2006; Heuner *et al.*, 2006). The overall operon structure is similar to that of *Pseudomonas* (Fig. 6.2B; Dasgupta *et al.*, 2003; Heuner and Steinert, 2003). Experimental data are available for several structural components of the flagellum, including flagellin, the flagellar ATPase FliI, the secretion system component FlhB, the motor components MotA1B1 and the cap protein FliD; together these data support the known general features of these structural proteins (Heuner *et al.*, 1995; Merriam *et al.*, 1997; Dietrich *et al.*, 2001; Molofsky *et al.*, 2005; Brüggemann *et al.*, 2006). Experiments revealed that *motA1* and *motA1B1* mutants of *L. pneumophila* are non-motile. Interestingly, in *motA1B1* mutants of *L. pneumophila* the flagellum is not stable and therefore flagellated mutant cells are rarely detected (Molofsky *et al.*, 2005; K. Heuner, unpublished). Since *L. pneumophila* exhibits a second MotAB encoding operon (*motA2B2*), the experiments suggest that *motA2B2* is not expressed and/or cannot complement the *motA1B1* mutant under conditions tested so far. The MotA1B1 proteins are homologous to the MotAB proteins of *Pseudomonas*, whereas the MotA2B2 proteins are more homologous to the MotAB proteins of *Nitrosomonas*, *Nitrosospira*, *Ralstonia* and *Methylobacillus*. Yet, the function of the MotA2B2 proteins in *Legionella* is unknown. The recently (*in silico*) identified homologue of MotY indicates that the flagellum of *L. pneumophila* seems to exhibit a sodium-driven motor composed of at least MotA1B1 and MotY.

A number of regulator proteins involved in flagellum expression have already been identified in different bacteria (Aldridge and Hughes, 2002). In *L. pneumophila*, the master regulator is FleQ, an activator protein of the alternative sigma factor σ^{54}, which acts together with σ^{54} and σ^{28} to control expression of the flagellar regulon (details, see below). Most of bacteria also encode chemotaxis proteins. However, in *L. pneumophila*, so far no homologues of chemotaxis genes (*cheWA*, *cheZYBR*, *cheB*) have been identified, with the exception of CheY-like proteins. Their absence may explain why chemotaxis by *Legionella* has not been described (Rodgers *et al.*, 1980; Heuner and Steinert, 2003). But since specific conditions are needed to observe swarming or chemotaxis activities, it also seems possible that the conditions needed to induce these activities have not yet been identified.

The flagellar regulon

Regulation of flagellum expression

Expression of the flagellum varies during the life cycle of *Legionella* and depends on its growth phase. *Legionella* switches from a non-flagellated replicating form to a flagellated non-replicating form (Rowbotham, 1986). Furthermore, flagellum expression is an early transmissive trait that starts when *Legionella* enter stationary phase, a period in broth cultures that corresponds to the end of the infection cycle in host cells (Robotham, 1986; Byrne and Swanson, 1998; Heuner *et al.*, 1999; Brüggemann *et al.*, 2006). The same pattern occurs when bacteria replicate within macrophages (Hammer and Swanson, 1999) and amoebae host cells. For example, in Fig. 6.1C and D bacterial flagellin was detected during the intracellular replication period using an anti-FlaA antiserum. Whereas no flagellin could be detected at early times (Fig. 6.1C), the bacteria are well stained at later stages of infection before lysis of the host cells (Fig. 6.1D).

Other kinds of switches also influence the expression of the flagellum. Growth on SMH agar leads to a conversion of virulent *L. pneumophila* to a non-flagellated and non-virulent form. There are conflicting reports about the ability of this phase variation to revert (McDade and Shepard, 1979; Elliott and Johnson, 1982; Catrenich and Johnson, 1988). Another phase variation has been described that is reversible and also leads to a non-virulent, non-flagellated form of *L. pneumophila*. In this case, the phase variable expression of an LPS epitope and the flagellum corresponds to the excision of a 30-kb genetic element. In the non-flagellated cell form, the element replicates as a high-copy plasmid (Lüneberg *et al.*, 1998, 2000, 2001).

Since *L. pneumophila* can replicate within host cells in very different aquatic habitats, it must modulate gene expression to respond to different environmental factors, such as temperature, osmolarity, viscosity and nutrition. Therefore, it is not surprising that a variety of conditions also influence expression of the flagellum (Fig. 6.3). Increased temperatures negatively regulate expression of the flagellum. Flagellin expression is induced by high osmolarity or low nutrient supply, whereas expression is repressed during exponential growth, increased temperatures, high viscosity, and abundant nutrients, specifically amino acids (Ott *et al.*, 1991; Mauchline *et al.*, 1992; Heuner *et al.*, 1995; Byrne and Swanson, 1998; Heuner *et al.*, 1999). In addition, the results shown in Fig. 6.3 demonstrate that *flaA* expression is repressed by the addition of amino acids (serine, threonine) while the addition of glucose had no effect. These results nicely demonstrate the fact that indeed amino acids are the major energy source of *L. pneumophila* (see chapter 11), since motility is generally suppressed under high nutrition conditions. However, the ability of *Legionella* to swarm or swim on agar plates has not yet been reported (Rodgers *et al.*, 1980; Heuner and Steinert, 2003). In the following section, we discuss the molecular mechanisms and regulators of flagellum expression.

The cascade of flagellin gene regulation

As mentioned above, genes that are predicted to belong to the flagellar regulon have been identified experimentally by molecular genetic and *in silico* methods (Fig. 6.2B). To obtain information about the cascade of flagellar gene regulation, the regulators involved were identified and specific mutant strains were generated and then analysed for flagellar gene expression (Heuner *et al.*, 1995, 1997, 2000, 2002a,b; Jacobi *et al.*, 2004; Molofsky *et al.*, 2005; Brüggemann *et al.*, 2006). Results of those studies allowed the construction of a hierarchical cascade of the flagellar regulon (Fig. 6.4).

FlaA (a class IV gene) encodes the flagellin and is directly regulated by the alternative σ^{28} factor (FliA). FliA was cloned by its ability to bind to the *E. coli* core RNA polymerase and direct transcription initiation from the *flaA*-specific promoter that was cloned upstream of a *luxAB* reporter gene. FliA of *L. pneumophila* partially restores the swarming ability of an *E. coli fliA* mutant strain. A *fliA* mutant of *L. pneumophila* does not express the flagellin and therefore is non-flagellated (Heuner *et al.*, 1997, 2002a). Later it was discovered that *L. pneumophila* also need FliA to express some of its virulence traits (see below). An additional gene contributes to *flaA* expression, named *flaR*, which encodes a protein belonging to the LysR family (Schell, 1993) of transcriptional regulators (Heuner *et al.*, 2000).

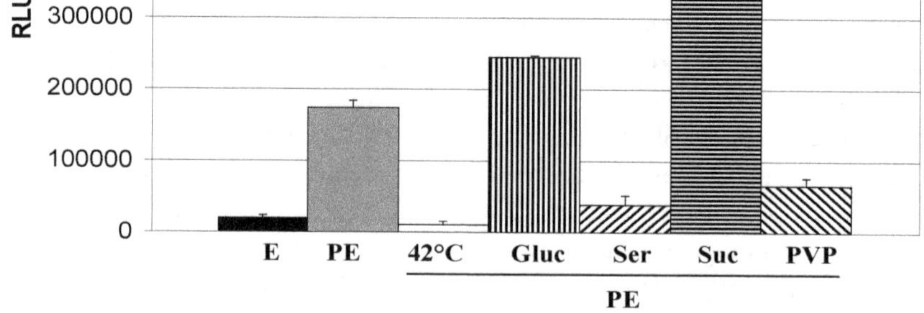

Figure 6.3 Modulation of flagellin expression in *L. pneumophila* Corby by different environmental factors. In general, *L. pneumophila* Corby (pKH23, p*fla-luxAB* fusion) was grown at 37°C in supplemented YEB medium. Luciferase activity is given in relative light units (RLU). E, exponential phase; PE, post-exponential phase; PVP, polyvinylpyrrolidone. (Heuner *et al.*, 1999; modified.)

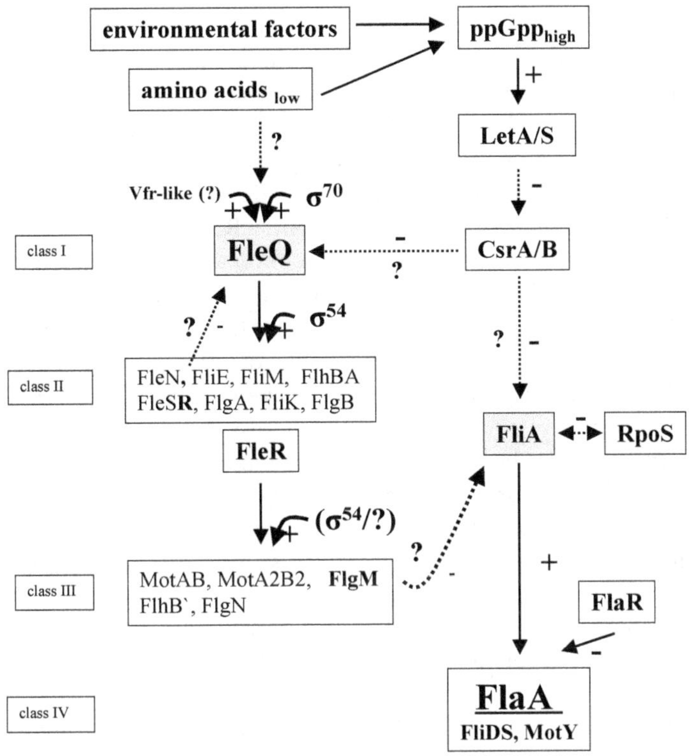

Figure 6.4 The proposed cascade of the flagellar regulon of *L. pneumophila*. The cascade was generated using data received by analysing flagellar mutant strains (strain Corby) and by transcriptome analysis (strains Paris and Lens). FleQ is the master regulator of the flagellar genes of classes II and III. FleR is the putative regulator of class III genes. FliA seems to be transcribed nearly independently of FleQ and is the direct regulator of class IV genes (for details, see text). Dotted bars represent an unknown mode of regulation (direct or indirect) and '?' indicates a putative mode of action. +, positive regulation; -, negative regulation. CsrA, carbon storage regulator A; FlaA, flagellin; FlaR, transcriptional regulator (LysR family); FleQ and FleR, σ54 activator protein; FlgM, anti-σ28 factor; FliA, alternative σ28 factor; LetA and LetS, two-component system; RpoN, alternative σ54 factor; RpoS, alternative σ38 factor; Vfr-like, putative cyclic AMP receptor protein homolog-like protein. (Brüggemann *et al.*, 2006, Heuner and Steinert, 2003; modified.)

FlaR binds to the *flaA* promoter. Therefore it seems possible that *L. pneumophila* relies on FlaR to repress FlaA expression during periods when FliA is needed to express virulence-associated traits, but FlaA expression would interfere with intracellular replication by consuming metabolic resources. However, further research is needed to verify this hypothesis.

The FliA-dependent expressed class IV genes are involved in assembly of the flagellum (*flaA*, *fliD*) and in motility (*motY*). So far little is known about *fliA* gene expression itself, but it is likely that RpoS and CsrA are involved (see below). *Salmonella typhimurium* control the activity of FliA by a post-transcriptional mechanism. By analogy, FlgM, an anti-σ^{28} factor, may bind to FliA to prevent activation of FliA-dependent gene expression (Kutsukake *et al.*, 1994; Aldridge *et al.*, 2006). Additionally, FlgN (the chaperone protein of FlgKL) may activate *flgM* mRNA in the absence of FlgKL. In other bacteria, once the flagellum hook-basal body structure is built, FlgM is exported and repression of FliA relieved so that the flagellum can be completed.

FleQ is the master regulator of the flagellar regulon (Jacobi *et al.*, 2004). FleQ is a σ^{54} activator protein which, together with the σ^{54} factor (RpoN), positively regulates the expression of class II genes. Class II genes encode proteins of the basal body and hook structure. In both *fleQ* and *rpoN* mutant strains, the flagellin gene is still transcribed due to the activity of FliA. However, the amount of FlaA in both mutants is reduced compared with the wild type strain, whereas *fliA* gene expression is only slightly affected by a mutation in *fleQ* of strain Corby (Jacobi *et al.*, 2004 and K. Heuner, unpublished). On the other hand, results of microarray analysis of *fliA* expression in *L. pneumophila* Paris revealed that *fliA* may be part of the FleQ regulon (Brüggemann *et al.*, 2006, and C. Albert-Weissenberger, C. Buchrieser and K. Heuner, unpublished). Together theses results indicate that certain aspects of *fliA* regulation may vary in different strains. Nevertheless, because of the predicted absence of a basal body, bacteria that lack FleQ and RpoN cannot assemble the expressed flagellin into a flagellar filament. Therefore, flagellin is degraded and its translation is repressed.

Within the class II genes there is another σ^{54} activator protein, FleR (Jacobi *et al.*, 2004), which seems to be a positive regulator of class III genes. But this hypothesis has to be tested by construction and analysis of a specific *fleR* mutant strain of *L. pneumophila*. However, FleS is the sensor protein of the putative FleSR two-component system, and a *fleS* mutant strain of *L. pneumophila* (strain AA100) was identified in a signature-tagged mutagenesis analysis. In particular, the *fleS* mutant strain was shown to be less invasive for host cells, but its flagellation was not investigated (Polesky *et al.*, 2001).

The class III genes encode proteins involved in the formation of the motor and regulation of further flagellum assembly. So far, details of FleQ expression are not known, but primer extension analysis indicated that the promoter seems to depend on sigma factor RpoD (σ^{70}). Furthermore, a Vfr-like (putative *E. coli* cAMP receptor protein homolog) DNA consensus sequence was identified upstream of its transcriptional start site (Jacobi *et al.*, 2004). Experiments are ongoing to define further details of this cascade by analysing different mutant strains by transcriptome analysis and phenotypic characterization. A link between FleQ- and FliA-dependent expression of the flagellum may be mediated through the anti-sigma factor FlgM, which is present in each of the *Legionella* genomes sequenced so far. It is also possible that LetA and/or CsrA are involved in this process (see below).

The regulatory cascade for the flagellum is comparable to the cascade of *Pseudomonas aeruginosa* (Dasgupta *et al.*, 2002, 2003; Jyot *et al.*, 2002). In both bacteria, FleQ is the master regulator, and FleQ interacts with the σ^{54} factor to induce the expression of class II and class III genes. Regulation of *fleQ* expression seems to be σ^{70} factor-dependent and another regulatory protein (Vfr-like) is also involved. Furthermore, in both bacteria FliA, which induces expression of class IV genes, seems to be regulated independently of FleQ. On the other hand, there are also some differences between the species. Preliminary results from a DNA microarray study of *fleR* expression in wild type and a *fleQ* mutant strain of *L. pneumophila* Paris in our lab suggest that regulation via FleR may be different to that found in *Pseudomonas* (C. Albert-Weissenberger, K. Heuner and C. Buchrieser, unpublished). However, further experiments are needed to clarify this.

Some additional regulators contribute to optimal expression of the flagellin (Fig. 6.4), but, for the most part, the mechanisms involved are unknown and are proposed to be indirect. Hence, in all these studies, the expression of FliA and/or FlaA was investigated, but the expression of FleQ was not. Nevertheless, we will briefly present the results available so far: RelA, a guanine 3′,5′-bispyrophosphate (ppGpp) synthetase, and its product ppGpp influences *flaA* expression. Accumulation of ppGpp seems to trigger the initiation of flagellar biosynthesis (Hammer and Swanson, 1999). In a *relA* mutant strain, the flagellum is still expressed, but *flaA* expression is reduced (Zusman *et al.*, 2002); the contribution of the second (ppGpp) synthetase, SpoT, remains to be addressed. The two-component regulator protein LetA (and LetS) positively influences *flaA* expression and motility. A more modest contribution is made by LetE, a regulatory protein of unknown function (Bachman and Swanson, 2004b, Gal-Mor and Segal, 2003). The homologueof the carbon storage regulator protein A (CsrA) of *L. pneumophila* negatively influences the expression of *flaA* and motility, probably by influencing *fliA* expression or by binding to flagellar gene transcripts (Fettes *et al.*, 2001; Molofsky and Swanson 2003;

Forsbach-Birk *et al.*, 2004). Furthermore, the alternative sigma factor RpoS also positively influences the expression of *fliA* and thus also *flaA* (Bachman and Swanson, 2001, 2004a, Chapter 4). An *in silico* screen of the genome sequence of *L. pneumophila* for open reading frames encoding proteins with sequence similarities to members of the LuxR family of transcriptional regulators identified three proteins (LpnR2, LpnR3 and LetA) that may influence *flaA* expression (Lebeau *et al.*, 2004). But until now there are no further details reported regarding the LpnR2- or LpnR3-dependent flagellin gene expression.

The FliA regulon

As mentioned above, FliA is not only the direct regulator of flagellin expression, but also contributes to expression of virulence traits of *L. pneumophila* (see below). Therefore, it is of interest to identify all FliA-dependent genes of *L. pneumophila*, the so-called FliA regulon, to determine which of these factors promote virulence. FliA recognizes a specific nucleotide sequence (σ^{28} consensus sequence, TAAA-N15-TCCGATAA) in the promoter region (Heuner *et al.*, 1995, 1997). This knowledge was then used to screen the genome of *L. pneumophila* strain Paris for putative σ^{28}-dependent promoters. Nearly 90 genes were identified; 30 genes exhibit very good matches, having a very similar consensus sequence plus a spacer distance equal to that found in the *flaA* promoter (C. Buchrieser, unpublished). However, in a recent transcriptome based analysis, only 10 genes were identified that appear to belong to the FliA regulon of intracellular replicating bacteria (Brüggemann *et al.*, 2006). Five of these ten genes were not found in the previous '*in silico*' screen. Only three of the ten genes identified belong to the flagellar regulon (*flaA*, *fliD*, *motY*). Of the seven other genes, three encode for proteins exhibiting a signal peptide (*lpp0972*, *lpp1290*, *lpp2260*), four encode proteins of unknown function (*lpp1841*, *lpp2260*, *lpp2282*, *lpp2998*), two for EnhA-like proteins (*lpp0972*, *lpp1290*) and one gene (*lpp0952*) encodes a putative regulator protein. Further analysis of these genes may help to elucidate the role of FliA for expression of virulence traits in *L. pneumophila*.

Contribution of the flagellar system to the virulence of *Legionella*

Role of the flagellum

Early it was shown that motile transmissive bacteria are more infectious for amoebae then the non-motile replicative phase, and it was hypothesized that flagellum expression and virulence are genetically linked (Rowbotham, 1986; Pruckler *et al.*, 1995; Bosshardt *et al.*, 1997; Gao *et al.*, 1997; Byrne and Swanson, 1998; Hammer and Swanson, 1999; Heuner *et al.*, 2002a). Furthermore, flagellated bacteria have been detected inside the human lung of Legionnaires' disease patients (Chandler *et al.*, 1980b). Later Tn10 insertion mutants were described which are defective in flagellum expression and compromised in their ability to infect host cells. But some of these mutants multiply as well as the wild-type strain (Pruckler *et al.*, 1995). By generating and analysing a specific *flaA* mutant strain, the ability of flagellin to enhance the invasion capacity and thus the infectivity of *L. pneumophila* was discovered. In contrast, intracellular replication was not significantly affected (Dietrich *et al.*, 2001). *L. pneumophila motAB* mutant strains are less flagellated and have severely reduced motility compared to wild type. Once again, a *motAB* mutant strain is less invasive but its intracellular multiplication is not significantly affected. Thus, motility but not flagellin appears to promote contact with host cells and increase infectivity (Molofsky *et al.*, 2005).

Legionella is able to form biofilms, and biofilm formation is believed to be a virulence factor for various bacteria. But the flagellum of *L. pneumophila* does not contribute to biofilm formation under the conditions tested so far (Mampel *et al.*, 2006). We will also mention here that flagellin does affect host resistance to Legionnaires' disease (Hawn *et al.*, 2003). Furthermore, loss of either bacterial flagellin or the macrophage cytosolic Naip5 (Birc1e) receptor makes mice susceptible to *L. pneumophila* infection, in part by impairing a pro-inflammatory cell death pathway of mouse macrophages (Swanson and Molofsky, 2005; Molofsky *et al.*, 2006; Amer *et al.*, 2006; Ren *et al.*, 2006). However, addition of recombinant IFN-γ to the infected macrophages restricted replication of the *flaA* mutant within macrophages (Akamine *et al.*, 2007). Furthermore, the flagellin of *L. pneumophila* is strongly immunogenic and capable of inducing a protective immunity in a A/J mouse model (Ricci *et al.*, 2005). For details, see Chapter 5.

Role of the flagellar regulon

As mentioned above, the flagellum and motility contributes to the invasion capacity of *L. pneumophila* into host cells, but it seems not to be necessary for intracellular replication or biofilm formation. However, motility probably equips *Legionella* to reach the next host cell for further rounds of multiplication. But which regulators are involved in the proposed link between flagellum and virulence traits expression in *L. pneumophila*? When we inactivated the regulator of *flaA* expression, the *fliA* gene, this mutant strain does not replicate in *Dictyostelium discoideum*, whereas the *flaA* mutant or wild type strain does. In addition, the *flaA* mutant strain is less haemolytic than the wild-type strain (Heuner *et al.*, 2002a, Heuner *et al.*, 2006). These results clearly show that the flagellar regulon is involved in the virulence of *L. pneumophila*. It further suggests that in addition to *flaA* more genes regulated by FliA are needed for full fitness and virulence of this bacterium; together these are now called the FliA regulon (see above). At the same time, another group showed that a *fliA* mutant strain of *L. pneumophila* is non-cytotoxic and less infectious for macrophages. The same group also reported that FliA is involved in lysosome evasion and regulates binding to the dye crystal violet (Hammer *et al.*, 2002; Molofsky *et al.*, 2005). Recently, another group showed that a *fliA* mutant is reduced in its ability to form biofilms (Mampel *et al.*, 2006). Taken together, all these results clearly confirm the hypothesis that the flagellar regulon is linked to the expression of virulence traits in *L. pneumophila* and that the FliA sigma factor plays a major role in this linkage.

On the other hand, the FliA regulon is not the only pathway involved in expression of flagellar regulon-dependent virulence traits in *L. pneumophila*. The FleQ/RpoN regulon and the FleSR two component system also affect *L. pneumophila* virulence (Polesky *et al.*, 2001; Jacobi *et al.*, 2004, Heuner *et al.*, 2006). In particular, a FleS mutant strain is less invasive for host cells, and *fleQ* and *rpoN* mutant strains of *L. pneumophila* Corby exhibit less haemolytic activity against red blood cells than does the wild-type strain. Similar results were obtained by analysis of *fleQ* and *rpoN* mutants of strain Paris (C. Albert-Weissenberger and K. Heuner, unpublished). Therefore, the FleQ regulon is also involved in the expression of virulence traits of *L. pneumophila*.

Regulation of virulence factors in *Legionella*

The infectious, transmissive phase of *L. pneumophila*

Virulence of *L. pneumophila* seems to be multifactorial, involving a complex interplay between the bacteria and the host cell. The virulence factors acquired for replication in protozoan host cells also enable *L. pneumophila* to infect human cells, and regulation of the virulence gene expression in response to the growth phase is essential to ensure a successful infection. The life cycle of *L. pneumophila* consists of at least two distinguishable phenotypes: post-exponential phase bacteria escape from the host cell and are flagellated, highly motile, cytotoxic and virulent; replicative bacteria that reside in the host vacuole are sodium resistant, unflagellated, and non-cytotoxic (Byrne and Swanson, 1998; Molofsky and Swanson, 2004). For details, see Chapter 4.

The global regulatory genes *csrA*, *letA*/*letS*/*letE* and *rpoS* control maintenance and induction of either the replicative-focused genes or the genes necessary for the post-exponential phenotype (Fig. 6.5). The small RNA-binding protein CsrA represses the expression of transmissive traits in *L. pneumophila* (Molofsky and Swanson, 2003). By contributing to *csrA* RNA stability, Hfq is involved in maintenance of the exponential phase. Expression of *hfq* in exponential phase is induced by RpoS. In stationary phase, *hfq* transcription is repressed either directly or indirectly by LetA (McNealy *et al.*, 2005). Besides, during nutrient deprivation (presumably signalled by ppGpp) the two-component regulator LetA/S

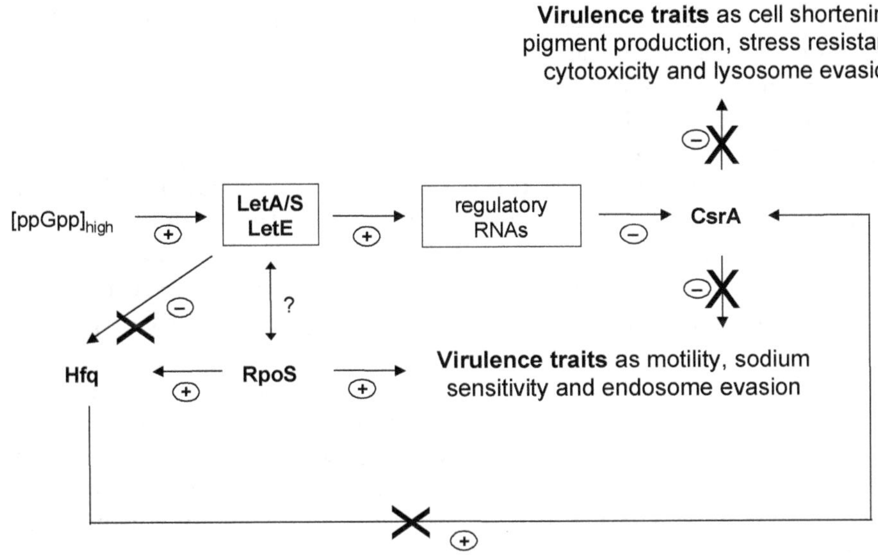

Figure 6.5 Model for induction of virulence traits specific for the transmissive phase in *L. pneumophila*. Synthesis of ppGpp caused by nutrition starvation probably acts via the two-component system LetA/S to mediate transcription of regulatory RNAs. These regulatory RNAs are thought to bind to CsrA and consequently antagonize its activity. In exponential phase, RpoS induces the expression of Hfq, which contributes to the stability of *csrA* RNA. But when cells respond to signals of nutrient deprivation, LetA turns off *hfq* transcription. Thus, repression of transmissive phase virulence traits by CsrA is abolished. Also involved in the expression of some transmissive phase virulence traits is RpoS. (Adapted from Molofsky and Swanson, 2003; Jacobi *et al.*, 2004; McNealy *et al.*, 2005.)

inhibits CsrA activity, probably mediated by not yet identified regulatory RNAs (Hammer and Swanson 1999, Molofsky and Swanson 2003). This transition is enhanced by LetE (Bachman and Swanson, 2004b). The *letE* gene promoter region exhibits a RpoN consensus sequence and seems to be expressed in a RpoN-dependent manner (Brüggemann *et al.*, 2006; C. Albert-Weissenberger, C. Buchrieser and K. Heuner, unpublished).

RpoS is also involved in the regulation of transmissive phase traits expression in *Legionella*, but its role is somewhat controversial. RpoS appears to be necessary for intracellular survival and is involved in regulating the replicative phase (Bachman and Swanson, 2001; 2004a; Abu-Zant *et al.*, 2006). On the other hand, RpoS expression increases during stationary phase and coordinates the expression of several transmissive phase traits of *L. pneumophila*, but not stress resistance (Bachman and Swanson, 2001; 2004a; Abu-Zant *et al.*, 2006). By analysing *relA* mutant bacteria, Abu-Zant and colleagues were not able to identify a link between the stringent response and RpoS. Therefore, further experiments are necessary to clarify the contributions of the stringent response and RpoS to the intracellular life cycle.

Temperature not only influences motility (as discussed above), but also the expression of virulence traits of *L. pneumophila*. *L. pneumophila* serogroup 1 multiplies within *Acanthamoeba palestinensis* at 35°C but not at 20°C. At 20°C, the amoebae instead digest most of the *Legionella* (Anand *et al.*, 1983). Similarly, in *A. castellanii L. pneumophila* Philadelphia 1 grow intracellularly at 37°C but not at room temperature (Moffat and Tompkins, 1992). However, there are conflicting reports about the influence of temperature on the virulence of *L. pneumophila*. When grown at 37°C in chemostat culture, *Legionella* are significantly more virulent than those grown at 24°C, as judged by aerosol infection of guinea pigs (Mauchline *et al.*, 1994). In a guinea pig model of lung infection, Edelstein and colleagues showed that bacteria grown in buffered yeast extract broth in batch culture at 25°C were more virulent then those grown at 41°C (Edelstein *et al.*, 1987). However, different strains and different routes of infection were used in the studies, which may explain the differences observed. Furthermore, since neither study analysed bacteria cultured to the transmissive virulent phase, it is difficult to interpret these results. Presumably genetic differences between different *L. pneumophila* strains, e.g. the existence of the Lvh T4ASS (Samrakandi *et al.*, 2002), also influences the ability to infect host cells at different temperatures. In fact, Ridenour and colleagues found that the *lvh*-encoded T4ASS positively influences the infectivity of *L. pneumophila* grown at lower temperatures (Ridenour *et al.*, 2003).

Regulation of secretion systems

Pathogenicity of *L. pneumophila* depends greatly on the ability to secrete virulence factors that are displayed on the bacterial cell surface, secreted into the extracellular milieu or injected directly into the host cell. Mainly associated with the virulence of *L. pneumophila* are the Dot/Icm T4BSS and the Lsp T2SS, but also the twin-arginine translocation pathway (Tat) and the *lvh* encoded T4ASS. For details, see Chapters 8–10.

The Lsp secretion system promotes the ability of *L. pneumophila* to infect protozoan and macrophage hosts, to grow in the mammalian lung and to grow at low temperatures (Rossier and Cianciotto, 2001; Soderberg *et al.*, 2004, DebRoy *et al.*, 2006b). It is involved in secretion of various proteins including many virulence-associated proteins (Hales and Shuman, 1999; Chapter 8). The T2SS of *L. pneumophila* is dependent on the *pilBCD* locus that is also involved in the biogenesis of type IV pili and on the *Legionella* secretion pathway, encoded by the *lspFGHIJK* and *lspDE* loci. Expression of *pilBCD* is influenced by temperature, increasing at temperatures below 30°C. However, irrespective of temperature, the *pilD* expression and T2SS-dependent hydrolytic activity increases from exponential to stationary phase (Liles *et al.*, 1998; Liles, *et al.*, 1999; Flieger *et al.*, 2001). The regulation of *pilD* has similarity with that of the *L. pneumophila* flagellin (*flaA*) gene. Like expression of *flaA*, *pilD* expression is regulated not only by temperature but also growth phase- (Ott *et al.*, 1991; Pope *et al.*, 1996; Heuner *et al.*, 1997; Heuner *et al.*, 1999; Heuner *et al.*, 2002; Brüggemann *et al.*, 2006). When type II-dependent enzymatic activities were monitored, *L. pneumophila rpoS* and *letA* mutants exhibited a significant reduction of secreted phospholipase A, glycerophospholipid:cholesterol acyltransferase, protease, phosphatase, decreased p-NPPC hydrolase activities and increased lysophospholipase A and lipase activities. In *letA* and *rpoS* mutant strains, the *pla*C mRNA level was reduced compared to the wild-type (Flieger *et al.*, 2002; Broich *et al.*, 2006).

Whether LetA and RpoS influence the hydrolytic enzyme activities directly, via transcriptional regulation of structural gene, or indirectly, by controlling the expression of activators, cofactors, or components of the secretion apparatus, still needs to be elucidated by future studies. Although the activities of these hydrolytic enzymes depend on the T2SS, the type II-dependent RNase activity was not affected, indicating that LetA and RpoS do not regulate the type II machinery directly (Broich *et al.*, 2006).

L. pneumophila requires the Dot/Icm-T4SS to replicate in both protozoa and human macrophage hosts (Andrews *et al.*, 1998; Edelstein *et al.*, 1999; Segal and Shuman, 1999). To establish a replicative niche after invading alveolar macrophages, this secretion system is proposed to translocate bacterial protein substrates into eukaryotic host cells during infection (Segal and Shuman, 1999; Dumenil and Isberg, 2001). The apparatus also mediates an anti-apototic signalling cascade in human macrophages (Abu-Zant *et al.*, 2007). Vincent and colleagues have conducted a comprehensive biochemical and genetic examination of the Dot/Icm secretion system (Vincent *et al.*, 2006b; Chapter 9). At present, a large number of substrates of the Dot/Icm-T4SS have been identified and their activities partially analysed.

Regulation of the genes encoding the Dot/Icm-T4SS seems to be complex and is probably directed by several regulators. Although some studies have been performed, there is still only a diffuse picture of Dot/Icm-T4SS regulation. The sigma factor RpoS, the ppGpp synthetase RelA, and the two-component regulatory system LetA/S affect expression of genes encoding the

Dot/Icm secretion system (Zusman *et al.*, 2002; Gal-Mor and Segal, 2003; Lynch *et al.*, 2003; Shi *et al.*, 2006). CpxR also plays a role in regulation of various *dot/icm* genes (Gal-Mor *et al.*, 2002; Gal-Mor and Segal, 2003; Vincent *et al.*, 2006a). The promoter region of some of the *icm* genes exhibit an RpoD-dependent -10 promoter consensus sequence and a conserved CpxR binding site ([T]GTAAA-box). Nevertheless, a *cpxR* mutant showed no growth defect inside *A. castellanii* and HL-60-derived human macrophages (Gal-Mor and Segal, 2003), indicating that other regulators must also contribute to expression of the Dot/Icm-T4SS. Indeed, the *fir* genes of *Legionella*, encoding chaperones of IcmQ, exhibit both a CpxR-binding site (TGTAAANAANNNGAAAG) and a PmrA-binding site (CTTAAGNTTNNCTTAAG/TNNT) that positively influences the expression of the *fir* genes (Feldman and Segal, 2007). The PrmA binding site was not present in the *icmR* (*fir*) gene of *L. pneumophila*. A similar PmrA box (cTTAATaTT) was identified in the promoter region of several Dot/Icm substrates and genes coregulated with these effector genes (Ceg genes) (Zusmann *et al.*, 2007).

Expression of almost all of the reported substrates of the Dot/Icm secretion system, such as RalF, SidC and members of the SidE family, are induced in stationary phase (Nagai *et al.*, 2002; Conover *et al.*, 2003; Luo and Isberg, 2004; Bardill *et al.*, 2005, Brüggemann *et al.*, 2006; Zusman *et al.*, 2007). LetA is necessary for full transcriptional expression of the *ralF* gene and RpoS for the *sidG* gene (Shi *et al.*, 2006; Zusman *et al.*, 2007). These observations are consistent with the enhanced virulence of *L. pneumophila* in stationary phase. The response regulator PmrA is also involved in the expression of Dot/Icm-translocated substrates. A mutant strain of *pmrA* does not replicate within *A. castellanii* (Zusman *et al.*, 2007). In contrast, SidJ expression is constant during the entire cell cycle of *L. pneumophila* (Liu and Luo, 2007). Similarly, LidA expression is only slightly induced in stationary phase bacteria, and the protein is secreted into the host cell throughout the replication cycle (Derre and Isberg, 2005). Expression of at least some Dot/Icm secretion system substrates, such as SidJ, RalF, and SdeC, is independent of a functional Dot/Icm transporter (Vincent *et al.*, 2006a; Liu and Luo, 2007). The expression pattern of the substrates presumably reflects the temporal requirements of each effector protein during the cell cycle.

Regulation of other virulence factors

A large number of factors that contribute to virulence have been described, including *katA*, *katB*, *prpCD*, *milA*, *iraA*, *pmiA*, *nudA*, *lpnE*, *lvgA*, *rtxA*, the *ccm* locus, *ligA*, *feoAB*, *rcp* and *laiA* (Bandyopadhyay and Steinman, 1998; Stone *et al.*, 1999; Bandyopadhyay and Steinman, 2000; Fettes *et al.*, 2000; Harb and Abu Kwaik, 2000; Viswanathan *et al.*, 2000; Robey *et al.*, 2001; Cirillo *et al.*, 2002; Robey and Cianciotto, 2002; Edelstein *et al.*, 2003; Chang *et al.*, 2005; Edelstein *et al.*, 2005; Miyake *et al.*, 2005; Newton *et al.*, 2006). However, little is known about the mechanisms that regulate their expression. Some of these factors are discussed below.

The macrophage infectivity potentiator (Mip)

Mip belongs to the enzyme family of FK-506 binding proteins that exhibit peptidyl prolyl cis/trans isomerase (PPIase) activity. It promotes the presence of a p-NPPC hydrolase activity in culture supernatants and is involved in the infection of eukaryotic host cells (Wintermeyer *et al.*, 1995; Cianciotto, 2001; Helbig *et al.*, 2001; Helbig *et al.*, 2003; Kohler *et al.*, 2003; DebRoy *et al.*, 2006a, Wagner *et al.*, 2007). Although the Mip protein is constitutively expressed in both free-living and intracellularly replicating *L. pneumophila* (Kohler *et al.*, 2000), the transcriptional activity of the *mip* promoter is repressed after invasion of the monocytic cell line MonoMac 6, and then regains activity after 24 h of intracellular replication (Wieland *et al.*, 2002). Thus, its pattern of expression is typical of flagellin and other factors thought to be dedicated to transmission but dispensable for replication (Molofsky and Swanson, 2004). Secretion of Mip into the supernatant apparently is positively regulated by LetA, since a *letA* mutation dramatically decreases the secretion of Mip but has no effect on membrane-associated Mip (Shi *et al.*, 2006).

The major cell-associated phospholipase A (PlaB)

PlaB contributes to the cytotoxicity of *L. pneumophila*, as it possesses phospholipase A and lysophospholipase A activity. Although the *plaB* mutant multiplies like wild type in U937 macrophages and in *A. castellanii* amoeba (Flieger *et al.*, 2004), *plaB* probably promotes dissemination of *L. pneumophila* in tissues (K. Heuner, E. Schunder, and F. Higa, unpublished). *plaB* is expressed especially before entry into the late logarithmic growth phase, and PlaB is responsible for the major lipolytic activity present during host cell infection. LetA and RpoS presumably promote PlaB expression or activity, since the cell-associated phospholipase A and lysophospholipase A activities are significantly reduced in *letA* and *rpoS* mutants (Broich *et al.*, 2006).

The major secreted zinc metalloproteinase A (MspA)

MspA is encoded by the *proA* gene. Although Blander and colleagues (1990) initially reported that MspA is not a virulence factor in guinea pigs, later studies showed that a *proA* mutant was attenuated in a guinea pig model. Furthermore, the *proA* gene was identified in a screen for virulence genes that used signature-tagged mutagenesis (Moffat *et al.*, 1994; Edelstein *et al.*, 1999). The *proA* gene is expressed during intracellular growth in amoebae (Moffat *et al.*, 1994). Likewise, MspA is produced in the lungs of guinea pigs during infection (Williams *et al.*, 1987; Conlan *et al.*, 1988). Purified protease can cause lesions like those of Legionnaires' disease in guinea pig and human lung (Baskerville *et al.*, 1986; Conlan *et al.*, 1986). Furthermore, guinea pigs immunized with MspA develop a protective immunity against lethal aerosol challenge with *L. pneumophila* (Blander and Horwitz, 1989, 1991), confirming its expression during infection and its contribution to disease. MspA presumably alters human phagocyte function (Sahney *et al.*, 2001). Overall, the data indicate that MspA may play a role in the pathogenesis of human Legionnaires' disease. LetA and RpoS may positively regulate the activity of secreted MspA (Broich *et al.*, 2006). As an aside, MspA itself is necessary for activation of PlaC (Banerji *et al.*, 2005).

The global stress protein (GspA)

GspA is needed for expression of full stress resistance of *L. pneumophila*. The sigma factors σ^{70} (RpoD) and σ^{32} (RpoH) seem to be involved in *gspA* expression (Abu Kwaik and Engelberg, 1994; Abu Kwaik *et al.*, 1997). The *gspA* gene is induced during intracellular infection and by heat shock, oxidative stress, acid- and osmotic shock.

The heat shock protein (Hsp60)

Another protein involved in stress response is the 60 kDa heat shock protein (Hsp60), encoded by the *htpAB* operon (Hoffman *et al.*, 1990). Hsp60 is necessary for full invasion of *L. pneumophila*. The protein is found in both the cytoplasm and the cell envelope, and it is also expressed and secreted during infection (Fernandez *et al.*, 1996; Garduno *et al.*, 1998a, b). Furthermore, the Dot/Icm secretion system seems to be involved in its secretion (Chong *et al.*, 2006). Hsp60 may also alter the pathogen-host interaction by interfering with signalling pathways, as suggested by the developmental changes initiated by a *Saccharomyces cerevisiae* that express *L. pneumophila* Hsp60. Upstream of *htpA*, an RpoH-like promoter element was identified (Hoffman *et al.*, 1990), consistent with the observation that the *htpAB* promoter seems to be under the control of RpoH. The authors proposed that RpoH is essential for *Legionella*, because they were unsuccessful in generating a *rpoH* mutant strain (Ewann and Hoffman, 2006). However, further research is necessary to elucidate the role of RpoH in *L. pneumophila*.

Conclusions

Thirty years after the identification of *L. pneumophila* as the causative agent of Legionnaires' disease, various factors and mechanisms responsible for the pathogenicity of *L. pneumophila* have been identified. But there is still little information about the mechanisms that regulate the virulence of *L. pneumophila*. We are well on the way to understanding the mechanisms of gene regulation of the flagellum, or the flagellar regulon, in more detail. However, there are still open questions regarding the link between the expression of the virulent phenotype and the expression of the flagellum. Except for the flagellin, further virulence factors regulated by the flagellar regulon (e.g. by FliA) are still waiting to be investigated. With the identification and characterization of FliA and LetA, we will probably be able to answer many remaining questions in the near future. The completion of the genome sequences of four different *L. pneumophila* strains which now provides the opportunity to perform whole genome microarray-based studies will be especially helpful in efforts to identify and characterize both additional regulators and their target genes.

The flagellum and the flagellar regulon remains a promising field to study not only *Legionella* virulence and protein secretion (T3SS), but also to investigate pathogen-host interactions, since flagellin recognition by Toll-like receptor 5 or Naip5 receptor activates the innate immune system. Together all these aspects indicate that the flagellar system is a rich area for further research in the *L. pneumophila* pathogenesis field.

Acknowledgements

Own work was supported by grants from the Deutsche Forschungsgemeinschaft (grants GRK 587/2, HE2845/2–1/2, HE2845/4–1 and HE2845/5–1) and by the Bavarian Research Foundation (Bayerische Forschungsstiftung).

References

Abu Kwaik, Y., and Engleberg, N.C. (1994). Cloning and molecular characterization of a *Legionella pneumophila* gene induced by intracellular infection and by various *in vitro* stress conditions. Mol. Microbiol. *13*, 243–251.

Abu Kwaik, Y., Gao, L.Y., Harb, O.S., and Stone, B.J. (1997). Transcriptional regulation of the macrophage-induced gene (*gspA*) of *Legionella pneumophila* and phenotypic characterization of a null mutant. Mol. Microbiol. *24*, 629–642.

Abu-Zant, A., Asare, R., Graham, J.E., and Abu-Kwaik, Y. (2006). Role for RpoS but not RelA of *Legionella pneumophila* in modulation of phagosome biogenesis and adaptation to the phagosomal microenvironment. Infect. Immun. *74*, 3021–3026.

Abu-Zant, A., Jones, S., Asare, R., Suttles, J., Price, C., Graham, J., and Abu-Kwaik, Y. (2007). Anti-apoptotic signalling by the Dot/Icm secretion system of *L. pneumophila*. Cell. Microbiol. 9, 246–264.

Aizawa, S.I. (2001). Bacterial flagella and type III secretion systems. FEMS Microbiol. Lett. *202*, 157–164.

Akamine, M., Higa, F., Haranaga, S., Tateyama, M., Mori, N., Heuner, K., and Fujita, J. (2007). Interferon gamma reverses the evasion of *Birc1e*/Naip5 gene mediated murine macrophage immunity by *Legionella pneumophila* mutant lacking flagellin. Microbiol. Immunol. *51*, 279–287.

Aldridge, P., and Hughes, K.T. (2002). Regulation of flagellar assembly. Curr. Opin. Microbiol. 5, 160–165.

Aldridge, P.D., Karlinsey, J.E., Aldridge, C., Birchall, C., Thompson, D., Yagasaki, J., and Hughes, K.T. (2006). The flagellar-specific transcription factor, sigma28, is the Type III secretion chaperone for the flagellar-specific anti-sigma28 factor FlgM. Genes Dev. *20*, 2315–2326.

Allison, C., Coleman, N., Jones, P.L., and Hughes, C. (1992). Ability of *Proteus mirabilis* to invade human urothelial cells is coupled to motility and swarming differentiation. Infect. Immun. *60*, 4740–4746.

Amer, A., Franchi, L., Kanneganti, T.D., Body-Malapel, M., Ozoren, N., Brady, G., Meshinchi, S., Jagirdar, R., Gewirtz, A., Akira, S., and Nunez, G. (2006). Regulation of *Legionella* phagosome maturation and infection through flagellin and host Ipaf. J. Biol. Chem. *281*, 35217–35223.

Anand, C.M., Skinner, A.R., Malic, A., and Kurtz, J.B. (1983). Interaction of *L. pneumophila* and a free living amoeba (*Acanthamoeba palestinensis*). J. Hyg. (Lond). *91*, 167–178.

Andrews, H.L., Vogel J.P., and Isberg, R.R. (1998). Identification of linked *Legionella pneumophila* genes essential for intracellular growth and evasion of the endocytic pathway. Infect. Immun. *66*, 950–958.

Bachman, M.A., and Swanson, M.S. (2001). RpoS co-operates with other factors to induce *Legionella pneumophila* virulence in the stationary phase. Mol. Microbiol. *40*, 1201–1214.

Bachman, M.A, and Swanson, M.S. (2004a). Genetic evidence that *Legionella pneumophila* RpoS modulates expression of the transmission phenotype in both the exponential phase and the stationary phase. Infect. Immun. *72*, 2468–2476.

Bachman, M.A, and Swanson, M.S. (2004b). The LetE protein enhances expression of multiple LetA/LetS-dependent transmission traits by *Legionella pneumophila*. Infect. Immun. *72*, 3284–3293.

Baker, M.D., Wolanin, P.M., and Stock, J.B. (2006). Signal transduction in bacterial chemotaxis. Bioessays. *28*, 9–22.

Bandyopadhyay, P., and Steinman, H.M. (1998). *Legionella pneumophila* catalase-peroxidases: cloning of the katB gene and studies of KatB function. J. Bacteriol. *180*, 5369–5374.

Bandyopadhyay, P., and Steinman, H.M. (2000). Catalase-peroxidases of *Legionella pneumophila*: cloning of the katA gene and studies of KatA function. J. Bacteriol. *182*, 6679–6686.

Banerji, S., Bewersdorff, M., Hermes, B., Cianciotto, N.P., and Flieger, A. (2005). Characterization of the major secreted zinc metalloprotease- dependent glycerophospholipid:cholesterol acyltransferase, PlaC, of *Legionella pneumophila*. Infect. Immun. *73*, 2899–2909.

Bangsborg, J.M., Hindersson, P., Shand, G., and Hoiby, N. (1995). The *Legionella micdadei* flagellin: expression in *Escherichia coli* K 12 and DNA sequence of the gene. APMIS. *103*, 869–877.

Bardill, J. P., Miller, J.L., and Vogel, J.P. (2005). IcmS-dependent translocation of SdeA into macrophages by the *Legionella pneumophila* type IV secretion system. Mol. Microbiol. *56*, 90–103.

Bardy, S.L., Ng, S.Y., and Jarrell, K.F. (2003). Prokaryotic motility structures. Microbiology *149*, 295–304.

Baskerville, A., Conlan, J.W., Ashworth, L.A., and Dowsett, A.B. (1986). Pulmonary damage caused by a protease from *Legionella pneumophila*. Br. J. Exp. Pathol. *67*, 527–536.

Blair, D.F. (1995). How bacteria sense and swim. Annu. Rev. Microbiol. *49*, 489–522.

Blander, S.J., Horwitz, M.A. (1989). Vaccination with the major secretory protein of *Legionella pneumophila* induces cell-mediated and protective immunity in a guinea pig model of Legionnaires' disease. J. Exp. Med. *169*, 691–705.

Blander, S.J., Horwitz, M.A. (1991). Vaccination with the major secretory protein of *Legionella* induces humoral and cell-mediated immune responses and protective immunity across different serogroups of *Legionella pneumophila* and different species of *Legionella*. J. Immunol. *147*, 285–291.

Blander, S.J., Szeto, L., Shuman, H.A., and Horwitz, M.A. (1990). An immunoprotective molecule, the major secretory protein of *Legionella pneumophila*, is not a virulence factor in a guinea pig model of Legionnaires' disease. J. Clin. Invest. *86*, 817–824.

Bornstein, N., Marmet, D., Dumaine, M.H., Surgot, M., and Fleurette, J. (1991). Detection of flagella in 278 *Legionella* strains by latex reagent sensitized with antiflagellum immunoglobulins. J. Clin. Microbiol. *29*, 953–956.

Bosshardt, S.C., Benson R.F., and Fields B.S. (1997). Flagella are a positive predictor for virulence in *Legionella*. Microb. Pathog. *23*, 107–112.

Broich, M., Rydzewski, K., McNealy, T.L., Marre, R., and Flieger, A. (2006). The global regulatory proteins LetA and RpoS control phospholipase A, lysophospholipase A, acyltransferase, and other hydrolytic activities *of Legionella pneumophila* JR32. J. Bacteriol. *188*, 1218–26.

Brüggemann, H., Hagman, A., Jules, M., Sismeiro, O., Dillies, M-A., Gouyette, C., Kunst, F., Steinert, M., Heuner, K., Coppée, J-Y., and Buchrieser, C. (2006). Virulence strategies for infecting phagocytes deduced from the transcriptional program of *Legionella pneumophila*. Cell. Microbiol. *8*, 1228–1240.

Byrne, B., and Swanson, M.S. (1998). Expression of *Legionella pneumophila* virulence traits in response to growth conditions. Infect. Immun. *66*, 3029–3034.

Catrenich, C.E., Johnson, W. (1988). Virulence conversion of *Legionella pneumophila*: a one-way phenomenon. Infect. Immun. *56*, 3121–3125.

Cazalet, C., Rusniok, C., Bruggemann, H., Zidane, N., Magnier, A., Ma, L., Tichit, M., Jarraud, S., Bouchier, C., Vandenesch, F., Kunst, F., Etienne, J., Glaser, P., and Buchrieser, C. (2004). Evidence in the *Legionella pneumophila* genome for exploitation of host cell functions and high genome plasticity. Nat. Genet. *36*, 1165–1173.

Chandler, F.W., Roth, I. L., Callaway, C.S., Bump, J.L., Thomason, B.M., Weaver, R.E. (1980a). Flagella on Legionnaires' disease bacteria: ultrastructural observations. Ann. Intern. Med. *93*, 711–714.

Chandler, F.W., Thomason, B.M., Hebert, G.A. (1980b). Flagella on Legionnaires' disease bacteria in the human lung. Ann. Intern. Med. *93*, 715–716.

Chang, B., Kura, F., Amemura-Maekawa, J., Koizumi, N., and Watanabe H. (2005). Identification of a novel adhesion molecule involved in the virulence of *Legionella pneumophila*. Infect. Immun. *73*, 4272–4280.

Chien, M., Morozova, I., Shi, S., Sheng, H., Chen, J., Gomez, S.M., Asamani, G., Hill, K., Nuara, J., Feder, M., Rineer, J., Greenberg, J.J., Steshenko, V., Park, S.H., Zhao, B., Teplitskaya, E., Edwards, J.R., Pampou, S., Georghiou, A., Chou, I.C., Iannuccilli, W., Ulz, M.E., Kim, D.H., Geringer-Sameth, A., Goldsberry, C., Morozov, P., Fischer, S.G., Segal, G., Qu, X., Rzhetsky, A., Zhang, P., Cayanis, E., De Jong, P.J., Ju, J., Kalachikov, S., Shuman, H.A., and Russo, J.J. (2004). The genomic sequence of the accidental pathogen *Legionella pneumophila*. Science. *305*, 1966–1968.

Chilcott, G.S., and Hughes, K.T. (2000). Coupling of flagellar gene expression to flagellar assembly in *Salmonella enterica* serovar typhimurium and *Escherichia coli*. Microbiol. Mol. Biol. Rev. *64*, 694–708.

Chong, A., Riveroll, A., Allan, D.S., Garduno, E., and Garduno, R.A. (2006). The Hsp60 chaperonin of *Legionella pneumophila*: An intriguing player in infection of host cells. In *Legionella*: State of the art 30 years after its recognition, N.P. Cianciotto, Y. Abu Kwaik, P.H. Edelstein, B.S. Fields, D.F. Geary, T.G. Harrison, C.A. Joseph, R.M. Ratcliff, J.E. Stout, and M.S. Swanson, eds. (Washington, DC: ASM Press), pp. 255–260.

Cianciotto, N.P. (2001). Pathogenicity of *Legionella pneumophila*. Int. J. Med. Microbiol. *291*, 331–343.

Cirillo, S.L.G., Yan, L., Littman, M., Samrakandi, M.M., and Cirillo, J.D. (2002). Role of the *Legionella pneumophila* rtxA gene in amoebae. Microbiology *148*, 1667–1677.

Conlan, J.W., Williams, A., and Ashworth, L.A. (1986). Separation of *Legionella pneumophila* proteases and purification of a protease which produces lesions like those of Legionnaires' disease in guinea pig lung. J. Gen. Microbiol. *132*, 1565–1574.

Conlan, J.W., Williams, A., and Ashworth, L.A. (1988). *In vivo* production of a tissue-destructive protease by *Legionella pneumophila* in the lungs of experimentally infected guinea-pigs. J. Gen. Microbiol. *134*, 143–149.

Conover, G.M., Derre, I., Vogel, J.P., and Isberg, R.R. (2003). The *Legionella pneumophila* LidA protein: a translocated substrate of the Dot/Icm system associated with maintenance of bacterial integrity. Mol. Microbiol. *48*, 305–321.

Dasgupta, N., Ferrell, E.P., Kanack, K.J., West, S.E., Ramphal, R. (2002). *fleQ*, the gene encoding the major flagellar regulator of *Pseudomonas aeruginosa*, is sigma70 dependent and is downregulated by Vfr, a homologueof *Escherichia coli* cyclic AMP receptor protein. J. Bacteriol. *184*, 5240–5250.

Dasgupta, N., Wolfgang, M.C., Goodman, A. L., Arora, S.K., Jyot, J., Lory, S., and Ramphal, R. (2003). A four-tiered transcriptional regulatory circuit controls flagellar biogenesis in *Pseudomonas aeruginosa*. Mol. Microbiol. *50*, 809–824.

DebRoy, S., Aragon, V., Kurtz, S., and Cianciotto, N.P. (2006a). *Legionella pneumophila* Mip, a surface-exposed peptidylproline cis-trans-isomerase, promotes the presence of Phospholipase C-like activity in culture supernatants. Infect. Immun. *74*, 5151–5160.

DebRoy, S., Dao, J., Söderberg, M., Rossier, O., and Cianciotto, N.P. (2006b). *Legionella pneumophila* type II secretome reveals unique exoproteins and a chitinase that promotes bacterial persistence in the lung. PNAS. *103*, 19146–19151.

Derre, I., and Isberg R.R. (2005). LidA, a translocated substrate of the *Legionella pneumophila* type IV secretion system, interferes with the early secretory pathway. Infect. Immun. *73*, 4370–4380.

Dietrich, C., Heuner, K., Brand, B. C., Steinert, M., Hacker, J. (2001). The flagellum of *Legionella pneumophila* positively affects the early phase of infection of eukaryotic host cells. Infect. Immun. *64*, 2116–2122.

Drake, D., and Montie, T.C. (1988). Flagella, motility and invasive virulence of *Pseudomonas aeruginosa*. J. Gen. Microbiol. *134*, 43–52.

Dumenil, G., and Isberg R.R. (2001). The *Legionella pneumophila* IcmR protein exhibits chaperone activity for IcmQ by preventing its participation in high-molecular-weight complexes. Mol. Microbiol. *40*, 1113–1127.

Edelstein, P.H., Beer, K.B., and DeBoynton, E.D. (1987). Influence of growth temperature on virulence of *Legionella pneumophila*. Infect. Immun. *55*, 2701–2705.

Edelstein, P.H., Edelstein, M.A., Higa, F., and Falkow, S. (1999). Discovery of virulence genes of *Legionella pneumophila* by using signature tagged mutagenesis in a guinea pig pneumonia model. Proc. Natl. Acad. Sci. USA *96*, 8190–8195.

Edelstein, P.H., Hu, B., Higa, F., and Edelstein, M. A. (2003). *lvgA*, a novel *Legionella pneumophila* virulence factor. Infect. Immun. *71*, 2394–2403.

Edelstein, P.H., Hu, B., Shinzato, T., Edelstein, M.C., Xu, W., and Bessman, M. (2005). *Legionella pneumophila* NudA is a nudix hydrolase and virulence factor. Infect. Immun. *73*, 6567–6576.

Elliott, J.A., and Johnson, W. (1981). Immunological and biochemical relationships among flagella isolated from *Legionella pneumophila* serogroups 1, 2, and 3. Infect. Immun. *33*, 602–10.

Elliot, J.A., and Johnson, W. (1982). Virulence conversion of *Legionella pneumophila* serogroup 1 by passage in guinea pigs and embryonated eggs. Infect. Immun. *35*, 943–947.

Ewann, F., and Hoffman, P.S. (2006). Modulation of *rpoH* expression using an antisense strategy. In *Legionella*: State of the art 30 years after its recognition, N.P. Cianciotto, Y. Abu Kwaik, P.H. Edelstein, B.S. Fields, D.F. Geary, T.G. Harrison, C.A. Joseph, R.M. Ratcliff, J.E. Stout, and M.S. Swanson, eds. (Washington, DC: ASM Press), pp. 336–338.

Feldman, M., and Segal, G. (2007). A pair of highly conserved two-component systems participates in the regulation of the hyper variable FIR proteins in different *Legionella* species. J. Bacteriol. 2; [Epub ahead of print].

Fernandez, R.C., Logan, S.M., Lee, S.H., and Hoffman, P.S. (1996). Elevated levels of *Legionella pneumophila* stress protein Hsp60 early in infection of human monocytes and L929 cells correlate with virulence. Infect. Immun. *64*, 1968–1976.

Fettes, P.S., Forsbach-Birk, V., Lynch, D., and Marre, R. (2001). Overexpresssion of a *Legionella pneumophila* homologue of the *E. coli* regulator *csrA* affects cell size, flagellation, and pigmentation. Int. J. Med. Microbiol. *291*, 353–360.

Fettes, P.S., Susa, M., Hacker, J., and Marre, R. (2000). Characterization of the *Legionella pneumophila* gene *ligA*. Int. J. Med. Microbiol. *290*, 239–250.

Flieger, A., Gong, S., Faigle, M., Northoff, H., and Neumeister, B. (2001). *In vitro* secretion kinetics of proteins from *Legionella pneumophila* in comparison to proteins from non-pneumophila species. Microbiol. *147*, 3127–3134.

Flieger, A., Neumeister, B., and Cianciotto, N.P. (2002). Characterization of the gene encoding the major secreted lysophospholipase A of *Legionella pneumophila* and its role in detoxification of lysophosphatidylcholine. Infect. Immun. *70*, 6094–6106.

Flieger, A., Rydzewski, K., Banerji, S., Broich, M., and Heuner, K. (2004). Cloning and characterization of the gene encoding the major cell-associated phospholipase A of *Legionella pneumophila*, *plaB*, exhibiting hemolytic activity. Infect. Immun. *72*, 2648–2658.

Forsbach-Birk, V., McNealy, T., Shi, C., Lynch, D., and Marre, R. (2004). Reduced expression of the global regulator protein CsrA in *Legionella pneumophila* affects virulence-associated regulators and growth in *Acanthamoeba castellanii*. Int. J. Med. Microbiol. *294*, 15–25.

Gal-Mor, O., and Segal, G. (2003). The *Legionella pneumophila* GacA homologue(LetA) is involved in the regulation of *icm* virulence genes and is required for intracellular multiplication in *Acanthamoeba castellanii*. Microb. Pathog. *34*, 187–194.

Gal-Mor, O., Zusman, T., and Segal, G. (2002). Analysis of DNA regulatory elements required for expression of the *Legionella pneumophila icm* and *dot* virulence genes. J. Bacteriol. *184*, 3823–3833.

Gao, L.Y., Harb, O.S., and Kwaik, Y.A. (1997). Utilization of similar mechanisms by *Legionella pneumophila* to parasitize two evolutionary distant Host Cells, mammalian macriophages and protozoa. Infect. Immun. *65*, 4738–4746.

Garduno, R.A., Faulkner, G., Trevors, M.A., Vats, N., and Hoffman, P.S. (1998a). Immunolocalization of Hsp60 in *Legionella pneumophila*. J. Bacteriol. *180*, 505–513.

Garduno, R.A., Garduno, E., and Hoffman, P.S. (1998b). Surface-associated *hsp60* chaperonin of *Legionella pneumophila* mediates invasion in a HeLa cell model. Infect. Immun. *66*, 4602–4610.

Gewirtz, A.T., Navas, T.A., Lyons, S., Godowski, P.J., and Madara, J.L. (2001a). Cutting edge: bacterial flagellin activates basolaterally expressed TLR5 to induce epithelial proinflammatory gene expression. J. Immunol. *167*, 1882–1885.

Gewirtz, A.T., Simon, P.O. Jr, Schmitt, C.K., Taylor, L.J., Hagedorn, C.H., O'Brien, A.D., Neish, A.S., and Madara, J.L. (2001b). *Salmonella typhimurium* translocates flagellin across intestinal epithelia, inducing a proinflammatory response. J. Clin. Invest. *107*, 99–109.

Ghelardi, E., Celandroni, F., Salvetti, S., Beecher, D.J., Gominet, M., Lereclus, D., Wong, A.C., and Senesi, S. (2002). Requirement of *flhA* for swarming differentiation, flagellin export, and secretion of virulence-associated proteins in *Bacillus thuringiensis*. J. Bacteriol. *184*, 6424–6433.

Glöckner, G., Albert-Weissenberger, C., Weinmann E., Jacobi, S., Schunder, E., Steinert, M., Hacker, J., and Heuner, K. (2007) Identification and characterization of a new conjugation/type IVA secretion system (*trb/tra*) of *L. pneumophila* Corby localized on two mobile genomic islands. doi:10.1016/j.ijmm.2007.07.012

Goon, S., Ewing, C.P., Lorenzo, M., Pattarini, D., Majam, G., and Guerry, P. (2006). A sigma28-regulated nonflagella gene contributes to virulence of *Campylobacter jejuni* 81–176. Infect. Immun. *74*, 769–772.

Grant, C.C., Konkel, M.E., Cieplak, W. Jr, and Tompkins, L.S. (1993). Role of flagella in adherence, internalization, and translocation of *Campylobacter jejuni* in nonpolarized and polarized epithelial cell cultures. Infect. Immun. *61*, 1764–1771.

Hales, L.M., and Shuman H.A. (1999). *Legionella pneumophila* contains a type II general secretion pathway required for growth in amoebae as well as for secretion of the Msp protease. Infect Immun *67*, 3662–3666.

Hammer, B.K., and Swanson, M.S. (1999). Co-ordination of *Legionella pneumophila* virulence with entry into stationary phase by ppGpp. Mol. Microbiol. *33*, 721–731.

Hammer, B.K., Tateda, E.S., and Swanson, M.S. (2002). A two-component regulator induces the transmission phenotype of stationary-phase *Legionella pneumophila*. Mol. Microbiol. *44*, 107–118.

Harb, O.S., and Abu Kwaik, Y. (2000). Characterization of a macrophage-specific infectivity locus (*milA*) of *Legionella pneumophila*. Infect. Immun. *68*, 368–376.

Hawn, T.R., Verbon, A., Lettinga, K.D., Zhao, L.P., Li, S.S., Laws, R.J., Skerrett, S.J., Beutler, B., Schroeder, L., Nachman, A., Ozinsky, A., Smith, K.D., and Aderem, A. (2003). A common dominant TLR5 stop codon polymorphism abolishes flagellin signaling and is associated with susceptibility to legionnaires' disease. J. Exp. Med. *198*, 1563–1572.

Hayashi, F., Smith, K.D., Ozinsky, A., Hawn, T.R., Yi, E.C., Goodlett, D.R., Eng, J.K., Akira, S., Underhill, D.M., AND Aderem, A. (2001). The innate immune response to bacterial flagellin is mediated by Toll-like receptor 5. Nature. *410*, 1099–1103.

Helbig, J.H., Konig, B., Knospe, H., Bubert, B., Yu, C.,Luck, C.P., Riboldi-Tunnicliffe, A., Hilgenfeld, R., Jacobs, E., Hacker, J., and Fischer, G. (2003). The PPIase active site of *Legionella pneumophila* Mip protein is involved in the infection of eukaryotic host cells. Biol. Chem. *384*, 125–137.

Helbig, J.H., Luck, P.C., Steinert, M., Jacobs, E., and Witt, M. (2001). Immunolocalization of the Mip protein of intracellularly and extracellularly grown *Legionella pneumophila*. Lett. Appl. Microbiol. *32*, 83–88.

Heuner, K., Bender-Beck, L., Brand, B.C., Lück, P.C., Mann, K.H., Marre, R., Ott, M., and Hacker, J. (1995). Cloning and genetic characterization of the flagellum subunit gene (*flaA*) of *Legionella pneumophila* serogroup 1. Infect. Immun. *63*, 2499–2507.

Heuner, K., Brand, B.C., and Hacker, J. (1999). The expression of the flagellum of *Legionella pneumophila* is modulated by different environmental factors. FEMS Microbiol. Lett. *175*, 69–77.

Heuner, K., Dietrich, C., Hacker J. (2002a). Influence of the alternative σ28 factor on virulence and flagellum expression of *L. pneumophila*. Infect. Immun. *70*, 1604–1608.

Heuner, K., Dietrich, C., Steinert, M., Göbel, U.B., and Hacker, J. (2000). Cloning and characterization of a *Legionella pneumophila*-specific gene encoding a member of the LysR family of transcriptional regulators. Mol. Gen. Genet. *264*, 204–211.

Heuner, K., Hacker, J., and Brand, B.C. (1997). The alternative sigma factor σ28 of *Legionella pneumophila* restores flagellation and motility to an *Escherichia coli fliA* mutant. J. Bacteriol. *179*, 17–23.

Heuner, K., Jacobi, S., Albert, C., Steinert, M., Brüggemann, H., Buchrieser, C. (2006) Gene Expression and Virulence in *Legionella*: The Flagellar Regulon. In *Legionella*: State of the art 30 years after its recognition, N.P. Cianciotto, Y. Abu Kwaik, P.H. Edelstein, B.S. Fields, D.F. Geary, T.G. Harrison, C.A. Joseph, R.M. Ratcliff, J.E. Stout, and M.S. Swanson, eds. (Washington, DC: ASM Press), pp. 327–332.

Heuner, K., and Steinert; M. (2003). The flagellum of *Legionella pneumophila* and its link to the expression of the virulent phenotype. Int. J. Med. Microbiol. *293*, 133–145.

Heuner, K., Steinert, M., Dietrich, C., Fischer, G., Köhler, R., Hacker, J. (2002b) Function and Expression of *Legionella* surface factors. In: *Legionella*- Proceedings of the 5th International Symposium. Eds., Marre *et al.* (Washington, DC: ASM Press), pp. 43–48.

Hoffman, P.S., Houston, L., and Butler, C.A. (1990). *Legionella pneumophila htpAB* heat shock operon: nucleotide sequence and expression of the 60-kilodalton antigen in *L. pneumophila*-infected HeLa cells. Infect. Immun. *58*, 3380–3387.

Jacobi, S., Hacker, J., Schade, R., Heuner, K. (2004). Characterization of the alternative sigma factor σ54 and the transcriptional regulator FleQ of *Legionella pneumophila*, which are both involved in the regulation cascade of flagellar gene expression. J. Bacteriol. *186*, 2540–2547.

Jyot, J., Dasgupta, N., and Ramphal, R. (2002). FleQ, the major flagellar gene regulator in *Pseudomonas aeruginosa*, binds to enhancer sites located either upstream or atypically downstream of the RpoN binding site. J. Bacteriol. *184*, 5251–5260.

Kirov, S.M., Castrisios, M., and Shaw, J.G. (2004). *Aeromonas* flagella (polar and lateral) are enterocyte adhesins that contribute to biofilm formation on surfaces. Infect. Immun *72*, 1939–1945.

Kohler, R., Bubert, A., Goebel, W., Steinert, M., Hacker, J., and Bubert, B. (2000). Expression and use of the green fluorescent protein as a reporter system in *Legionella pneumophila*. Mol. Gen. Genet. *262*, 1060–1069.

Kohler, R., Fanghanel, J., Konig, B., Luneberg, E., Frosch, M., Rahfeld, J.U., Hilgenfeld, R., Fischer, G., Hacker, J., and Steinert, M. (2003). Biochemical and functional analyses of the Mip protein: influence of the N-terminal half and of peptidylprolyl isomerase activity on the virulence of *Legionella pneumophila*. Infect. Immun. *71*, 4389–4397.

Konkel, M.E., Klena, J.D., Rivera-Amill, V., Monteville, M.R., Biswas, D., Raphael, B., and Mickelson, J. (2004). Secretion of virulence proteins from *Campylobacter jejuni* is dependent on a functional flagellar export apparatus. J. Bacteriol. *186*, 3296–3303.

Kutsukake, K., Iyoda, S., Ohnishi, K., and Iino, T. (1994). Genetic and molecular analyses of the interaction between the flagellum-specific sigma and anti-sigma factors *in Salmonella typhimurium*. EMBO J. *13*, 4568–4576.

Lebeau, I., Lammertyn, E., De Buck, E., Maes, L., Geukens, N., Van Mellaert, L., and Anne, J. (2004). Novel transcriptional regulators of *Legionella pneumophila* that affect replication in *Acanthamoeba castellanii*. Arch. Microbiol. *181*, 362–370.

Liles, M.R., Edelstein, P.H., and Cianciotto, N.P. (1999). The prepilin peptidase is required for protein secretion by and the virulence of the intracellular pathogen *Legionella pneumophila*. Mol. Microbiol. *31*, 959–970.

Liles, M.R., Viswanathan, V.K., and Cianciotto, N.P. (1998). Identification and temperature regulation of *Legionella pneumophila* genes involved in type IV pilus biogenesis and type II protein secretion. Infect. Immun. *66*, 1776–1782.

Liu, S. L., Ezaki, T., Miura, H., Matsui, K., Yabuuchi, E. (1988). Intact motility as a *Salmonella typhi* invasion-related factor. Infect. Immun. *56*, 1967–1973.

Liu, Y., and Luo Z.Q. (2007). The *Legionella pneumophila* effector SidJ is required for efficient recruitment of endoplasmic reticulum proteins to the bacterial phagosome. Infect. Immun. *75*, 592–603.

Lüneberg, E., Mayer, B., Daryab, N., Kooistra, O., Zahringer, U., Rohde, M., Swanson, J., Frosch, M. (2001). Chromosomal insertion and excision of a 30 kb unstable genetic element is responsible for phase variation of lipopolysaccharide and other virulence determinants in *Legionella pneumophila*. Mol. Microbiol. *39*, 1259–1271.

Lüneberg, E., Zahringer, U., Knirel, Y. A., Steinmann, D., Hartmann, M., Steinmetz, I., Rohde, M., Kohl, J., and Frosch, M. (1998). Phase-variable expression of lipopolysaccharide contributes to the virulence of *Legionella pneumophila*. J. Exp. Med. *188*, 49–60.

Lüneberg, E., Zetzmann, N., Alber, D., Knirel, Y. A., Kooistra, O., Zahringer, U., and Frosch, M. (2000). Cloning and functional characterization of a 30 kb gene locus required for lipopolysaccharide biosynthesis in *Legionella pneumophila*. Int. J. Med. Microbiol. *290*, 37–49.

Luo, Z.Q., and Isberg R.R. (2004). Multiple substrates of the *Legionella pneumophila* Dot/Icm system identified by interbacterial protein transfer. Proc. Natl. Acad. Sci. USA. *101*, 841–846.

Lynch, D., Fieser, N., Gloggler, K., Forsbach-Birk, V., and Marre, V. (2003). The response regulator LetA regulates the stationary-phase stress response in *Legionella pneumophila* and is required for efficient infection of *Acanthamoeba castellanii*. FEMS Microbiol. Lett. *219*, 241–248.

Macnab, R.M. (1999). The bacterial flagellum: reversible rotary propellor and type III export apparatus. J. Bacteriol. *181*, 7149–7153.

Macnab, R.M. (2003). How bacteria assmble flagella. Annu. Rev. Micobiol *57*, 77–100.

Macnab, R.M. (2004). Type III flagellar protein export and flagellar assembly. Biochim. Biophys. Acta. *1694*, 207–217.

Majander, K., Anton, L., Antikainen, J., Lang, H., Brummer, M., Korhonen, T.K., and Westerlund-Wikstrom, B. (2005). Extracellular secretion of polypeptides using a modified *Escherichia coli* flagellar secretion apparatus. Nat. Biotechnol. *23*, 475–481.

Mampel, J., Spirig, T., Weber, S.S., Haagensen, J.A., Molin, S., and Hilbi, H. (2006). Planktonic replication is essential for biofilm formation by *Legionella pneumophila* in a complex medium under static and dynamic flow conditions. Appl. Environ. Microbiol. *72*, 2885–28895.

Manson, M.D., Armitage, J.P., Hoch, J.A., and Macnab, R.M. (1998). Bacterial locomotion and signal transduction. J. Bacteriol. *180*, 1009–1022.

Mauchline, W.S., Araujo, R., Wait, R., Dowsett, A.B., Dennis, P.J., and Keevil, C.W. (1992). Physiology and morphology of *Legionella pneumophila* in continuous culture at low oxygen concentration. J. Gen. Microbiol. *138*, 2371–2380.

Mauchline, W.S., James, B.W., Fitzgeorge, R.B., Dennis, P.J., and Keevil, C.W. (1994). Growth temperature reversibly modulates the virulence of *Legionella pneumophila*. Infect. Immun. *62*, 2995–2997.

McCarter, L.L. (2006). Regulation of flagella. Curr. Opin. Microbiol. *9*, 180–186.

McDade, J.E., and Shepard, C.C. (1979). Virulent to avirulent conversion of Legionnaires' disease bacterium (*Legionella pneumophila*)--its effect on isolation techniques. J. Infect. Dis. *139*, 707–711.

McNealy, T.L., Forsbach-Birk, V., Shi, C., and Marre, R. (2005). The Hfq homologuein *Legionella pneumophila* demonstrates regulation by LetA and RpoS and interacts with the global regulator CsrA. J. Bacteriol. *187*, 1527–1532.

Merriam, J.J., Mathur, R., Maxfield-Boumil, R., and Isberg, R.R. (1997). Analysis of the *Legionella pneumophila fliI* gene: intracellular growth of a defined mutant defective for flagellum biosynthesis. Infect. Immun. *65*, 2497–2501.

Minamino T, and Macnab, R.M. (1999). Components of the *Salmonella* flagellar export apparatus and classification of export substrates. J. Bacteriol. *181*, 1388–1394.

Minamino T, and Namba, K. (2004). Self-assembly and type III protein export of the bacterial flagellum. J. Mol. Microbiol. Biotechnol. *7*, 5–17.

Miyake, M., Watanabe, T., Koike, H., Molmeret, M., Imai, Y., and Abu Kwaik, Y. (2005). Characterization of *Legionella pneumophila pmiA*, a gene essential for infectivity of protozoa and macrophages. Infect. Immun. *73*, 6272–6282.

Mobley, H.L., Belas, R., Lockatell, V., Chippendale, G., Trifillis, A.L., Johnson, D.E., and Warren, J.W. (1996). Construction of a flagellum-negative mutant of *Proteus mirabilis*: effect on internalization by human renal epithelial cells and virulence in a mouse model of ascending urinary tract infection. Infect. Immun. *64*, 5332–5340.

Moffat, J.F., Edelstein, P.H., Regula, D.P. Jr, Cirillo, J.D., Tompkins, L.S. (1994). Effects of an isogenic Zn-metalloprotease-deficient mutant of *Legionella pneumophila* in a guinea-pig pneumonia model. Mol. Microbiol. *12*, 693–705.

Moffat, J.F., and Tompkins, L.S. (1992). A quantitative model of intracellular growth of *Legionella pneumophila* in *Acanthamoeba castellanii*. Infect. Immun. *60*, 296–301.

Molofsky, A.B., Byrne, B.G., Whitfield, N.N., Madigan, C.A., Fuse, E.T., Tateda, K., and Swanson, M.S. (2006). Cytosolic recognition of flagellin by mouse macrophages restricts *Legionella pneumophila* infection. J. Exp. Med. *17*, 1093–1104.

Molofsky, A.B., Shetron-Rama, L.M., and Swanson, M.S. (2005). Components of the *Legionella pneumophila* flagellar regulon contribute to multiple virulence traits, including lysosome avoidance and macrophage death. Infect. Immun. *73*, 5720–5734.

Molofsky, A.B., and Swanson, M.S. (2003). *Legionella pneumophila* CsrA is a pivotal repressor of transmission traits and activator of replication. Mol. Microbiol. *50*, 445–461.

Molofsky, A.B., and Swanson, M.S. (2004). Differentiate to thrive: lessons from the *Legionella pneumophila* life cycle. Mol. Microbiol. *53*, 29–40.

Nagai, H., Kagan, J.C., Zhu, X., Kahn, R.A., and Roy, C.R. (2002). A bacterial guanine nucleotide exchange factor activates ARF on *Legionella* phagosomes.' Science *295*, 679–682.

Newton, H. J., Sansom, F.M., Bennett-Wood, V., and Hartland, E. (2006). Identification of *Legionella pneumophila*-specific genes by genomic subtractive hybridization with *Legionella micdadei* and identification of *lpnE*, a gene required for efficient host cell entry. Infect. Immun. *74*, 1683–1691.

Orrison, L.H., Cherry, W.B., Tyndall, R.L., Fliermans, C.B., Gough, S.B., Lambert, M.A., McDougal, L.K., Bibb, W.F., and Brenner, D.J. (1983). *Legionella oakridgensis*: unusual new species isolated from cooling tower water. Appl. Environ. Microbiol. *45*, 536–545.

O'Toole, G., Kaplan, H.B., and Kolter, R. (2000). Biofilm formation as microbial development. Annu. Rev. Microbiol. *54*, 49–79.

Ott, M., Messner, P., Heesemann, J., Marre, R., and Hacker, J. (1991). Temperature-dependent expression of flagella in *Legionella*. J. Gen. Microbiol. *137*, 1955–1961.

Pallen, J.M., Penn, C.W., and CHaudhuri, R.R. (2005). Bacterial flagellar diversity in the post-genomic era. TRENDS in Microbiol. *13*, 143–149.

Polesky, A.H., Ross, J.T., Falkow, S., and Tompkins, L.S. (2001). Identification of *Legionella pneumophila* genes important for infection of amoebas by signature-tagged mutagenesis. Infect. Immun. *69*, 977–987.

Pope, C., O'Connell, W., and Cianciotto, N.P. (1996). *Legionella pneumophila* mutants that are defective for iron acquisition and assimilation and intracellular infection. Infect. Immun. *64*, 629–636.

Pruckler, J.M., Benson, R.F., Moyenuddin, M., Martin, T., and Fields, B.S. (1995). Association of flagellum expression and intracellular growth of *Legionella pneumophila*. Infect. Immun. *63*, 4928–4932.

Ren, T., Zamboni, D.S., Roy, C.R., Dietrich, W.F., and Vance, R.E. (2006). Flagellin deficient *Legionella* mutants evade caspase-1- and Naip5-mediated macrophage immunity. PLoS Pathog. *2*, e18.

Ricci, M.L., Torosantucci, A., Scaturro, M., Chiani, P., Baldassarri, L., and Pastoris, M.C. (2005). Induction of protective immunity by *Legionella pneumophila* flagellum in an A/J mouse model. Vaccine. *23*, 4811–4820.

Ridenour, D.A., Cirillo, S.L., Feng, S., Samrakandi, M.M., and Cirillo, J. (2003). Identification of a gene that affects the efficiency of host cell infection by *Legionella pneumophila* in a temperature-dependent fashion. Infect. Immun. *71*, 6256–6263.

Robey, M., and Cianciotto, N.P. (2002). *Legionella pneumophila feoAB* promotes ferrous iron uptake and intracellular infection. Infect. Immun. *70*, 5659–5669.

Robey, M., O'Connell, W., and Cianciotto, N.P. (2001). Identification of *Legionella pneumophila rcp*, a *pagP*-like gene that confers resistance to cationic antimicrobial peptides and promotes intracellular infection. Infect. Immun. *69*, 4276–4286.

Rodgers, F.G., Greaves, P.W., Macrae, A.D., Lewis, M. J. (1980). Electron microscopic evidence of flagella and pili on *Legionella pneumophila*. J. Clin. Pathol. *33*, 1184–1188.

Rossier, O., and Cianciotto, N.P. (2001). Type II protein secretion is a subset of the PilD-dependent processes that facilitate intracellular infection by *Legionella pneumophila*. Infect. Immun. *69*, 2092–2098.

Rowbotham, T.J. (1986). Current view on the relationship between amoebae, legionellae and man. Isr. J. Med. Sci. *22*, 678–689.

Sahney, N.N., Summersgill, J.T., Ramirez, J.A., Miller, R.D. (2001). Inhibition of oxidative burst and chemotaxis in human phagocytes by *Legionella pneumophila* zinc metalloprotease. J. Med. Microbiol. *50*, 517–525.

Samrakandi, M.M., Cirillo, S.L., Ridenour, D.A., Bermudez, L.E., and Cirillo, J.D. (2002). Genetic and phenotypic differences between *Legionella pneumophila* strains. J. Clin. Microbiol. *40*, 1352–1362.

Schell, M.A. (1993). Molecular biology of the LysR family of transcriptional regulators. Annu. Rev. Microbiol. *47*, 597–626.

Segal, G., and Shuman, H.A. (1999). *Legionella pneumophila* utilizes the same genes to multiply within *Acanthamoeba castellanii* and human macrophages. Infect. Immun. *67*, 2117–2124.

Shi, C., Forsbach-Birk, V., Marre, R., and McNealy, T.L. (2006). The *Legionella pneumophila* global regulatory protein LetA affects DotA and Mip. Int. J. Med. Microbiol. *296*, 15–24.

Shimizu, R., Taguchi, F., Marutani, M., Mukaihara, T., Inagaki, Y., Toyoda, K., Shiraishi, T., and Ichinose, Y. (2003). The Delta *fliD* mutant of *Pseudomonas syringae* pv. tabaci, which secretes flagellin monomers, induces a strong hypersensitive reaction (HR) in non-host tomato cells. Mol. Genet. Genomics. *269*, 21–30.

Smith, K.D., Andersen-Nissen, E., Hayashi, F., Strobe, K., Bergman, M.A., Barrett, S.L., Cookson, B.T., and Aderem, A. (2003). Toll-like receptor 5 recognizes a conserved site on flagellin required for protofilament formation and bacterial motility. Nat. Immunol. *4*, 1247–1253.

Smith, K.D., and Ozinsky, A. (2002). Toll-like receptor-5 and the innate immune response to bacterial flagellin. Curr. Top. Microbiol. Immunol. *270*, 93–108.

Soderberg, M.A., Rossier, O., and Cianciotto, N.P. (2004). The type II protein secretion system of *Legionella pneumophila* promotes growth at low temperatures. J. Bacteriol. *186*, 3712–3720.

Stone, B.J., Brier, A., and Abu-Kwaik, Y. (1999). The *Legionella pneumophila prp* locus; required during infection of macrophages and amoebae. Microb. Pathog. 27, 369–376.

Swanson, M.S., and Hammer, B.K. (2000). *Legionella pneumophila* pathogenesis: a fateful journey from amoebae to macrophages. Annu. Rev. Microbiol. *54*, 567–613.

Swanson, M.S., and Molofsky A.B. (2005). Autophagy and inflammatory cell death, partners of innate immunity. Autophagy. *1*, 174–176.

Szurmant, H., and Ordal, G.W. (2004). Diversity in chemotaxis mechanisms among the bacteria and archaea. Microbiol. Mol. Biol. Rev. *68*, 301–319.

Vincent, C.D., Buscher, B.A, Friedman, J.R., Williams, L.A., Bardill, P., and Vogel, J.P. (2006a). Identification of non-*dot/icm* suppressors of the *Legionella pneumophila ΔdotL* lethality phenotype. J. Bacteriol. *188*, 8231–8243.

Vincent, C.D., Friedman, J.R., Jeong, K.C., Miller, J.L., and Vogel, J.P. (2006b). Identification of the core transmembrane complex of the *Legionella* Dot/Icm type IV secretion system. Mol. Microbiol. *62*, 1278–1291.

Viswanathan, V.K., Edelstein, P.H., Pope, C.D., and Cianciotto, N.P. (2000). The *Legionella pneumophila iraAB* locus is required for iron assimilation, intracellular infection, and virulence. Infect. Immun. *68*, 1069–1079.

Wagner, C., Khan, A.S., Kamphausen, T., Schmausser, B., Unal, C., Lorenz, U., Fischer, G., Hacker, J., Steinert, M. (2007). Collagen binding protein Mip enables *Legionella pneumophila* to transmigrate through a barrier of NCI-H292 lung epithelial cells and extracellular matrix. Cell. Microbiol. *9*, 450–462.

Wieland, H., Faigle, M., Lang, F., Northoff, H., and Neumeister, B. (2002). Regulation of the *Legionella mip*-promotor during infection of human monocytes. FEMS Microbiol. Lett. *212*, 127–132.

Williams, A., Baskerville, A., Dowsett, A.B., and Conlan, J.W. (1987). Immunocytochemical demonstration of the association between *Legionella pneumophila*, its tissue-destructive protease, and pulmonary lesions in experimental legionnaires' disease. J. Pathol. *153*,:257–264.

Wintermeyer, E., Ludwig, B., Steinert, M., Schmidt, B., Fischer, G., and Hacker, J. (1995). Influence of site specifically altered Mip proteins on intracellular survival of *Legionella pneumophila* in eukaryotic cells. Infect. Immun. *63*, 4576–4583.

Yao, R., Burr, D.H., Doig, P., Trust, T.J., Niu, H., and Guerry, P. (1994). Isolation of motile and non-motile insertional mutants of *Campylobacter jejuni*: the role of motility in adherence and invasion of eukaryotic cells. Mol. Microbiol. *14*, 883–893.

Young, G.M., Badger, J.L., and Miller, V.L. (2000). Motility is required to initiate host cell invasion by *Yersinia enterocolitica*. Infect. Immun. *68*, 4323–4326.

Young, G.M., Schmiel, D.H., and Miller, V.L. (1999). A new pathway for the secretion of virulence factors by bacteria: the flagellar export apparatus functions as a protein-secretion system. Proc. Natl. Acad. Sci. USA. *96*, 6456–6461.

Young, B.M., and Young, G.M. (2002). YplA is exported by the Ysc, Ysa, and flagellar type III secretion systems of *Yersinia enterocolitica*. J. Bacteriol. *184*, 1324–1334.

Zusman, T., Aloni, G., Halperin, E., Kotzer, H., Degtyar, E., Feldman, M., and Segal, G. (2007). The response regulator PmrA is a major regulator of the icm/dot type IV secretion system in *Legionella pneumophila* and *Coxiella burnetii*. Mol. Microbiol. *63*, 1508–1523.

Zusman, T., Gal-Mor, O., and Segal, G. (2002). Characterization of a *Legionella pneumophila relA* insertion mutant and toles of RelA and RpoS in virulence gene expression. J. Bacteriol. *184*, 67–75.

Genomics and Transcriptomics of *Legionella pneumophila*: Insights into the Lifestyle of an Intracellular Pathogen

7

Christel Cazalet, Matthieu Jules and Carmen Buchrieser*

Abstract

The last years have seen a giant step in genomics of *Legionella pneumophila*. The determination and publication of the complete genome sequences of three clinical *L. pneumophila* isolates in 2004 now paves the way for major breakthroughs in understanding the biology of *L. pneumophila* in particular and *Legionella* in general. Here we describe the major features deduced from genome sequence analyses of *L. pneumophila* and highlight the characteristic features and common traits of the *L. pneumophila* genomes. Emphasis is given to putative virulence and *Legionella* life cycle related functions. In depth comparative genome analysis using DNA arrays to study the gene content of 180 *Legionella* strains revealed the high genome plasticity and frequent horizontal gene transfer identified through whole genome comparisons. Further insight in the *L. pneumophila* life cycle was recently gained by investigating the intracellular gene expression profile of *L. pneumophila* in *Acanthamoeba castellanii*, its natural host. This study shows that *L. pneumophila* exhibits a biphasic life cycle *in vitro* as well as *in vivo* and defines transmissive and replicative traits according to gene expression profiles. These specific traits are discussed, and future perspectives in *Legionella* genomics are presented.

Introduction

Pathogens that are able to enter and multiply within human cells are responsible for multiple diseases and millions of deaths worldwide. Thus, the challenge is to elucidate those pathogen-specific and cell biological mechanisms involved in intracellular growth and spread. Many different techniques, such as molecular genetics, tissue culture systems, high-resolution microscopy, *in vivo* infection models, and recently also *in vivo* imaging techniques have been applied to the study of the mechanisms of intracellular pathogenesis. Since the publication of the first bacterial genome sequence in 1995 (Fleischmann *et al.*, 1995), a tremendous increase in genomic information has substantially altered our view on bacterial pathogenesis and has led to the application of many different genomics and post genomics approaches in microbial research. Today over 340 complete bacterial genome sequences are available in public databases, representing hundreds of species, allowing in depth comparative and functional genomic studies. Comparative genomics aims to identify genetic differences across entire genomes, to correlate those differences with biological functions, and to gain insight into selective evolutionary pressures and patterns of gene transfer or loss, particularly with respect to virulence in pathogens. One of the most obvious conclusions, which resulted from analyses of hundreds of genome sequences and microarray-based comparative genomics, is the great amount of diversity that exists even among different genomes from the same bacterial species.

In *Legionella* research, the genomics era started only recently. However, the situation has changed rapidly as the complete genome sequences of three clinical isolates were determined and published in 2004 (Cazalet *et al.*, 2004; Chien *et al.*, 2004). The sequenced strains are the endemic

*Corresponding author.

strain Paris and the epidemic strain Lens, responsible for two major outbreaks in France in 2001 and 2004 (Cazalet *et al.*, 2004), and strain Philadelphia-1, isolated from the first recognized outbreak of the disease in 1976 (Chien *et al.*, 2004). Recently, the genome of a fourth isolate, that of *L. pneumophila* strain Corby, a virulent strain isolated from a human legionellosis case (Jepras *et al.*, 1985), was determined (Steinert *et al.*, 2007). Knowledge of these sequences is now the basis for major breakthroughs in understanding the biology of *L. pneumophila*, as it gives new insight into the bacterium's life style and its way of adapting to the host. As described below in detail, the analysis of these genome sequences revealed several specific features of *L. pneumophila*, some of which are undoubtedly related to its intracellular life, and allowed generation of hypothesis on how *L. pneumophila* subverts host functions to its advantage.

In depth comparative genomics can now be undertaken to help to unravel the molecular basis of the pathogenesis, evolution and phenotypic differences among different strains. Furthermore, additional *Legionella* sequences are in the pipeline and will soon allow comparative genomics of pathogenic and 'non-pathogenic' (found in the environment but not human infection) strains. This approach should reveal genetic differences important for the host range, pathogenic lifestyle and disease manifestations.

Genome analysis of *L. pneumophila*

General features and organization of the *L. pneumophila* genome

The *L. pneumophila* genomes comprise 3,503,610 base pairs (bp) (Paris), 3,345,687 bp (Lens) 3,397,754 bp (Philadelphia-1) and 3,576,461 bp (Corby) with an average GC content of 38% and ~3000 predicted protein-coding genes in each genome (Cazalet *et al.*, 2004; Chien *et al.*, 2004; Steinert *et al.*, 2007). The sequence and the annotation of the *L. pneumophila* Paris, Lens and Philadelphia-1 genomes are accessible at http://genolist.pasteur.fr/LegioList. Throughout the text *lpp*, *lpl*, *lpg* indicate predicted coding sequences of *L. pneumophila* strain Paris, Lens and Philadelphia, respectively. A high proportion of the predicted genes (17–21%) are unique to the genus *Legionella*. Analysis of the *L. pneumophila* genomes allowed determination of common features as well as particular features of *Legionellae*. One common feature is that about 2400 genes have an orthologous gene in all *L. pneumophila* genomes, constituting the common backbone of this species. Furthermore, a quite high conservation of the order and the relative orientation of these orthologous genes was identified. Principally, the genomes contain three large plasticity zones, where the synteny is disrupted: a 260 kb inversion in strain Lens with respect to strains Paris and Philadelphia-1; a 130 kb fragment that is only partly similar and that is inserted in a different genomic location in strains Paris and Philadelphia-1 but is truncated in strain Lens; and the chromosomal region carrying the Lvh T4SS (Fig. 7.1). However, despite this considerable number of common genes, many differences in gene content are also present, some of which are without doubt related to the ability of *L. pneumophila* to replicate intracellularly within protozoa and to cause disease in humans.

Specific features of the *L. pneumophila* genomes – eukaryotic-like proteins

In contrast to their relatively recent recognition as human pathogens (McDade *et al.*, 1977), *Legionella* species have probably evolved as parasites of various protozoa for hundreds of millions of years. In 1980, Rowbotham described the ability of *L. pneumophila* to multiply intracellularly within freshwater protozoa (Rowbotham, 1980). This observation has led to a new percept in microbiology: bacteria may parasitize protozoa and can utilize the same process to infect human cells. Since then, *L. pneumophila* has been described to multiply in many species of protozoa, an interaction that is central to the pathogenesis and ecology of *L. pneumophila*. There is today much evidence suggesting that protozoa play a major role in the emergence of *L. pneumophila* as a pathogen of human alveolar macrophages. This tight co-evolution of *Legionella* with protozoa is reflected in the genome sequence. Within its genome *L. pneumophila* encodes a striking abundance of eukaryotic-like proteins. This 'eukaryotic' signature seem to reflect the origin of the DNA, and some of those are probably able

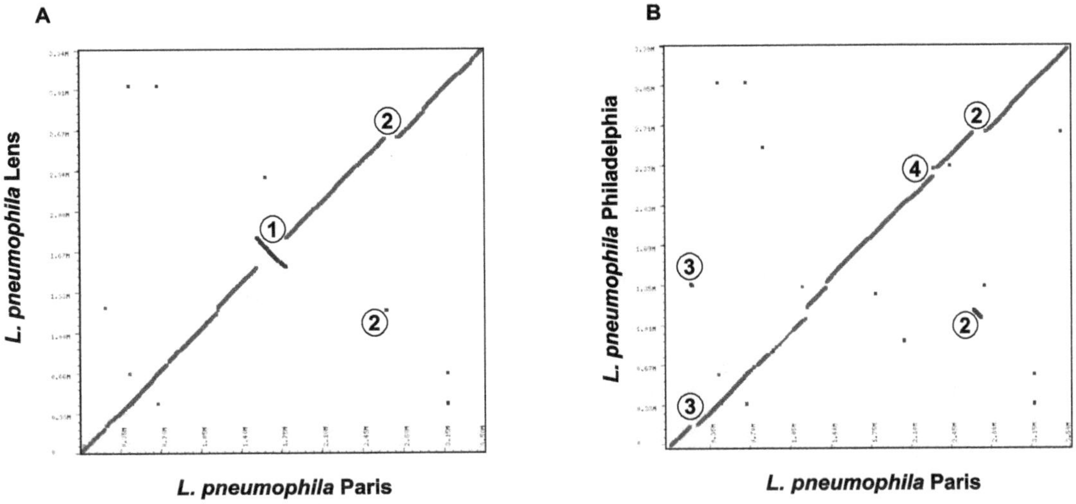

Figure 7.1 Synteny plot of the chromosomes of strains Paris and Lens (A) and strains Paris and Philadelphia-1 (B). Inversions between the genomic sequences are represented in blue. Genome-wide synteny is disrupted by a 260 kb inversion 1 and a 130 kb plasticity zone in strain Paris 2 as well as by some smaller clusters. The insertion 2 is also present in strain Philadelphia but in a different chromosomal location as well as the lvh secretion system 3. Synteny is also disrupted by an insertion of a 65 kb pathogenicity island in strain Philadelphia-1 4. Plots were created with the mummer software package developed by Tigr (http://www.tigr.org).

to interfere in the different steps of the infectious cycle by mimicking functions of eukaryotic proteins. However, others are also likely to mediate protein-protein interactions or enzymatic reactions inside the bacterial cell. Whether these eukaryotic-like proteins of *L. pneumophila* evolved by lateral gene transfer from eukaryotic hosts to *L. pneumophila*, by convergent evolution, or by a combination of these two processes is not known yet.

By a systematic search in the genome sequence of *L. pneumophila* strain Paris, 105 genes were identified as having their best hits against eukaryotic genomes, including both housekeeping genes and genes that could play a role in the virulence of *L. pneumophila*. Detailed analyses lead to the identification of 30 eukaryotic-like genes (i.e. at least 20% amino acid identity over more than a third of the length of the most similar protein). A search against Pfam and Prosite databases identified 38 additional genes containing motifs that are present mostly in eukaryotes (ankyrin, sel-1, sec7, F-box, U-box) where they mediate protein-protein interactions. When comparing the three sequenced genomes, it becomes evident that most of these proteins should fulfil important functions, as nearly all are conserved in the three sequenced genomes (Tables 7.1 and 7.2) (Cazalet *et al.*, 2004; de Felipe *et al.*, 2005; Brüggemann *et al.*, 2006a). Some of these proteins are predicted to divert eukaryotic regulatory pathways or to be secreted into eukaryotic cells, making them strong candidates for directing *L. pneumophila* invasion, trafficking in the host cell, or modulating or evading host cell functions.

An example is RalF, a protein that contains a eukaryotic Sec7 domain. RalF, a substrate of the Dot/Icm T4SS, is required for the regulatory protein ARF to associate with the *L. pneumophila* phagosome and is thought to contribute to ER recruitment (Nagai *et al.*, 2002; Chapter 10). Other examples are the three eukaryotic-like serine/threonine protein kinases (STPK) present in each of the sequenced *L. pneumophila* strains. Eukaryotic-like STPK have been identified in several prokaryotic genomes where they regulate various activities such as developmental processes (Dasgupta *et al.*, 2006), secondary metabolism (Umeyama T, 2002), stress response (Neu *et al.*, 2002) or virulence. For example, *Mycobacterium tuberculosis*, which like *L. pneumophila* blocks phagosome lysosome fusion, produces eleven such eukaryote-like STPKs (Cole *et al.*, 1998).

Table 7.1 Proteins with the highest similarity score to eukaryotic proteins, their distribution and gene expression in the three sequenced strains

Predicted product	*L. pneumophila* Paris	FC	*L. pneumophila* Lens	FC	*L. pneumophila* Philadelphia	FC	Best Hit BLASTp
purC	*lpp1647**	~	*lpl1640*	~	*lpg1675*	~	(AAR06292.1 *Nicotania tabacum*)
exoA exodeoxyribonuclease III	*lpp0702**	~	*lpl0684*	~	*lpg0648*	~	(EAA20230.1 *Plasmodium yoelii yoelii*)
RNA binding protein precursor	*lpp0321*	–2.5	–	–	*lpg0251*	–2.4	(AAL07519 *Solanum tubeosum*)
Pyruvate decarboxylase	*lpp1157**	3.4	*lpl1162*	11.8	*lpg1155*	5.8	(AAB16855.1 *Arabidopsis thaliana)*
Thiamine biosynthesis protein NMT-1	*lpp1522**	~	*lpl1461*		*lpg1565*		(AAC64375.1 *Botryotinia fuckeliana*)
nuoE NADH dehydrogenase I chain E	*lpp2832*	–3.3	*lpl2701*	-2.3	*lpg2785*	–3.1	(BAA25988.1 *Homo sapiens*)
hypersensitive induced response Protein	*plpp0050**	~	-	–	–	-	(AAN17462.1 *Hordeum vulgare subsp. Vulgare*)
Hypothetical protein	*lpp0634**	–2.6	*lpl0618*	–2.4	*lpg0584*	–1.9	(XP_306643.1 *Anopheles gambiae*)
DegP protease	*lpp0965*	~	*lpl0935*	~	*lpg0903*	~	(NP_189431.2 *Arabidopsis thaliana*)
Phytanoyl-coA dioxygenase	*lpp2748**	~	*lpl2621*	~	*lpg2694*	~	(XP_372144.1 *Homo sapiens*)
Sphingosine-1-phosphate lyase	*lpp2128*	n.d.	*lpl2102*	-4.2	*lpg2176*	–2.2	(NP_775139.1 *Rattus norvegicus*)
Glucoamylase	*lpp0489*	-1.9	*lpl0465*	-4.1	*lpg0422*	–2.4	(P42042 *Arxula adeninivorans*)
Cytokinin oxidase	*lpp0955*	2.5	*lpl0925*	7.7	*lpg0894*	3.7	(NP_484368.1 *Nostoc* sp.)
Phytanoyl coA dioxygenase	*lpp0578**	~	*lpl0554*	~	*lpg0515*	~	(EAA70100.1 *Gibberella zeae*)
Hypothetical protein	*lpp0379*	~	*lpl0354*	~	*lpg0301*	~	(CAD21525.1 *Taenia solium*)
Ectonucleoside triphosphate diphosphohydrolase (apyrase)	*lpp1033*	~	*lpl1000*	~	*lpg0971*	~	nucleoside phosphatase signature conserved (Q9MYU4 *Sus scrofa*)
6-Pyruvoyl-tetrahydropterin synthase	*lpp2923**	~	*lpl2777*	~	*lpg2865*	~	(NP_703938.1 *Plasmodium falciparum*)

Zinc metalloproteinase	*lpp3071**	–5.2	*lpl2927*	-	*lpg2999*	-2.1	(AAF56122.1 *Drosophila melanogaster*)
SAM-dependent methyltransferase	*lpp2134**	~	*lpl2109*	~	*lpg2182*	~	(BAC98835.1 *Bombyx mori*)
Ectonucleoside triphosphate diphosphohydrolase (apyrase)	*lpp1880*	~	*lpl1869*	~	*lpg1905*	~	(CAE70887.1 *Caenorhabditis briggsae*)
SAM dependent methyltransferase	*lpp2747*	4.4	*lpl2620*	4.1	*lpg2693*	n.d.	(EAA20288.1 *Plasmodium yoelii yoelii*)
Cytochrome P450	*lpp2468**	~	*lpl2326*		*lpg2403*		(NP_487786.1 *Nostoc sp.*)
Nuclear membrane binding protein	*lpp1824**	~	-	-	-	–	(NP_082559.1 *Mus musculus*)
Uracyl DNA glycosylase	*lpp1665*	~	*lpl1659*	~	*lpg1700*	~	(EAA36774.1 *Giardia lamblia*)
Chromosome condensation 1-like	*lpp1959**	~	*lpl1953*	~	*lpg1976*	~	Pattern of chromosome condensation regulator conserved
Hypothetical protein	*lpp0358*	~	*lpl0334*	~	*lpg0282*	~	(EAA20288.1 *Plasmodium yoelii yoelii*)
Ca^{2+}-transporting ATPase	*lpp1127**	2.0	*lpl1131*	1.7	*lpg1126*	1.6	(AAB81284.1 *Paramecium tetraurelia*)
Uridine kinase	*lpp1167*	~	*lpl1173*	–2.4	*lpg1165*	~	(AAM09314.2 *Mus musculus*)
Ser/thr proteine kinase domain	*lpp2626**	~	*lpl2481*	~	*lpg2556*	~	Conserved domain
Serine threonine protein kinase	*lpp1439**	~	*lpl1545*	~	*lpg1483*	~	Conserved domain

lpp, *lpl*, *lpg* indicate predicted coding sequences of *L. pneumophila* strain Paris, Lens and Philadelphia respectively; grey lines indicate genes which are also mentioned in Table 7.2 because of their conserved eukaryotic domains; protein accession numbers are given in parentheses. FC, fold change of expression ratios; negative FC indicates upregulated in exponential growth phase, positive FC upregulated in post-exponential growth phase; ~ not differentially expressed during growth; * the encoded protein contains a putative signal for recognition by the Dot/Icm T4SS.

Table 7.2 *L. pneumophila* genes identified that encode domains preferentially found within eukaryotic proteins and their gene expression

L. pneumophila Paris	FC	*L. pneumophila.* Lens	FC	*L. pneumophila* Philadelphia	FC	Motif identified
enhC (*lpp2692*)	n.d.	*enhC (lpl2564)*	7.3	*enhC (lpg2639)*	4.2	21 sel-1 domains
lidL(lpp1174) - EnhC paralogue	4.6	*lidL (lpl1180)* – EnhC paralogue	7.7	*lidL (lpg1172)* – EnhC paralogue	3.2	6 sel-1 domains
lpp1310 - EnhC paralogue	8.0	*lpl1307* – EnhC paralogue	20.7	*lpg1356* – EnhC paralogue	11.6	4 sel-1 domains
lpp2174 - EnhC paralogue	13.0	*lpl1303* – EnhC paralogue	37.5	*lpg2222* *– EnhC paralogue*	19.8	3 sel-1 domains
–		*lpl1059* – EnhC paralogue	1.8	*lpg1062* – EnhC paralogue	1.8	7 sel-1 domains
ralF (*lpp1932*)	16.6	*ralF lpl1919*	16.4	*ralF* (*lpp1950*)	9.2	sec7 domain
*lpp0267**	~	*lpl0262*	~	*lpg0208*	~	Ser/thr proteine kinase domain
*lpp2626**	~	*lpl2481*	~	*lpg2556*	~	Ser/thr proteine kinase domain
*lpp1439**	~	*lpl1545*	~	*lpg1483*	~	Ser/thr proteine kinase domain
lpp2065	2.8	*lp2055*	1.7	*lpg2214*	1.6	Ankyrin repeat
lpp0037	~	*lpl0038*	~	*lpg0038*	~	Ankyrin repeat
*plpp0098**	~	–	–	–	–	Ankyrin repeat
*lpp2058**	~	*lpl2048*	~	–	–	Ankyrin repeat
*lpp0750**	4.3	*lpl0732*	3.8	*lpg0695*	2.0	Ankyrin repeat
lpp2061	~	*lpl2051*	~	–	~	Ankyrin repeat
lpp2270	–6.7	*lpl2242*	–2.7	*lpg2322*	–5.5	Ankyrin repeat
lpp0503	~	*lpl0479*	~	*lpg0436*	~	Ankyrin repeat
*lpp1905**	~	–	–	–	–	Ankyrin repeat
*lpp1683**	2.2	*lpl1682*	1.7	*lpg1718*	2.0	Ankyrin repeat + SET domain
lpp2248	2.7	*lpl2219*	4.3	*lpg2300*	2.4	Ankyrin repeat
*lpp0202**	2.1	—	–	–	–	Ankyrin repeat
*lpp0469**	1.6	*lpl0445*	2.1	*lpg0403*	2.6	Ankyrin repeat
*lpp2517**	~	*lpl2370*	~	*lpg2452*	~	Ankyrin repeat
*lpp1100**	5.2	–	–	–	–	Ankyrin repeat
*lpp0126**	2.2	*lpl0111*	1.4	*lpg0112*	2.2	Ankyrin repeat
*lpp0356**	~	–	–	–	–	Ankyrin repeat
lpp2522	4.0	*lpl2375*	4.1	*lpg2456*	2.8	Ankyrin repeat
*lpp0547**	–2.8	*lpl0523*	–2.7	*lpg0483*	–2.0	Ankyrin repeat
-	–	*lpl1681*	26.3	–	–	Ankyrin repeat
-	–	*lpl2344**	22.0	–	–	Ankyrin repeat
-	–	*lpl2058*	4.2	*lpg2128*	2.2	Ankyrin repeat
-	–	–	–	*lpg0402**	2.1	Ankyrin repeat
-	–	–	–	*lpg2131**	3.2	Ankyrin repeat
lpp2082	2.1	*lpl2072*	n.d.	*lpg2144*	4.1	F-box domain + ankyrin repeat

Table 7.2 *continued*

L. pneumophila Paris	FC	*L. pneumophila.* Lens	FC	*L. pneumophila* Philadelphia	FC	Motif identified
*lpp2486**	~	–	–	–	–	F-box domain + coiled-coil
–	–	–	–	*lpg2224*	n.d.	F-box domain
*lpp0233**	~	*lpl0234*	~	*lpg0171*	~	F-box domain
*lpp2887**	4.1	–	–	*lpg2830*	3.6	2 U-box domains

FC, fold change of expression ratios; negative FC indicates upregulated in exponential growth phase, positive FC upregulated in post-exponential growth phase; ~ not differentially expressed during growth; * the encoded protein contains a putative signal for recognition by the Dot/Icm T4SS.

Given the function of STPK in other prokaryotes, it is likely that the *L. pneumophila* enzymes are also implicated in modulating trafficking in the host cell.

The most prominent family of proteins with eukaryotic-like domains are ankyrin repeat proteins. Ankyrin motifs are tandem-repeats of around 33 amino acids that represent one of the most common modular protein-protein interaction motif of eukaryotes (Mosavi *et al.*, 2002) involved in many cellular pathways. Recently it was suggested that the signal for substrate recognition by the Dot/Icm T4SS could be an hydrophobic residue or a proline at the C-terminal -3 or -4 position (Nagai *et al.*, 2005). Interestingly, 15 out of the 24 ankyrin proteins encoded by *L. pneumophila* fulfil this criteria and may indeed be secreted via the Dot/Icm T4SS (Table 7.2).

Additional eukaryotic-like proteins found in the *L. pneumophila* genomes are two ecto-nucleoside triphosphate diphosphohydrolases (ecto-NTPDases) implicated in entry and replication (Sansom *et al.*, 2007), a sphingosine-1-phosphate lyase probably involved in the modulation of the host cell cycle signalling, and a glucoamylase putatively implicated in glycogen degradation. Among the proteins with protein domains mainly found in eukaryotes, F-box and U-box containing proteins are notable. Ubiquitination is a fundamental post-translational protein modification in eukaryotes. It is a recognized signal for protein degradation; however, ubiquitination is also implicated in control of cell cycle progression, transcriptional regulation, signal transduction or recognition, and resistance to pathogens (Kerscher *et al.*, 2006; Welchman *et al.*, 2005). Furthermore, exploiting the host's ubiquitination system seems a general tactic among bacteria to control the timing of their effectors' action by programming them for degradation, to block specific intermediates in mammalian or plant innate immunity, or to target host proteins for degradation by mimicking specific ubiquitin/proteasome system components (Veiga and Cossart, 2005; Angot *et al.*, 2006; Munro *et al.*, 2007). The Ub ligase enzyme (E3) controls the specificity of substrate ubiquitination by recruiting the target protein. E3 Ub ligases constitute a large protein family comprising F-box and U-box type proteins.

Also present are proteins with predicted coiled-coil (CC) domains, many of which also show weak similarities to eukaryotic proteins. For example, strain Paris encodes 247 predicted proteins with these features, representing 7.6% of its gene content. CC domains mediate protein–protein interactions either for protein multimerization or macromolecular recognition. Thus, CC domains may target proteins to the appropriate localization in the eukaryotic host. Interestingly, all nine substrates (SidA-H, SdeC) of the Dot/Icm T4SS identified by Luo and Isberg (Luo and Isberg, 2004) as well as LidA (Conover *et al.*, 2003), LepA and LepB (Chen *et al.*, 2004) contain coiled-coil domains and are among the 105 proteins having a eukaryotic best hit.

Finally, to act in the eukaryotic cell, *L. pneumophila* must translocate these eukaryotic-like proteins to the host cytoplasm. Among the 66 genes listed in Tables 7.1 and 7.2, only EnhC, the three conserved EnhC homologues, the homologueof the DegP protease, and the two apyrases have a predicted signal peptide sequence. The 59 remaining proteins are candidate substrates for the T4SS, as demonstrated for VirF (a F-box protein) in *A. tumefaciens* (Vergunst *et al.*, 2000) or RalF in *L. pneumophila* (Nagai *et al.*, 2002). A screen for the presence of a hydrophobic residue or a proline at the C-terminal 3 or 4 positions as T4SS signal (Nagai *et al.*, 2005) identified such residues in 36 of the 59 proteins (Tables 7.1 and 7.2). This proposed signal for substrate recognition by the Dot/Icm T4SS was identified based on the study of the translocation mechanism of RalF in strain *L. pneumophila* Philadelphia-1 (Nagai *et al.*, 2005). Surprisingly, in strains *L. pneumophila* Paris and Lens, the C-terminal part of the RalF protein differs from the protein of strain Philadelphia-1, and no hydrophobic residue or proline was found at the C-terminal -3 and -4 positions in these two strains. This finding is intriguing: Perhaps the Dot/Icm T4SS also recognizes its substrates via hydrophobic residues at different positions, or by some other motif. Additional studies are needed to resolve how such a large number of proteins can behave as substrates for the Dot/Icm T4SS.

Distribution of eukaryotic-like proteins in completely sequenced prokaryotic genomes

Given the large number of eukaryotic-like proteins present in *L. pneumophila*, the question arises whether this feature is unique to *L. pneumophila*. The taxonomic breakdown of the best hits for all proteins encoded in completely sequenced prokaryotic genomes is available through the genome division of the Entrez retrieval system at http://www.ncbi.nlm.nih.gov/entrez/query.fcgi?db = genomeprj ('TaxMap' for each individual genome). Table 7.3 presents results for the three *L. pneumophila* strains and for 22 other microorganisms (12 intracellular pathogens, seven extracellular pathogens and three plant pathogens). Interestingly, eukaryotic-like genes are found mainly in intracellular bacteria. The only exception is *Helicobacter pylori*, with 2.2% of its proteins with highest similarity to eukaryotic proteins. With 6.8% and 4.9%, respectively, *Rickettsia bellii* and *Rickettsia felis* (Ogata *et al.*, 2005; Ogata *et al.*, 2006) contain the highest percentage of proteins showing similarity to eukaryotic proteins. Third is *L. pneumophila* with 3.5%. A motif search for six eukaryotic motifs (F-box, serine/threonine kinase domain, Sec7, ankyrin, SET, U-box) showed that, compared to *L. pneumophila*, none of 12 selected bacterial genomes encodes the same variety and number of proteins containing motifs found preferentially in eukaryotes (Albert-Weissenberger *et al.*, 2007). With the exception of the ankyrin repeat family, which is most abundant in *Wolbachia pipientis* (Wu *et al.*, 2004) and *R. felis* (Ogata *et al.*, 2005), and STPK domains, which are more abundant in *M. tuberculosis* (Cole *et al.*, 1998), *L. pneumophila* is the organism with the highest number of proteins with eukaryotic-like domains. Thus, the genome sequence reflects the co-evolution of *L. pneumophila* with its eukaryotic hosts, e.g. freshwater amoebae, and suggests that the different eukaryotic-like proteins are important for the intracellular life cycle of *L. pneumophila*.

Specific features of the *L. pneumophila* genomes – high genetic plasticity and diversity

A second specific feature of the *L. pneumophila* genome is its high diversity and plasticity. Plasmids, genomic and pathogenicity islands, IS-elements and transposases small DNA-pieces, termed 'islets', as well as integrative plasmids are present. Comparison of the four completely sequenced *L. pneumophila* strains identified about 300 genes (~10%) in each genome as strain specific, constituting the flexible gene pool.

Most of the strain-specific genes code proteins of unknown function, however some proteins may be related to niche adaptation or to specific virulence functions. Strain Paris seems to be a particularly well equipped strain, as its specific genetic complement contains many regulators (two CsrA homologues, 11 transcriptional regulators), five ankyrin-repeat proteins, one specific U-box domain protein and two eukaryotic-like proteins as well as several restriction modification genes (DNA modification methylases, endonucleases), a specific *o*-type cytochrome ubiquinol oxidase, and a type V secretion system (autotransporter). Strain Lens contains four specific transcriptional regulators and three specific proteins with eukaryotic domains, two of which are ankyrin repeat proteins. In addition to four specific transcriptional regulators, strain

Table 7.3. Distribution of best Blast P hits among different taxa (bacteria, eukaryota, archea, virus) for all proteins encoded by 23 selected microorganisms

Taxa	*L. pneumophila* Paris		*L. pneumophila* Lens		*L. pneumophila* *Philadelphia*		*Mycobacterium tuberculosis*		*Chlamydia trachomatis*	
Bacteria	2380	(79)	2287	(79.5)	2345	(79.7)	3445	(86.4)	846	(94.6)
Eukaryota	105	(3.5)	99	(3.4)	101	(3.4)	93	(2.3)	0	(0)
Archaea	15	(0.5)	14	(0.49)	11	(0.37)	18	(0.45)	0	(0)
Virus	3	(0.1)	3	(0.1)	4	(0.1)	16	(0.4)	0	(0)

Taxa	*Rickettsia felis*		*Francisella tularensis*		*Shigella flexneri*		*Salmonella typhimurium*		*Coxiella burnetii*	
Bacteria	1091	(77.9)	1348	(84)	4025	(96.2)	4213	(95.2)	1384	(68.6)
Eukaryota	68	(4.9)	26	(1.6)	0	(0)	3	(0.06)	55	(2.7)
Archaea	7	(0.5)	10	(0.6)	0	(0)	2	(0.04)	18	(0.9)
Virus	1	(0.07)	6	(0.37)	31	(0.74)	38	(0.86)	1	(0.05)

Taxa	*Rickettsia typhi*		*Yersinia pestis*		*Wolbachia pipientis*		*Rickettsia bellii*		*Listeria monocytogenes*	
Bacteria	724	(86.4)	3595	(0.93)	866	(72.5)	1047	(73.3)	2586	(90)
Eukaryota	25	(3)	13	(0.33)	43	(3.5)	98	(6.8)	10	(0.4)
Archaea	2	(0.2)	2	(0.05)	4	(0.3)	7	(0.5)	5	(0.2)
Virus	0	(0)	16	(0.4)	30	(2.5)	30	(2)	51	(1.8)

Taxa	*Pseudomonas aeruginosa*		*Helicobacter pylori*		*Haemophilus influenzae*		*Neiseria meningitidis*		*Streptococcus pneumoniae*	
Bacteria	5256	(94.4)	1272	(80.7)	1685	(94)	1728	(83.6)	1728	(82.5)
Eukaryota	21	(0.38)	36	(2.2)	0	(0)	10	(0.48)	13	(0.6)
Archaea	2	(0.03)	9	(0.57)	1	(0.05)	2	(0.09)	4	(0.19)
Virus	26	(0.47)	1	(0.06)	22	(1.2)	9	(0.4)	10	(0.48)

Taxa	*Escherichia coli* CFT073		*Bordetella pertussis*		*Synechocystis* sp PCC6803		*Xylella fastidiosa*		*Xanthomonas campestris*	
Bacteria	4648	(86.4)	3301	(96)	3002	(94.8)	1955	(70.7)	3807	(91)
Eukaryota	9	(0.17)	6	(0.17)	14	(0.44)	8	(0.29)	20	(0.47)
Archaea	1	(0.01)	7	(0.2)	28	(0.88)	3	(0.1)	4	(0.1)
Virus	58	(1)	8	(0.2)	1	(0.03)	59	(2.1)	31	(0.74)

The percentage of proteins having their best Blastp-hit in one of these different taxa is indicated between brackets. No color indicates intracellular pathogens, light grey corresponds to extracellular pathogens and dark grey to plant pathogens. This table is based on results available from http://www.ncbi.nlm.nih.gov/entrez/query.fcgi?db = genomeprj, TaxMap heading.

Philadelphia-1 codes two specific ankyrin repeat proteins, one F-box domain protein, and a 65 kb pathogenicity island that contains the putative virulence determinants *msrA1,2,3* and *magA* (Brassinga *et al.*, 2003). Furthermore, each of the sequenced genomes has a slightly different array of Sid proteins, candidate effectors of the Dot/Icm T4SS (Brüggemann *et al.*, 2006a).

Interestingly, all four *L. pneumophila* strains that have been sequenced contain chromosomal regions that can excise and be maintained as plasmids (Cazalet *et al.*, 2004; Chien *et al.*, 2004; Doleans *et al.*, 2006; Glöckner *et al.*, 2007; Steinert *et al.*, 2007). These elements contain a type IVA secretion system: the Lvh T4SS in strains Paris, Lens and Philadelphia-1, and two new type IVA secretion systems (*trb*/*tra*) in strain Corby (Glöckner *et al.*, 2007). The Lvh type IVA secretion system was reported to be involved in virulence-related phenotypes under conditions mimicking the spread of Legionnaires' disease from environmental niches (Bandyopadhyay *et al.*, 2007). Diversity in this system, as well as differences in copy number due to excision in a growth phase-dependent manner (Doleans *et al.*, 2006), might contribute to survival and spread of *L. pneumophila* in the environment. Interestingly, although the *lvh* region is highly conserved in the three sequenced genomes, the size and sequence of the DNA flanking the *lvh* locus are specific to each strain, suggesting that these regions were probably acquired by horizontal gene transfer. Both the *lvh* region and the newly described type IVA secretion system of strain Corby are inserted in the same tmRNA in strains Paris and Lens, but in strain Philadelphia 1 the *lvh* region is inserted in an Arg tRNA. These elements are probably of plasmid origin as they have a higher GC content (43%) than the rest of the chromosome (38%), and they carry several plasmid-related genes. A similar element that contributes to phase variation of strain OLDA, can also exist in a plasmid as well as in an integrated form (Luneberg *et al.*, 2001). These different elements probably play a role in the adaptation and versatility mechanisms of *L. pneumophila*.

Comparative analysis of the four genome sequences revealed also variations at the gene level. In strain Paris, the gene *rtxA* is fused with the *arpB* gene and to about 30 highly conserved repeats of 549 bp (Cazalet *et al.*, 2004). A similar structure is encoded by strain Lens; however, the latter strain contains two repeated motifs both of which differ from that of strain Paris (Cazalet *et al.*, 2004). When compared to strain Philadelphia-1, *arpB* is absent, and *rtxA* is also followed by tandem repeats, again exhibiting a different motif (Chien *et al.*, 2004). The recently sequenced strain Corby encodes a fifth repeated motif, not present in the other three strains (G. Glöckner, unpublished). RtxA was shown to be involved in many different steps of the *L. pneumophila* infectious cycle: entry, adherence, cytotoxicity, pore formation (Cirillo *et al.*, 2001) and intracellular trafficking in amoebae (Cirillo *et al.*, 2002). Thus, variations in the sequences and combinations of the different repeats may contribute to differences in the entry, adherence or cytotoxic capacities of *L. pneumophila*. However, the exact role of these repeats is still unknown.

Secretion machineries – many and varied

Analysis of the genome sequence revealed a rather large number and wide variety of secretion systems, underlining the importance of protein secretion for *L. pneumophila*. The Dot/Icm type-IV secretion system (Chapter 9) and the *lsp* type II secretion system (Chapter 8) contribute to virulence. The Dot/Icm secretion system (Berger and Isberg, 1993; Marra *et al.*, 1992; Vogel *et al.*, 1998) is required for intracellular growth in human macrophages as well as in amoebae and for intracellular trafficking (for reviews see Sexton and Vogel, 2002; Segal *et al.*, 2005; Chapter 9). This macromolecular complex is encoded by 26 genes located in two genomic regions. Its chromosomal location is conserved in strains Paris, Lens, Philadelphia 1 and Corby. Morozova and colleagues (Morozova *et al.*, 2004) showed by sequence comparison and DNA/DNA hybridization that the organization of this region is conserved among 18 *L. pneumophila* and 17 other *Legionella* species tested. Hybridization studies using a *Legionella* macroarray carrying these 26 genes further established their presence and conservation among 165 tested *L. pneumophila* strains of all serotypes and different origins (Cazalet *et al.*, 2006), underlining the importance of the Dot/Icm type IV system for *Legionella*. In the Dot/Icm system of *L. longbeachae* and *L. micdadei* the global organization is also conserved.

An interesting difference is the *icmR* gene. In *L. micdadei* it is replaced by two genes, *migA* and *migB*, that do not show any homology to *icmR*; in *L. longbeachae* it is replaced by a gene called *ligB* (Feldman and Segal, 2004). The *icmR* gene encodes the chaperone of IcmQ, a protein that forms pores in lipid membranes, and thus regulates the activity of IcmQ. IcmQ and IcmR are essential components of the Dot/Icm system (Dumenil and Isberg, 2001). A recent analysis of the *icmR-icmQ* region in 29 additional *Legionella* species identified a large hypervariable gene family named functional homologues of *icmR* (*fir*). Despite the highly variable sequence of the *fir* genes, in all investigated species they are located next to *icmQ* and encode a functionally similar protein that interacts with IcmQ. Thus, co-evolution between the IcmQ and its relevant

FIR protein seems to have occurred, possibly allowing adaptation to various protozoan hosts (Feldman *et al.*, 2005).

Different substrates of the Dot/Icm secretion system have been identified; these are discussed in detail in Chapter 10. A large number of candidate effector proteins, designated SidA-G (substrate of Icm/Dot transporter), were identified using a two-hybrid screen, and subsequent screening using a Cre/loxP-based system (Luo and Isberg, 2004). Interestingly, the Sid protein substrates have one or more paralogues present in the *L. pneumophila* genomes, and each of the three sequenced genomes has a slightly different array of Sid proteins (Brüggemann *et al.*, 2006a). Interestingly, many of the substrates show no sequence similarity to proteins from other organisms.

Another substrate of the Dot/Icm secretion system is SidM/DrrA, a guanosine nucleotide exchange factor that is translocated into host cells where it recruits the host GTPase Rab1 to *Legionella*-containing vacuoles (Machner and Isberg, 2006; Murata *et al.*, 2006; Chapter 10). Interestingly, the gene encoding SidM/DrrA is absent from strain Paris, again showing that each strain has a slightly different array of effector proteins. Perhaps this diversity indicates subtle differences in host adaptation; alternatively, factors that are not widely spread in the species may have been recently acquired by horizontal transmission. Additional substrates secreted via the Dot/Icm secretion system are VipA, VipD and VipF. VipA (Shohdy *et al.*, 2005), WipA, WipB, YlfA (Campodonico *et al.*, 2005; de Felipe *et al.*, 2005; Ninio *et al.*, 2005), which are discussed in detail in Chapter 10. The second type IV secretion system encoded by *L. pneumophila* is a type-IVA named Lvh and a new type IVA secretion system (*trb/tra*) in strain Corby discussed already before.

L. pneumophila encodes a type-II secretion (Lsp) system that is critical for infection of amoebae, macrophages, and mice (Chapter 8). It is highly conserved in the three sequenced genomes (protein identities of 96 to 100%) and also seems to be conserved throughout the genus *Legionella* (Rossier and Cianciotto, 2001). The Lsp secretion system promotes the ability of *L. pneumophila* to infect both protozoan and macrophage hosts, to grow in the mammalian lung and to grow at low temperatures (Soderberg *et al.*, 2004) (see Chapter 8).

L. pneumophila possesses several additional secretion systems, including a type I secretion system called *lss* that seems to be specific to *L. pneumophila* (Jacobi and Heuner, 2003); a sec apparatus, which is complete with the exception of *secG*; and a twin arginine translocation (TAT) pathway (De Buck *et al.*, 2004; Rossier and Cianciotto, 2005; Chapter 8). In contrast to the usual organization of *tatA*, *tatB* and *tatC* in an operon, only the *tatA* and *tatB* genes are co-transcribed, and *tatC* is in a different location on the *L. pneumophila* chromosome. This organization is conserved in strains Paris, Lens and Philadelphia 1, where the *tatAB* operon and the *tatC* gene are located 24 kb, 41 kb and 27 kb apart, respectively. However, despite the different distances between the *tatAB* and *tatC* genes, the flanking regions are homologous in the three strains. The Tat secretion system is implicated in growth under low iron conditions and within macrophages in an Lsp-independent manner (Rossier and Cianciotto, 2005). In addition to these conserved secretory pathways, strain Paris encodes a type V secretion system that is specific to this strain.

Regulation and signal transduction – clues to adaptation

Pseudomonas aeruginosa and *Listeria monocytogenes*, which are both able to adapt to the host environment and to multiply and survive in a wide variety of environments outside of a host, contain large sets of regulatory proteins with 8.4% and 7.3% of the genome, respectively, dedicated to regulatory proteins (Glaser *et al.*, 2001; Stover *et al.*, 2000). In contrast, *L. pneumophila* is an intracellular pathogen that does not seem to replicate as readily in the environment outside the host. Consistent with a less versatile lifestyle, the *L. pneumophila* regulatory repertoire is relatively modest. Genome analysis identified about 100 transcriptional regulators in each of the sequences. This represents only a little more than 3% of the predicted genes. Each *L. pneumophila* genome encodes six putative sigma factors, the homologues of RpoD (the major sigma factor), RpoH, RpoS, RpoN, FliA and the

ECF-type sigma factor RpoE. Furthermore one sigma 54 (RpoN) modulator, the sigma factor-dependent enhancers FleQ, FleR, and PilR, the anti-sigma factors FlgM and the anti-sigma factor B are encoded by *L. pneumophila* (Polesky *et al.*, 2001; Heuner and Steinert 2003; Jacobi *et al.*, 2004).

Although substantially lower than that of free-living bacteria like *E. coli* (35) or *P. aeruginosa* (100), the number of two-component systems (13 histidine kinases and 13 response regulators) is also conserved in all sequenced genomes. Three of these two component systems, LetAS, CpxAR and PmrAB, have been characterized in the most detail. LetAS, together with the enhancer protein LetE, the alternative sigma factor RpoS, FliA and perhaps the flagellar regulators RpoN and FleQ, promotes the expression of transmission traits (Hammer *et al.*, 2002; Heuner *et al.*, 2002; Molofsky and Swanson, 2004; Molofsky *et al.*, 2005; Chapter 6). CpxAR was implicated in *dot/icm* gene expression, as it activates *icmR* gene expression and, to a lesser extent, also that of the transcriptional units *icmV-dotA* and *icmWX* (Gal-Mor and Segal, 2003). The PmrA regulator appears to be a fundamental regulator of the Dot/Icm type IV secretion systems, since 13 of the genes coding for *L. pneumophila*-translocated substrates seem to be under control of the PmrA response regulator (Zusman *et al.*, 2007). Recently it was shown that the two two-component systems CpxAR and PmrAB both positively regulate the expression of the *fir* genes described above. Alignment of the regulatory regions of 29 *fir* genes revealed three regulatory groups; one contains a binding site for the CpxR response regulator previously shown to regulate the *L. pneumophila fir* gene (*icmR*) (Feldman *et al.*, 2005); the second one, which includes most of the *fir* genes, contains the CpxR binding site and an additional regulatory element that was identified as a PmrA binding site; and the third group contains only the PmrA binding site (Feldman and Segal, 2007). The group containing both binding sites was shown to be positively regulated by both PrmA and CpxR. Probably, changes in the regulation of the *fir* genes occurs as different *Legionella* species adapt to different protozoan hosts in the environment.

The two most prominent classes of transcriptional regulators in *L. pneumophila* are LysR-type and phage-like repressors. The abundance of the latter class should indicate that the genome is subject to continuous phage-mediated horizontal transfer. However, it is interesting that in none of the three genomes sequenced was a complete prophage identified. Regulators of the GGDEF/EAL family are also abundant in *L. pneumophila*. The sequenced strains encode 21 to 24 regulators of this type, similar to many other bacteria, like *Vibrio cholerae* (41), *P. aeruginosa* (33), *Wolinella succinogenes* (26), and *E. coli* (19). These regulators contain different N-terminal signalling domains and a C-terminal output domain, which is composed of two subdomains, GGDEF (DUF1) and EAL (DUF2) (Tal *et al.*, 1998). Those domains modulate the intracellular level of cyclic-di-GMP, a recently discovered secondary messenger in the bacterial cell that governs multicellular behaviour such as the switch from sessility (e.g. in a biofilm) to motility (e.g. swarming) in the pathogenic organisms *Salmonella enterica* serovar Typhimurium, the nosocomial pathogen *P. aeruginosa* and the commensal species *E. coli* (Simm *et al.*, 2004). Moreover, an EAL domain protein in *S. enterica* contributes to resistance to oxidative stress and the ability to kill macrophages (Hisert *et al.*, 2005). Thus, EAL/GGDEF proteins may have a global role in governing the developmental differentiation of *L. pneumophila*.

Comparative genomics to study diversity

Inter- and intraspecies comparison of complete genome sequences of pathogenic as well as non-pathogenic species has led to a wealth of new information on the genetic diversity of a large number of bacterial species. Based on this knowledge, complete genome sequences DNA microarrays were developed as powerful tools for genome comparisons between different strains of the same or closely related species, as well as for the comparison of the genomic DNA of unsequenced strains with sequenced reference strains. This technique has been extensively used to assess inter- and intraspecies variability at the genomic level in different prokaryotic organisms like *M. tuberculosis* (Behr *et al.*, 1999; Marmiesse *et al.*, 2004), *Helicobacter pylori* (Salama *et al.*, 2000), *Staphylococcus aureus* (Fitzgerald *et al.*, 2001; Trad *et al.*, 2004; Monecke *et al.*, 2007) *Yersinia pestis* and *Yersinia pseudotuberculosis* (Hinchliffe *et al.*, 2003), *Campylobacter jejuni* (Dorrell *et al.*, 2001), *Vibrio cholerae* (Dziejman *et al.*, 2005) and *Listeria monocytogenes* (Doumith *et al.*, 2004). Several studies combining genome sequencing and microarray-based comparative genomics showed that, although considerable diversity is present among different bacterial isolates of the same species, clonal expansion of highly virulent sub-populations of a bacterial pathogen may exist. We thus applied comparative genomic hybridization to *Legionella* to better understand the diversity within this genus and to assess correlations between functional traits, phylogeny and epidemiological characteristics of the strains.

Analysis of the *L. pneumophila* gene content using DNA arrays

Despite the abundance of many different species of *Legionella* in aquatic reservoirs, the vast majority of human disease is caused by strains of *L. pneumophila* serogroup 1 (Sg 1), the type responsible for 88.9% of the legionellosis cases worldwide. The only exception is Australia and New Zealand where *L. pneumophila* Sg1 accounts for 45.7% and *Legionella longbeachae* for 30.4% of the legionellosis cases (Yu *et al.*, 2002). To understand differences in the pathogenicity and the ecology of *Legionellae*, knowledge of the genetic diversity that exists among the different isolates is a prerequisite. Thus, DNA/DNA hybridization was undertaken on a selected strain collection to characterize and compare the gene content of these strains. Focused DNA arrays were developed that contain all genes variable among the three sequenced strains Paris, Lens and Philadelphia-1, genes encoding known and putative virulence factors, as well as genes identified as specific to *L. longbeachae*. (Cazalet *et al.*, 2006; C. Cazalet and C. Buchrieser, unpublished). The selected strain collection contained environmental and clinical strains implicated in either sporadic cases or epidemics isolated in different European countries (France, Germany, Austria, Sweden and Switzerland) as well as in the USA, Japan and Africa. Furthermore, to identify possible specificities of Sg 1 strains that may account for its higher virulence for humans, strains representing the 15 different *L. pneumophila* serogroups were incorporated into this collection. In all, the gene content of 180 *Legionella* isolates, including 165 *L. pneumophila* strains and 15 representatives of other *Legionella* non-*pneumophila* species, was characterized. The method used identified genes as present when the DNA identity was at least 80% as compared to the probe present on the array.

Global gene content analysis by hierarchical clustering

The large dataset obtained from 180 *Legionella* strains allowed the strains to be grouped according to shared genetic profiles. This approach allowed us to define four main lineages (I, II, III and IV) among the 165 *L. pneumophila* isolates tested and also to distinguish *L. pneumophila* strains from *Legionella* strains belonging to other species. All non-*pneumophila* species clustered separately into so-called lineage V. However, lineage V isolates are characterized by the absence of nearly all genes spotted on the membrane, indicating that a high diversity exists among the different *Legionella* species, as most of the genes of the flexible gene pool of *L. pneumophila* do not obviously belong to the gene pool of other *Legionella* species. Within the species *L. pneumophila*, each lineage can be subdivided according to the genetic profiles obtained by hybridization. Lineage I contains two groups (I.1 and I.2) and includes strains isolated in all different countries investigated as well as all French isolates that were previously identified as Paris strains according to an identical pulsed-field gel electrophoresis profile (Aurell *et al.*, 2003). Lineage II contains only the strains isolated during the epidemic in Lens/France in 2003/2004. Lineage III can be subdivided in five subgroups (III.1, III.2, III.3, III.4 and III.5). Finally, lineage IV, which includes the sequenced strain Philadelphia-1, can be subdivided into two subgroups (IV.1 and IV.2). Interestingly, clustering according to the genetic profiles of gene presence and absence was independent of geographical origin or epidemiological features of the isolate (clinical or environmental strains) (C. Cazalet and C. Buchrieser, unpublished).

Distribution and variability of known *L. pneumophila* plasmids and genomic islands

The plasmids of strains Paris and Lens show a heterogeneous distribution and can either be intact or truncated. The plasmid identified in strain Paris (pLPP) is present in all strains of lineage I.1 but seems never to be identical within strains belonging to other lineages. Interestingly all strains of lineage I.2 contain only parts of the plasmid, but the region carrying the two-component regulator system *lrpR-lskS* implicated in the virulence of *L. longbeachae* is always present. However, a region bordered by an integrase and encoding partition protein homologues ParAB, replication proteins RepAB as well as conjugation transfer proteins homologue to Tra proteins and also a homologueof CsrA is systematically absent from these strains. Perhaps this region no longer exists as a plasmid but has been integrated in the genome. Depending on the strains, two other variable regions contain mostly genes that might be implicated in antibiotic resistance, such as acyltransferases, homologues of beta-lactamase, penicillin-binding proteins, spectinomycin adenylyltransferase and certain transcriptional regulators.

The Lvh T4SS is unevenly distributed among different *Legionella* strains. It is present in only about half of the tested *L. pneumophila* strains. Furthermore, this cluster has a mosaic structure with respect to different isolates. As already seen from the genome comparisons of the completely sequenced genomes, the *lvh* region is highly conserved, but the flanking DNA regions are specific to each strain (Fig. 7.2) (Cazalet and Buchrieser, 2006). Similarly, the additional strains analysed by hybridization seem again to have different rearrangements in the chromosomal parts flanking the *lvh* region, as only parts of the flanking regions of the Paris, Lens or Philadelphia strain that are represented on the macroarray hybridized with these strains. However, in strains Paris and Lens, the regions that flank *lvh* contain a common phage integrase (*lpp0194-lpl0179*), a phage excisionase (*lpp0193-lpl0178*), and a *traA*-like gene. Strain Lens encodes an additional nine transposases in its *lvh* region, and strain Paris harbours a truncated integrase (*lpp0206*). Thus, the *lvh* region seems to be located in a plasticity zone that may have been acquired through different phages with different flanking regions. Alternatively, this region may be a hot spot of integration/deletion/rearrangement events, since it is bordered by different mobile elements that seem to constantly rearrange (C. Cazalet and C. Buchrieser, unpublished).

The 65 kb pathogenicity island that was initially described as specific for strain Philadelphia-1 (Brassinga *et al.*, 2003) is also present in some other *L. pneumophila* strains studied. More interestingly, this island is also conserved in the three tested strains of *L. anisa*, a species rarely implicated in human disease. Thus, horizontal gene transfer seems to take place not only among *L. pneumophila* strains but also among different *Legionella* species.

Distribution of known virulence genes and eukaryotic-like genes

Analysis of the distribution of genes that are known or are predicted to play a role in virulence reveals a very high level of conservation in all *L. pneumophila* isolates tested, regardless of their clinical or environmental origin or their Sg. An exception is *ralF*, (Nagai *et al.*, 2002) that is absent from all *L. pneumophila* Sg 2 strains tested. Analysis of the distribution of eukaryotic-like genes and those coding proteins with eukaryotic domains identified during sequence analysis (Cazalet *et al.*, 2004) showed that most of those genes that were conserved in the three sequenced strains are also conserved throughout the species *L. pneumophila*, further strengthening the hypothesis that they are required for intracellular growth and that they might play a role in virulence. The two genes encoding homologues of apyrase (*lpp1033* and *lpp1880*), the gene encoding an homologueof the eukaryotic glucoamylase (*lpp0489*), the gene homologous to eukaryotic sphingosine-1-phosphate lyase (*lpp2128*) as well as a gene encoding a putative serine/threonine protein kinase (*lpp0267*) are conserved in all analysed *L. pneumophila* strains (C. Cazalet and C. Buchrieser, unpublished).

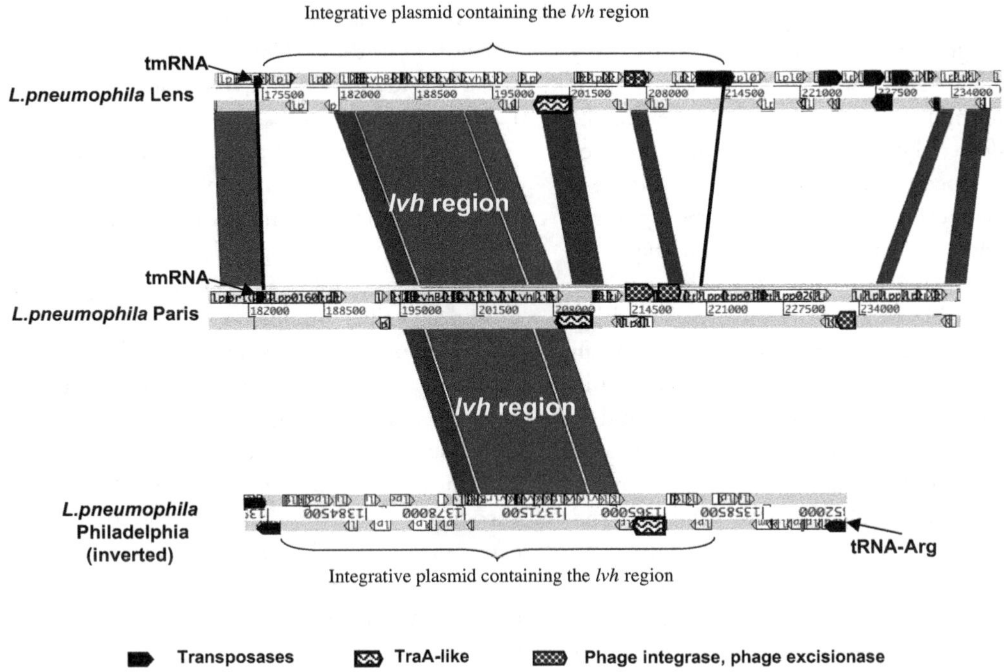

Figure 7.2 Schematic representation of the integrative multi copy plasmid carrying the lvh type IV system and comparison of this region in the three strains. Dark grey bars: Indicate homologous regions in the three genomes. Light grey arrow: Genes predicted in this region. Arrows and numbers below indicate the insertion position on the chromosome of each strain. The presence of many mobility genes like integrases, transposons and Tra-like genes is highlighted.

Interestingly, most of the known or putative virulence genes of *L. pneumophila* did not give any signal when hybridized with DNA from *Legionella* strains belonging to non-*pneumophila* species. This is due either to absence of the genes in certain cases or more probably to a high divergence of the corresponding orthologous genes in the non-*pneumophila* species. Indeed, a previous comparative sequence analysis of *dot/icm* genes in different *Legionella* species revealed that the sequence similarity of these genes ranges from 62% to 79% (Morozova *et al.*, 2004), a value below the detection limit of the hybridization conditions used (C. Cazalet and C. Buchrieser, unpublished).

Evolutionary thoughts

Different genomic approaches, such as whole genome sequencing and comparison and large scale gene content studies, applied to the genus *Legionella* provide for the first time a global view and a more complete knowledge of the gene distribution and the genetic content present in the gene pool of this genus. We have learned that *L. pneumophila* strains differ substantially in gene content and that these differences are mostly due to plasmids, pathogenicity islands and gene acquisition and loss. Most likely these differences confer *L. pneumophila* traits that provide selective advantages in the environment and the infected host.

Mechanisms leading to genetic diversity and genome plasticity of *L. pneumophila*

Genome reduction, gene duplication and horizontal gene transfer are the main driving forces for genome diversity in prokaryotes. Analysis and comparison of the *L. pneumophila* genomes did not reveal any signs of genome reduction. However, the comparison with sequence information from an isolate of a *Legionella*-like amoebal pathogen (LLAP), which are primarily obligate intracellular parasites of amoebae, might allow this question to be addressed in the future. In contrast, gene duplications and horizontal gene transfer are indeed important contributors to the evolution of *L. pneumophila*. Natural competence may also contribute to diversity. All sequenced genomes contain plasmids that may excise and integrate into the genome as well as several plasmids and genomic islands that might be transferred horizontally between strains of the same species and also between different *Legionella* species, as shown for the genomic island Trb-1 encoding a conjugation/type IV secretion system (*trb/tra*) (Glöckner *et al.*, 2007). Unstable plasmids capable of switching between an integrated chromosomal form and an excised circular form have also been reported in other microorganisms. In *Shigella flexneri* and in enteroinvasive *E. coli*, the integration of the large virulence plasmid is associated with silencing of plasmid-encoded virulence genes (Dorman and Porter, 1998). This mechanism might also play an important role in *L. pneumophila* as at least two different genetic islands are subject to such a mechanism. The first such locus described is a 30 kb element associated with virulence that was identified in *L. pneumophila* strain OLDA (Luneberg *et al.*, 2001). When the element is integrated in the chromosome, the strain exhibits a wild-type virulent phenotype, but excision and replication as a high-copy plasmid is associated with alteration of an LPS epitope and loss of virulence. Surprisingly, despite the wide distribution of the different mobile genetic elements identified in *L. pneumophila*, this 30 kb element seems to be specific to strain OLDA, as the probes specific to this island did not hybridize with any of the over 100 strains tested. The Lvh T4SS is another element that can be present in both an integrated and an excised form in each of the three sequenced strains.

The selective advantage and the mechanism of chromosomal integration and excision of these elements are not yet understood. For the 30 kb element described in strain OLDA, the results are paradoxical: Excision is enhanced under *in vivo* conditions in the guinea pig, yet excision causes loss of virulence. For the *lvh* gene cluster, switching between the integrated and the circular form was shown to be growth phase dependent (Doleans *et al.*, 2006). Integration and excision of different genomic regions might be one mechanism of genome plasticity that enables *L. pneumophila* to adapt to different environmental conditions. The identification of the exact mechanism leading to the excision and integration of these elements will help to understand the adaptation and versatility mechanisms *L. pneumophila* employs.

Interestingly, a feature shared by most of the mobile elements (plasmids of strains Paris and Lens, *lvh* gene cluster, 65 kb putative pathogenicity island originally identified in strain Philadelphia-1 and the cluster encoding several multidrug efflux pumps) is that they encode a paralogue of CsrA. This protein is described as a repressor of transmission traits and activator of replication, raising the question whether these elements are implicated in virulence. One hypothesis would be that the paralogue of CsrA encoded on the *lvh* mobile element controls the expression of the Lvh system and/or regulates the switch between integrated and circular forms of the plasmid, which is known to be growth phase dependent.

Origin of the eukaryotic-like proteins

Given the high number of eukaryotic-like proteins in *L. pneumophila*, the question arises 'What is the origin of these proteins?' Two hypotheses can be proposed: (i) they have been acquired by horizontal gene transfer or (ii) they evolved by convergent evolution. Most of the *L. pneumophila* eukaryotic-like genes have a GC content ranging from 32% to 48%, which differs significantly from that of the rest of the genome (38%) and therefore supports the hypothesis that these genes have been acquired through horizontal gene transfer. Another line of evidence supporting this hypothesis is the presence of a 11.3 kb gene cluster containing 11 genes with similarity to eukaryotic genes that is bordered by a Lys-tRNA (Cazalet and Buchrieser, 2006). Accordingly, the tRNA locus may be the integration site, a scenario reported previously for many genomic and pathogenicity islands in different bacterial species (Hacker and Kaper, 2000).

However, it is also feasible that some of these proteins have arisen in *L. pneumophila* through convergent evolution. In this case, they would not necessarily share primary amino acid similarity with eukaryotic proteins, and their putative function would consequently not be predicted by conventional annotation methods. A recent study clearly illustrates this hypothesis:

the AvrPtoB protein of *Pseudomonas syringae* is translocated into plant cells, where it inhibits immunity-associated programmed cell death (PCD) (Janjusevic *et al.*, 2006). The amino acid sequence of AvrPtoB does not show any homology to eukaryotic or bacterial proteins, but the 3D structure of the C-terminal domain of AvrPtoB shows remarkable homology to U-box components of eukaryotic E3 ubiquitin ligases. Further functional studies confirmed that AvrPtoB has ubiquitin ligase activity that is correlated with the anti-PCD activity of the protein (Abramovitch *et al.*, 2006; Janjusevic *et al.*, 2006). Thus, different mechanisms like convergent evolution and/or acquisition of eukaryotic genes may target the same eukaryotic pathway and allow pathogens to adapt to their eukaryotic host.

If *L. pneumophila* has acquired some of these genes by horizontal gene transfer from eukaryotes, where do they come from? Interaction between *L. pneumophila* and amoeba, or more generally freshwater protozoa, is considered to be central to the ecology but also the pathogenesis of *L. pneumophila*, as there is much evidence that protozoa played a major role in the emergence of *L. pneumophila* as a pathogen of human alveolar macrophages. It is possible that *L. pneumophila* acquired some of its eukaryotic-like genes from amoeba, which seem to be a privileged place for lateral gene transfer. This hypothesis is reinforced by the analysis of the best BLASTp-hits of the 113 eukaryotic-like genes identified in the genome of *L. pneumophila* strain Paris: 33% of these proteins have their best match to sequences of amoebae or ciliated protozoae. It is not likely that these genes had been acquired in humans, as person to person transmission of *Legionella* has never been observed and thus any genetic modifications selected in humans would not be conserved in the species. The fact that *L. pneumophila*, *R. bellii* and *Protochlamydia amoebophila*, three bacteria that live inside amoeba, are the organisms that contain the most eukaryotic-like genes, with respectively 3.5%, 6.8% and 8.6% of their genes having best-matches to eukaryotic genes, argues for acquisition of some of these genes from amoeba. It has also been suggested that unicellular organisms that feed on bacteria, including amoebae, might have incorporated prokaryotic DNA in their chromosome (Field *et al.*, 2000). In particular, the recent publication of the complete genome sequence of *Entamoeba histolytica* provides strong evidence for such horizontal gene transfer (Loftus *et al.*, 2005). A phylogenetic screen to identify relatively recent horizontal gene transfer from prokaryotes to eukaryotes in the *Entamoeba* genome identified 96 candidate genes, mostly encoding different metabolic enzymes (Loftus *et al.*, 2005).

Recent sequencing of the genome of *R. bellii*, an obligate intracellular pathogen that replicates within amoeba, revealed that many *R. bellii* ORFs (8.8%) exhibit a high level of sequence similarity to homologues found in *L. pneumophila* and the amoeba-endosymbiont *P. amoebophila* (Ogata *et al.*, 2006). This percentage is significantly higher than that seen for alpha-proteobacteria that do not live within amoeba like *Mesorhizobium loti* (0.9%) and *Pelagibacter ubique* (2.5%), suggesting that horizontal gene transfers among these different bacterial species may take place within amoeba. Thus, amoebae are likely a reservoir for horizontal gene transfer between amoeba-associated bacteria and between the bacteria and their host.

Functional genomics of *L. pneumophila*

Adaptation to the host environment and exploitation of host cell functions is the basis of the success of intracellular pathogens. Microarrays provide a powerful tool to study this process as they allow experimentalists to quantify all mRNAs encoded by the genome of a bacterial pathogen and their eukaryotic host cell. Such expression profiles serve as a monitor of the host cell environment as well as an indicator of the bacterial adaptation to its intracellular niche. In recent years there have been numerous reports of analyses of the host cell transcriptome after infection with bacteria such as uropathogenic *Escherichia coli* (Snyder *et al.*, 2004), *Listeria monocytogenes* (Chatterjee *et al.*, 2006; Joseph *et al.*, 2006), *Mycobacterium tuberculosis* (Schnappinger *et al.*, 2003) or *Salmonella enterica* sv. Typhimurium (Eriksson *et al.*, 2003). Most of these studies have been performed in *in vitro* infection models and have provided the first insights into the complexity of host–bacterial interactions.

A recent analysis of the interaction of *L. pneumophila* with its natural host *Acanthamoeba castellanii* revealed clues about this host–pathogen encounter (Brüggemann *et al.*, 2006b).

L. pneumophila gene expression *in vitro* and during infection of *Acanthamoeba castellanii* reflects the biphasic life cycle

In vitro studies in broth indicated that the life cycle of *L. pneumophila* consists of at least two phases, *i.e.* a replicative phase (RP) and a transmissive phase (TP). Bacteria in RP are not highly infectious, sodium resistant and not flagellated; in contrast TP bacteria are highly infectious, flagellated and highly motile (Molofsky and Swanson, 2004). The expression profile of *in vitro* (BYCE medium) grown cells in the exponential (E) growth phase as compared to the post-exponential (PE) clearly reflected a biphasic life cycle. Overall, about 650 genes were two-fold or higher upregulated in the E phase, and over 500 genes were upregulated in the PE phase. Differently expressed genes could be classified as replicative or transmissive traits according to their expression levels in E or in PE growth phase (Brüggemann *et al.*, 2006b).

Transcriptional analyses of *L. pneumophila* during infection of *A. castellanii* revealed that two distinct phases also exist *in vivo* (Brüggemann *et al.*, 2006b). Furthermore, the three strains investigated showed similar, if not identical results, with sets of RP genes (405) up-regulated at the earlier time points, and TP genes (393) up-regulated at the late time points. The RP genes comprised DNA and RNA polymerase complexes, transcription and translation elongation factors, ribosomal proteins, protein secretion and translocation systems (*secAB, secEF, secGY*), genes coding proteins involved in purine and pyrimidine metabolism, and other genes encoding metabolic factors. Among the identified TP genes, several regulators (e.g. *fliA, cpxR, rpoE*), virulence factors (e.g. *ralF, dotA, letE, enhA, sdeA, sdcA*) and flagella biosynthesis genes (e.g. *fliS, fliD, fliN*) were induced (Brüggemann *et al.*, 2006b). Thus, the transcriptome during RP shows that cell division progresses as opposed to the TP, which prepares *L. pneumophila* to escape its host cell and to invade a new one.

Intracellular metabolism of *L. pneumophila* as deduced from gene expression

Although *L. pneumophila* growing in minimal medium obtains carbon and energy from amino acids, and not sugars (George *et al.*, 1980; Tesh and Miller, 1981), it is an open question whether the bacterium not only scavenges amino acids but also carbohydrates from the host when growing inside the replicative vacuole (see Chapter 11). Supporting the amino acid diet, a *L. pneumophila* threonine transporter PhtA (phagosomal transporter) was recently identified as essential for differentiation and proliferation in macrophages (Sauer *et al.*, 2005, Chapter 11). Likewise, the transcriptional analyses showed that the catabolism and uptake systems of amino acid encoding genes (>30) are upregulated in early and late RP, in particular those of serine, threonine (*phtA*), alanine, glycine, tyrosine and histidine (Brüggemann *et al.*, 2006b). Moreover, several aminopeptidase and protease encoding genes are up-regulated during RP, which further suggest the acquisition of host amino acids (Fig. 7.3).

Listeria monocytogenes can use alternative carbon sources like phosphorylated glucose and glycerol during replication in epithelial cells (Joseph *et al.*, 2006). Similarly, *L. pneumophila* may use carbohydrate-derivatives as indicated by the upregulation of the genes coding for the Entner–Doudoroff pathway (*lpp0483–5, lpp0487*), a putative glucokinase (*lpp0486*), a sugar transporter (*lpp0488*) and an eukaryotic-like glucoamylase (*lpp0489*, Fig. 7.1; Brüggemann *et al.*, 2006b). The NADPH producing Entner–Doudoroff pathway is commonly used by bacteria for sugar and/or gluconate assimilation (Peekhaus and Conway, 1998) and was also found to be upregulated during intracellular growth of *Salmonella enterica* (Eriksson *et al.*, 2003). Moreover, the *L. pneumophila* eukaryotic-like glucoamylase is upregulated *in vivo* and not *in vitro*, which suggests the presence of an *in vivo*-dependent signal that may trigger the use of the host glycogen.

It is apparent that the switch from the RP to the TP is accompanied by a thorough change in its metabolic strategy. The central core metabolism and oxidative phosphorylation employed in the RP is shut down in the TP,

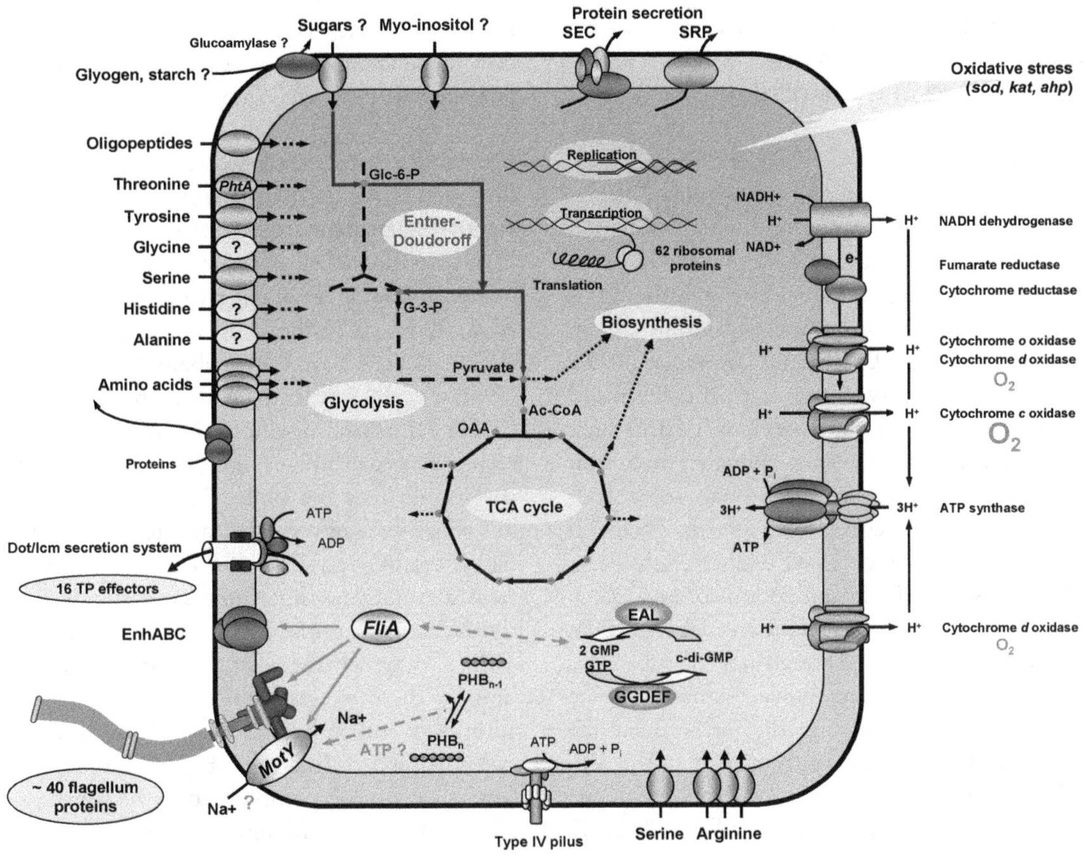

Figure 7.3 Overview of major characteristics of the biphasic life cycle of *L. pneumophila* as deduced from transcriptional analyses [reproduced with permission from (Brüggemann *et al.*, 2006b)]. The colour gradation of the background represents transcriptional changes from the replicative phase (RP, pink) to the transmissive phase (TP, cyan). Main replicative traits are: amino acid and oligopeptide transporters, proteases, sugar transporters, the glycolysis and the Entner-Doudoroff pathways, Sec- and SRP (signal recognition protein)-dependent protein secretion systems, oxidative stress response proteins, and the replication, transcription and translation machinery. Endoxidases of the energy-producing respiratory chain are cytochrome o, d and c oxidases, upregulated during the RP, while in TP an alternative cytochrome d oxidase is overexpressed. Important transmissive traits are: GGDEF/EAL proteins, serine and arginine transporters, type IV pilus biosynthesis, synthesis and degradation of polyhydoxybutyrate (PHB), flagellum formation, invasion-associated traits (e.g. EnhABC) and host cell-modulators secreted by the Dot/Icm secretion system.

growth is abolished, and maintenance functions are activated. *L. pneumophila* redirects its metabolism toward the fermentative and anaplerotic functions. For example, genes encoding malic enzyme (*lpp3043*), pyruvate decarboxylase (*lpp1157*), acetoacetate decarboxylase (*lpp0728*), and 3-hydroxybutyrate dehydrogenase (*lpp2264*) are strongly induced in the TP. In addition, when reaching the post-exponential growth phase, the ATP synthase (*atp* genes), c-type cytochrome (*ccmEFGH*), c-type cytochrome oxidase (*coxCAB*, *ctaG*), c-type cytochrome reductase (*petBA*) and NADH deshydrogenase (*nuo* genes) encoding genes are down-regulated (Fig. 7.1). In contrast the d-type cytochrome oxidase (*qxtAB*), which has a high affinity for O_2, is clearly up-regulated (Brüggemann *et al.*, 2006b). Furthermore, some metabolic functions of the TP seem to be involved in increasing the bacterium's fitness in the aquatic environment once released from the amoebal host. Some of those proteins might confer resistance, i.e. the multidrug efflux transporters or aminoglycoside 6-N-acetyltransferases, whereas others may catalyze the formation of compatible solutes or secondary compounds, such as choline monooxygenase, homospermidine synthase or the enzymes for polyhydroxybutyrate synthesis and its degradation (Brüggemann *et al.*, 2006b).

Dot/Icm substrates, motility and virulence genes are upregulated in transmissive phase

The life cycle switch is reflected in the transcriptional programme, as up-regulated genes include the flagellar operon, which shows a time-dependent cascade of gene expression in the TP. The strongest induced gene in the TP is *flaA*, encoding the structural flagellum filament protein flagellin (Byrne and Swanson 1998; Heuner *et al.*, 1999; Chapter 6). Furthermore, many substrates of the Dot/Icm T4BSS have been identified. These are thought to inhibit phagosome maturation, alter phagosome trafficking, or mediate egress from amoeba (Nagai *et al.*, 2002; Conover *et al.*, 2003; Chen *et al.*, 2004; Luo and Isberg, 2004; Campodonico *et al.*, 2005; Shohdy *et al.*, 2005; Machner and Isberg, 2006; Murata *et al.*, 2006). Though not all substrates of the Dot/Icm system showed significant gene expression changes during the intracellular life cycle, many are strongly upregulated in the TP: *ralF*, *sidB*-paralogue *sdbB*; *sidC* and paralogue *sdcA*; *sidG*; *sidE*-paralogues *sdeA*, *sdeB*, *sdeC*, *lpp1615* and *lpp1453*; *sidH*-paralogues *sdhB* and *lpp2886; vipD* and *vipE*. More moderately up-regulated in the TP are *lidA* (*lpp1002*), *sidF* (*lpp2637*); *sidE* (*lpp0304*) and *ceg10* (*lpp0360*) (Brüggemann *et al.*, 2006b). Many of these substrates are implicated in early steps of the intracellular life, thus indicating that, late in TP, *L. pneumophila* primes itself for the next invasion. The late response regulator PmrA was recently shown to control the expression of 13 substrates of the Dot/Icm T4BSS (Zusman *et al.*, 2007). However, no transcriptional correlation between these 13 targets during the cell cycle progression was observed. Thus, additional, yet unknown transcriptional regulators might contribute to their regulation. Importantly, many of the established and putative *in vivo* upregulated substrates of the Dot/Icm T4BSS (LidA, RalF, SdeA/LaiA, SdeB/LaiB, SidE/LaiD, SdhB and SdcA) did not show any differential regulation in *in vitro* growing cells (Brüggemann *et al.*, 2006b). The specific *in vivo* regulation of the expression of RalF, which interferes with vesicle trafficking (Nagai *et al.*, 2002), and LidA, which is implicated in the establishment of the replicative niche (Derre and Isberg, 2005), suggest that '*in vivo* -dependent' signals trigger adaptation of *L. pneumophila* to its environment.

Expression of eukaryotic-like proteins during growth in *Acanthamoeba castellanii*

The *L. pneumophila* genome analysis identified various homologues of eukaryotic genes (Cazalet *et al.*, 2004; de Felipe *et al.*, 2005; Brüggemann *et al.*, 2006a; Albert-Weissenberger *et al.*, 2007). Ten out of the 30 identified eukaryotic-like genes (Table 7.1) show significant expression changes related to growth phase. Five (strain Paris), and six (strains Lens and Philadelphia-1) eukaryotic-like genes were upregulated during the exponential phase of growth (Table 7.1). Among these were a putative sphingosine-1-phosphate lyase (*lpp2128*) and a putative glycoamylase (*lpp0849*). This is in agreement with their suggested function, as these enzymes might induce or retard autophagy or apoptosis in RP or scavenge carbohydrates during replication. Some of these factors are probably implicated in the subversion of host metabolic functions to provide carbon and energy to the growing bacterium.

Four genes are significantly upregulated in TP in all three strains studied: a SAM-dependent methyltransferase (*lpp2747*) and three coding for proteins having probably metabolic functions (*lpp0955*, *lpp1127*, *lpp1157*). Interestingly, *lpp0955* encodes a putative cytokinin oxidase. Cytokinin oxidases/dehydrogenases catalyse the irreversible degradation of cytokinins by cleavage and release of an adenine moiety. Cytokinins are phytohormones that play various roles in the development of plants by activating two-component phosphorelay pathways (Gray, 2004). Interestingly, *D. discoideum* produces a cytokinin (discadenine) (Taya *et al.*, 1978) that acts as an inhibitor of spore germination through its probable interaction with the histidine kinase DhkB, but it might also play a role in the initiation of sporulation (Wang *et al.*, 1999). Thus, although we cannot exclude a function of this *Legionella* enzyme in metabolism, it is tempting to assume that it might modulate the development of the host amoeba. In line with this idea, the *L. pneumophila* cytokinin oxidase encoding gene is up-regulated in TP (Table 7.1) (Jules and

Buchrieser, 2007). Furthermore, infection of *D. discoideum* with *L. pneumophila* leads to a strong inhibition (85%) of *Dictyostelium* differentiation into the multicellular fruiting body stage (Hägele *et al.*, 2000). Perhaps by degrading discadenine the *L. pneumophila* cytokinin oxidase arrests host cell differentiation.

When analysing the expression profiles of the proteins containing domains preferentially found in eukaryotes (Table 7.2), it becomes obvious that most of these proteins are regulated in a growth phase dependent manner. From the 31 proteins identified in strain Paris, 18 are expressed in a growth phase dependent way. Similarly, 18 of 28 present in strain Lens and 20 out of the 28 identified proteins in strain Philadelphia show growth phase dependent expression profiles (Table 7.2). With the exception of two ankyrin repeat proteins (*lpp2270/lpl2242/lpg2322* and *lpp0547/lpl0523/lpg0483*), all are strongly induced in late TP relative to RP. This suggests that these proteins are important in the late stages of the intracellular life cycle, and probably also in the early invasion events, as exemplified by the strong up-regulation in late TP of the EnhC protein and its paralogues. EnhC is important for entry, since, compared to wild-type bacteria, *enhC* mutants display significantly reduced entry into host cells (Cirillo *et al.*, 2000).

Strain specific genes induced during *in vivo* growth

As previously highlighted, a strong correlation of the transcriptional response of the core genome of the three-strains was observed during infection. When analysing the approximately 300 strain-specific genes of each isolate with respect to gene expression, it appeared that most of them are not differentially regulated during growth in *A. castellani* (Brüggemann *et al.*, 2006b). However, certain strain-specific genes were differentially expressed, indicating heterogeneity in the gene pool required for intracellular growth as well as specific adaptations to different intracellular environments. In RP, only few strain-specific genes of strains Lens and Philadelphia were upregulated. These comprise putative transposases (*lpl2032, lpl2869, lpl0184, lpg2114*) and outer membrane proteins (*lpl2148, lpl1942*). However, strain Paris showed more diversity, with the upregulation of a o-type cytochrome ubiquinol oxidase (*cyoCD/lpp0296–7, cyoA/lpp0295*) and a putative autotransporter or type V secretion system (*lpp0779*). The first enzyme might improve environmental adaptation and survival of the bacterium, whereas the autotransporter may be implicated either in better biofilm formation in the environment or confer an advantage in adhesion and invasion during infection.

In contrast, during TP several strain-specific genes probably important for virulence were highly upregulated. In strains Paris and Lens in particular specific eukaryotic-like and eukaryotic domain containing proteins (*lpp2168, lpl1158, lpl2330, lpl0063, lpl2477, lpp1100, lpp2344, lpl1681*) and GGDEF/EAL regulators (*lpp2477, lpl2826*) were induced. In addition, in each of the strains, up-regulation of strain specific genes putatively involved in signal transduction (*lpp0300, lpl2476*) or transcriptional regulation (*lpl1048, lpl1926, lpg2524, lpl2105, lpl2107*) was observed. Strain Philadelphia-1 contains a specific pathogenicity island (Brassinga *et al.*, 2003). Interestingly, genes encoded on this island showed a strong TP upregulation (from 5-fold to 10-fold). An example is the *msrA* locus (*lpg2098/msrA3, lpg2099/msrA2, lpg2111/msrA1*) encoding putative methionine sulfoxide reductases and the gene encoding the 24 kDa macrophage-induced major protein (*lpg2112*; Miyamoto *et al.*, 1993). The methionine sulfoxide reductase (Msr) is an antioxidant repair enzyme that catalyzes the reduction of methionine sulfoxide (Dhandayuthapani *et al.*, 2001) residues in proteins to methionine. It was shown to be involved in adherence of *Mycoplasma genitalium* to erythrocytes, possibly by protecting mycoplasma protein structures from oxidative damage or through alternate virulence-related pathways (Dhandayuthapani *et al.*, 2001). Altogether, these data suggest that each strain has acquired genes allowing specific adaptation to different preferential hosts and/or to increase its fitness in the environment.

Specific traits induced *in vivo* in *A. castellanii* but not *in vitro* in broth culture

A way to learn about factors that may be important for intracellular growth is to investigate the differences in the expression profiles of *in vitro* and *in vivo* grown *L. pneumophila*. This comparison revealed a strong correspondence between the transcription profiles of the three *L. pneumophila* strains cultured in BYCE broth (*in vitro*) or in *A. castellanii* (*in vivo*). Indeed, over 80% of the *in vivo* RP genes were upregulated in the *in vitro* RP and almost 80% of the *in vivo* TP genes showed upregulation in the *in vitro* PE phase. However, important differences can be observed, in particular with regard to certain virulence traits that are highly upregulated in TP *in vivo* but that are not differentially regulated in the post-exponential growth phase *in vitro*.

Members of this group are virulence genes such as the effector genes *ralF, sdeAB, sdhB, lidA* and *sdcA*. Similar gene expression differences were observed for periplasmic amino acid-binding proteins (*lpp0036, lpp0102, lpp1512*), fimbrial synthesis (*lpp0682-lpp0686, lpp1480*), the regulators PaiA (*lpp1118*), RpoE, and the two-component system *lpp2523/lpp2524*. Furthermore, the genes coding the type IV pili (*pilE/lpp0681,lpp0682–6,pilBCD/lpp1479–81*), which displayed a up to 20-fold up-regulation *in vivo* but were not differentially regulated during *in vitro* growth are worth mentioning (Brüggemann *et al.*, 2006b). This may indicate variability in the *in vitro* or *in vivo* DNA uptake and competence capacity of *L. pneumophila*, but may also be related to the role of type IV pili in adherence to eukaryotic cells (Stone and Abu Kwaik, 1998).

Another important class of proteins that is regulated differently *in vitro* and *in vivo* are GGDEF/EAL domain containing proteins, which represent a very prominent family of regulators in *L. pneumophila* (Cazalet *et al.*, 2004). Eleven out of eighteen GGDEF/EAL protein-encoding genes of strain Paris were induced in *in vivo* TP but not differentially regulated in broth.

Taken together, the analysis of expression profiles of three different *L. pneumophila* strains during its life cycle in broth and in *A. castellanii* provides insights into the pathobiology of this unique host-parasite interaction. In addition to pointing out limitations of the standard medium used to culture *L. pneumophila*, the *in vivo* analysis allowed a more subtle differentiation of life cycle phases and identification of many previously uncharacterized candidate genes for Dot/Icm-dependent and -independent virulence traits.

Future perspectives in *Legionella* genomics

Sequence analysis and comparative genomics of the species *L. pneumophila* and the genus *Legionella* revealed many particular features of the *L. pneumophila* genomes that are probably related to their capacity to subvert host functions. The association between the pathogen and its hosts is reflected in its gene content, which has been shaped through co-evolution. Deciphering the functions of the many eukaryotic-like proteins identified will lead to important insight into the *L. pneumophila* intracellular life. Furthermore, high genetic diversity among *L. pneumophila* strains and other *Legionella* species is particularly concentrated within specific plasticity regions defined in the *L. pneumophila* genomes. In the future, complete genome comparison of pathogenic and environmental members of the genus *Legionella* should bring more insight into virulence functions important for disease in humans and into the evolution of *Legionellae*. Thus, further sequencing efforts are needed for in depth comparative genomics.

Additional sequence information is becoming available as a fourth *L. pneumophila* isolate, strain Corby, was just sequenced (Glöckner *et al.*, 2007). The ongoing sequencing projects aimed at determining the complete genome sequence of *Legionella hackeliae, Legionella micdadei* and of a *Legionella*-like amoebal pathogen (LLAP) by the University of Jena and Würzburg and of that of an isolate of *L. longbeachae* by the Institut Pasteur, Paris, will be the driving force for understanding the functions of the many factors encoded by the genome, whether involved or not in virulence, and to understand strain specific differences in niche adaptation and virulence.

Web recourses

http://genolist.pasteur.fr/LegioList/ LegioList World wide web server, which is an interactive genome analysis tool where you find the completely annotated genome sequences of strains *L. pneumophila* Paris, Lens and Philadelphia 1.

http://genome3.cpmc.columbia.edu/~legion/ Web page of the *Legionella* Genome project at the Columbia University/USA that contains the data for strain *L. pneumophila* Philadelphia-1.

http://genodb.pasteur.fr/cgi-bin/WebObjects/GenoList/ GenoList is an integrated environment for the analysis of microbial genomes. GenoList provides a complete dataset of DNA and protein sequences derived from selected bacteria, linked to the relevant annotations and functional assignments It currently contains genome data from 103 bacteria including the three completely sequenced *Legionella* genomes.

http://www.ewgli.org/ web page of the European Working Group for *Legionella* Infections (EWGLI). The aim is to improve knowledge and information on the epidemiological and microbiological (clinical and environmental) aspects of Legionnaires' disease.

http://www.cdc.gov/ncidod/dbmd/diseaseinfo/legionellosis_g.htm web page of the CDC (Centre of Disease Control) giving information on the epidemiological and microbiological aspects of Legionnaires' disease.

http://www.ncbi.nlm.nih.gov/entrez/query.fcgi?db = genomeprj/ The NCBI Entrez Genome Project database is a searchable collection of complete and incomplete (in-progress) large-scale sequencing, assembly, annotation, and mapping projects for cellular organisms. The database is organized into organism-specific overviews that function as portals from which all projects in the database pertaining to that organism can be browsed and retrieved

Acknowledgements

We would like to thank many of our colleagues who have contributed in different ways to this research. This work received financial support from the Institut Pasteur, the AFSSET (Agence française de sécurité sanitaire de l'environnement et du travail) Project N° ARCL-2005–001 and the Network of Excellence 'Europathogenomics' LSHB-CT-2005–512061. M. Jules received support from PTR 185 (Institut Pasteur –Bio-rad) and AFSSET Project N° ARCL-2005–002 and C. Cazalet is holder of a fellowship jointly financed by the Centre National de la Recherche, France and VeoliaWater – Anjou Recherche.

References

Abramovitch, R.B., Janjusevic, R., Stebbins, C.E., and Martin, G.B. (2006). Type III effector AvrPtoB requires intrinsic E3 ubiquitin ligase activity to suppress plant cell death and immunity. Proc. Natl. Acad. Sci. USA *103*, 2851–2856.

Albert-Weissenberger, C., Cazalet, C., and Buchrieser, C. (2007). *Legionella pneumophila* – a human pathogen that co-evolved with freshwater protozoa. Cell Mol Life Sci *64*, 432–448.

Angot, A., Peeters, N., Lechner, E., Vailleau, F., Baud, C., Gentzbittel, L., Sartorel, E., Genschik, P., Boucher, C., and Genin, S. (2006). *Ralstonia solanacearum* requires F-box-like domain-containing type III effectors to promote disease on several host plants. Proc. Natl. Acad. Sci. USA *103*, 14620–14625.

Aurell, H., Etienne, J., Forey, F., Reyrolle, M., Girardo, P., Farge, P., Decludt, B., Vandenesch, F., and Jarraud, S. (2003). *Legionella pneumophila* Serogroup 1 Strain Paris: Endemic Distribution Throughout France. J. Clin. Microbiol. *41*, 3320–3322.

Bandyopadhyay, P., Liu, S., Gabbai, C.B., Venitelli, Z., and Steinman, H.M. (2007). Environmental Mimics and the Lvh Type IVA Secretion System Contribute to Virulence-related Phenotypes of *Legionella pneumophila*. Infect. Immun. *75*, 723–735.

Behr, M.A., Wilson, M.A., Gill, W.P., Salamon, H., Schoolnik, G.K., Rane, S., and Small, P.M. (1999). Comparative genomics of BCG vaccines by whole-genome DNA microarray. Science *284*, 1520–1523.

Berger, K.H., and Isberg, R.R. (1993). Two distinct defects in intracellular growth complemented by a single genetic locus in *Legionella pneumophila*. Mol. Microbiol. *7*, 7–19.

Brassinga, A.K., Hiltz, M.F., Sisson, G.R., Morash, M.G., Hill, N., Garduno, E., Edelstein, P.H., Garduno, R.A., and Hoffman, P.S. (2003). A 65-Kilobase Pathogenicity Island Is Unique to Philadelphia-1 Strains of *Legionella pneumophila*. J. Bacteriol. *185*, 4630–3637.

Brüggemann, H., Cazalet, C., and Buchrieser, C. (2006a). Adaptation of *Legionella pneumophila* to the host environment: role of protein secretion, effectors and eukaryotic-like proteins. Curr. Opin. Microbiol. *9*, 86–94.

Brüggemann, H., Hagman, A., Jules, M., Sismeiro, O., Dillies, M., Gouyette, C., Kunst, F., Steinert, M., Heuner, K., Coppée, J., and Buchrieser, C. (2006b). Virulence strategies for infecting phagocytes deduced from the *in vivo* transcriptional program of *Legionella pneumophila*. Cell. Microbiol. *8*, 1228–1240.

Byrne, B., and Swanson, M.S. (1998). Expression of *Legionella pneumophila* virulence traits in response to growth conditions. Infect. Immun. *66*, 3029–3034.

Campodonico, E.M., Chesnel, L., and Roy, C.R. (2005). A yeast genetic system for the identification and characterization of substrate proteins transferred into host cells by the *Legionella pneumophila* Dot/Icm system. Mol. Microbiol. *56*, 918–933.

Cazalet, C., and Buchrieser, C. (2006). Genomics of *Legionella pneumophila*. In Pathogenomics: Genome analysis of pathogenic microbes; U. Dobrindt and

H. Hacker, eds. (Weinheim, Ger: Wiley-VCH). pp. 315–338

Cazalet, C., Jarraud, S., Ghavi-Helm, Y., Allignet, J., Kunst, F., Etienne, J., Glaser, P., and Buchrieser, C. (2006). Genetic diversity of the species *Legionella pneumophila*. In *Legionella*: State of the art 30 years after its recognition, N.P. Cianciotto, Y. Abu Kwaik, P.H. Edelstein, B.S. Fields, D.F. Geary, T.G. Harrison, C.A. Joseph, R.M. Ratcliff, J.E. Stout, and M.S. Swanson, eds. (Washington, DC: ASM Press), pp. 355–358.

Cazalet, C., Rusniok, C., Bruggemann, H., Zidane, N., Magnier, A., Ma, L., Tichit, M., Jarraud, S., Bouchier, C., Vandenesch, F., Kunst, F., Etienne, J., Glaser, P., and Buchrieser, C. (2004). Evidence in the *Legionella pneumophila* genome for exploitation of host cell functions and high genome plasticity. Nat. Genet. *36*, 1165–1173.

Chatterjee, S.S., Hossain, H., Otten, S., Kuenne, C., Kuchmina, K., Machata, S., Domann, E., Chakraborty, T., and Hain, T. (2006). Intracellular gene expression profile of *Listeria monocytogenes*. Infect. Immun. *74*, 1323–1338.

Chen, J., de Felipe, K.S., Clarke, M., Lu, H., Anderson, O.R., Segal, G., and Shuman, H.A. (2004). *Legionella* effectors that promote nonlytic release from protozoa. Science. *303*, 1358–1361.

Chen, J., Reyes, M., Clarke, M., and Shuman, H.A. (2007). Host cell-dependent secretion and translocation of the LepA and LepB effectors of *Legionella pneumophila*. Cell. Microbiol. 9 (7), 1660–1671

Chien, M., Morozova, I., Shi, S., Sheng, H., Chen, J., Gomez, S.M., Asamani, G., Hill, K., Nuara, J., Feder, M., Rineer, J., Greenberg, J.J., Steshenko, V., Park, S.H., Zhao, B., Teplitskaya, E., Edwards, J.R., Pampou, S., Georghiou, A., Chou, I.C., Iannuccilli, W., Ulz, M.E., Kim, D.H., Geringer-Sameth, A., Goldsberry, C., Morozov, P., Fischer, S.G., Segal, G., Qu, X., Rzhetsky, A., Zhang, P., Cayanis, E., De Jong, P.J., Ju, J., Kalachikov, S., Shuman, H.A., and Russo, J.J. (2004). The genomic sequence of the accidental pathogen *Legionella pneumophila*. Science. *305*, 1966–1968.

Cirillo, S.L., Bermudez, L.E., El-Etr, S.H., Duhamel, G.E., and Cirillo, J.D. (2001). *Legionella pneumophila* entry gene *rtxA* is involved in virulence. Infect. Immun. *69*, 508–517.

Cirillo, S.L., Lum, J., and Cirillo, J.D. (2000). Identification of novel loci involved in entry by *Legionella pneumophila*. Microbiology. *146*, 1345–1359.

Cirillo, S.L., Yan, L., Littman, M., Samrakandi, M.M., and Cirillo, J.D. (2002). Role of the *Legionella pneumophila rtxA* gene in amoebae. Microbiology. *148*, 1667–1677.

Cole, S.T., Brosch, R., Parkhill, J., Garnier, T., Churcher, C., Harris, D., Gordon, S.V., Eiglmeier, K., Gas, S., Barry, C.E.,Tekaia, F., Badcock, K., Basham, D., Brown, D., Chillingworth, T., Connor, R., Davies, R., Devlin, K., Feltwell, T., Gentles, S., Hamlin, N., Holroyd, S., Hornsby, T., Jagels, K., Krogh, A., McLean, J., Moule, S., Murphy, L., Oliver, K., Osborne, J., Quail, M.A., Rajandream, M.A., Rogers, J., Rutter, S., Seeger, K., Skelton, J., Squares, R., Squares, S., Sulston, J.E., Taylor, K., Whitehead, S., and Barrell, B.G. (1998). Deciphering the biology of *Mycobacterium tuberculosis* from the complete genome sequence. Nature; *393*, 537–544.

Conover, G.M., Derre, I., Vogel, J.P., and Isberg, R.R. (2003). The *Legionella pneumophila* LidA protein: a translocated substrate of the Dot/Icm system associated with maintenance of bacterial integrity. Mol. Microbiol. *48*, 305–321.

Dasgupta, A., Datta, P., Kundu, M., and Basu, J. (2006). The serine/threonine kinase PknB of *Mycobacterium tuberculosis* phosphorylates PBPA, a penicillin-binding protein required for cell division. Microbiology. *152*, 493–504.

De Buck, E., Lebeau, I., Maes, L., Geukens, N., Meyen, E., Van Mellaert, L., Anne, J., and Lammertyn, E. (2004). A putative twin-arginine translocation pathway in *Legionella pneumophila*. Biochem. Biophys. Res. Commun. *317*, 654–661.

de Felipe, K.S., Pampou, S., Jovanovic, O.S., Pericone, C.D., Ye, S.F., Kalachikov, S., and Shuman, H.A. (2005). Evidence for acquisition of *Legionella* type IV secretion substrates via interdomain horizontal gene transfer. J. Bacteriol. *187*, 7716–7726.

Derre, I., and Isberg, R.R. (2005). LidA, a translocated substrate of the *Legionella pneumophila* type IV secretion system, interferes with the early secretory pathway. Infect. Immun. *73*, 4370–4380.

Dhandayuthapani, S., Blaylock, M.W., Bebear, C.M., Rasmussen, W.G., and Baseman, J.B. (2001). Peptide methionine sulfoxide reductase (MsrA) is a virulence determinant in *Mycoplasma genitalium*. J. Bacteriol. *183*, 5645–5650.

Doleans, A., Akermi, M., Ginevra, C., Cazalet, C., Kay, E., Schneider, D., Buchrieser, C., Atlan, D., Vandenesch, F., Etienne, J., and Jarraud, S. (2006). Growth phase-dependent mobility of the *lvh*-encoding region in *Legionella* strain Paris. Microbiology. *152*, 3561–3568.

Dorman, C.J., and Porter, M.E. (1998). The *Shigella* virulence gene regulatory cascade: a paradigm of bacterial gene control mechanisms. Mol. Microbiol. *29*, 677–684.

Dorrell, N., Mangan, J.A., Laing, K.G., Hinds, J., Linton, D., Al-Ghusein, H., Barrell, B.G., Parkhill, J., Stoker, N.G., Karlyshev, A.V., Butcher, P.D., and Wren, B.W. (2001). Whole genome comparison of *Campylobacter jejuni* human isolates using a low-cost microarray reveals extensive genetic diversity. Genome. Res. *11*, 1706–1715.

Doumith, M., Cazalet, C., Simoes, N., Frangeul, L., Jaquet, C., Kunst, F., Martin, P., Cossart, P., Glaser, P., and Buchrieser, C. (2004). New aspects regarding evolution and virulence of *Listeria monocytogenes* revealed by comparative genomics. Infect. Immun. *72*, 1072–1083.

Dumenil, G., and Isberg, R.R. (2001). The *Legionella pneumophila* IcmR protein exhibits chaperone activity for IcmQ by preventing its participation in high-molecular-weight complexes. Mol. Microbiol. *40*, 1113–1127.

Dziejman, M., Serruto, D., Tam, V. C., Sturtevant, D., Diraphat, P., Faruque, S.M., Rahman, M.H.,

Heidelberg, J.F., Decker, J., Li, L., Montgomery, K.T., Grills, G., Kucherlapati, R., and Mekalanos, J.J. (2005). Genomic characterization of non-O1, non-O139 *Vibrio cholerae* reveals genes for a type III secretion system. Proc. Natl. Acad. Sci. USA. *102*, 3465–3470.

Eriksson, S., Lucchini, S., Thompson, A., Rhen, M., and Hinton, J. C. (2003). Unravelling the biology of macrophage infection by gene expression profiling of intracellular *Salmonella enterica*. Mol. Microbiol. *47*, 103–118.

Feldman, M., and Segal, G. (2004). A specific genomic location within the *icm/dot* pathogenesis region of different *Legionella* species encodes functionally similar but nonhomologous virulence proteins. Infect. Immun. *72*, 4503–4511.

Feldman, M., and Segal, G. (2007). A pair of highly conserved two-component systems participates in the regulation of the hypervariable FIR proteins in different *Legionella* species. J. Bacteriol. *189*, 3382–3391.

Feldman, M., Zusman, T., Hagag, S., and Segal, G. (2005). Coevolution between nonhomologous but functionally similar proteins and their conserved partners in the *Legionella* pathogenesis system. Proc. Natl. Acad. Sci. USA. *102*, 12206–12211.

Field, J., Rosenthal, B., and Samuelson, J. (2000). Early lateral transfer of genes encoding malic enzyme, acetyl-CoA synthetase and alcohol dehydrogenases from anaerobic prokaryotes to *Entamoeba histolytica*. Mol. Microbiol. *38*, 446–455.

Fitzgerald, J.R., Sturdevant, D.E., Mackie, S.M., Gill, S.R., and Musser, J.M. (2001). Evolutionary genomics of *Staphylococcus aureus*: insights into the origin of methicillin-resistant strains and the toxic shock syndrome epidemic. Proc. Natl. Acad. Sci. USA. *98*, 8821–8826.

Fleischmann, R.D., Adams, M.D., White, O., Clayton, R.A., Kirkness, E.F., Kerlavage, A.R., Bult, C.J., Tomb, J.F., Dougherty, B.A., Merrick, J.M., *et al.* (1995). Whole-genome random sequencing and assembly of *Haemophilus influenzae* Rd. Science. *269*, 496–512.

Gal-Mor, O., and Segal, G. (2003). Identification of CpxR as a positive regulator of icm and dot virulence genes of Legionella pneumophila. J. Bacteriol. *185*, 4908–4919.

George, J.R., Pine, L., Reeves, M.W., and Harrell, W.K. (1980). Amino acid requirements of *Legionella pneumophila*. J Clin Microbiol *11*, 286–291.

Glaser, P., Frangeul, L., Buchrieser, C., Rusniok, C., Amend, A., Baquero, F., Berche, P., Bloecker, H., Brandt, P., Chakraborty, T., Charbit, A., Chetouani, F., Couvé, E., de Daruvar, A., Dehoux, P., Domann, E., Domínguez-Bernal, G., Duchaud, E., Durant, L., Dussurget, O., Entian, K.D., Fsihi, H., García-del Portillo, F., Garrido, P., Gautier, L., Goebel, W., Gómez-López, N., Hain, T., Hauf, J., Jackson, D., Jones, L.M., Kaerst, U., Kreft, J., Kuhn, M., Kunst, F., Kurapkat, G., Madueno, E.Maitournam, A., Vicente, J.M., Ng, E., Nedjari, H., Nordsiek, G., Novella, S., de Pablos, B., Pérez-Diaz, J.C., Purcell, R., Remmel, B., Rose, M., Schlueter, T., Simoes, N., Tierrez, A., Vázquez-Boland, J.A., Voss, H., Wehland, J., and Cossart, P. (2001). Comparative genomics of *Listeria* species. Science. *294*, 849–852.

Glöckner, G., Albert-Weissenberger, C., Weinmann, E., Jacobi, S., Schunder, E., Steinert, M., Hacker, J., and Heuner, K. (2007). Identification and characterization of a new conjugation type IVA secretion system (trb/tra) of *L. pneumophila* Corby localized on two mobile genomic islands. Int. J. Med. Microbiol. doi:10.1016/j.ijmm.2007.07.012

Gray, W. M. (2004). Hormonal regulation of plant growth and development. PLoS. Biol. 2, E311.

Hacker, J., and Kaper, J.B. (2000). Pathogenicity islands and the evolution of microbes. Annu. Rev. Microbiol. *54*, 641–679.

Hägele, S., Kohler, R., Merkert, H., Schleicher, M., Hacker, J., and Steinert, M. (2000). *Dictyostelium discoideum*: a new host model system for intracellular pathogens of the genus *Legionella*. Cell. Microbiol. *2*, 135–171.

Hammer, B.K., Tateda, E.S., and Swanson, M.S. (2002). A two-component regulator induces the transmission phenotype of stationary-phase *Legionella pneumophila*. Mol. Microbiol. *44*, 107–118.

Heuner, K., Brand, B.C, Hacker, J. (1999). The expression of the flagellum of *Legionella pneumophila* is modulated by different environmental factors. FEMS. Microbiol. Lett. *175*:69–77.

Heuner, K., Dietrich, C., Skriwan, C., Steinert, M., Hacker, J. (2002). Influence of the alternative sigma(28) factor on virulence and flagellum expression of *Legionella pneumophila*. Infect. Immun. *70*:1604–1608.

Heuner, K., Steinert, M. (2003). The flagellum of *Legionella pneumophila* and its link to the expression of the virulent phenotype. Int. J. Med. Microbiol. *293*:133–43.

Hinchliffe, S.J., Isherwood, K.E., Stabler, R.A., Prentice, M.B., Rakin, A., Nichols, R.A., Oyston, P.C., Hinds, J., Titball, R.W., and Wren, B.W. (2003). Application of DNA microarrays to study the evolutionary genomics of *Yersinia pestis* and *Yersinia pseudotuberculosis*. Genome. Res. *13*, 2018–2029.

Hisert, K.B., MacCoss, M., Shiloh, M.U., Darwin, K.H., Singh, S., Jones, R.A., Ehrt, S., Zhang, Z., Gaffney, B.L., Gandotra, S., Holden, D.W., Murray, D., and Nathan, C. (2005). A glutamate-alanine-leucine (EAL) domain protein of *Salmonella* controls bacterial survival in mice, antioxidant defense and killing of macrophages: role of cyclic diGMP. Mol. Microbiol. *56*, 1234–1245.

Jacobi, S., and Heuner, K. (2003). Description of a putative type I secretion system in *Legionella pneumophila*. Int. J. Med. Microbiol. *293*, 349–358.

Jacobi, S., Schade, R., Heuner, K. (2004). Characterization of the alternative sigma factor sigma54 and the transcriptional regulator FleQ of *Legionella pneumophila*, which are both involved in the regulation cascade of flagellar gene expression. J. Bacteriol. *186*:2540–2547.

Janjusevic, R., Abramovitch, R.B., Martin, G.B., and Stebbins, C.E. (2006). A bacterial inhibitor of host programmed cell death defenses is an E3 ubiquitin ligase. Science. *311*, 222–226.

Jepras, R.I., Fitzgeorge, R.B., and Baskerville, A. (1985). A comparison of virulence of two strains of *Legionella pneumophila* based on experimental aerosol infection of guinea pigs. J. Hyg. *95*, 29–38.

Joseph, B., Przybilla, K., Stuhler, C., Schauer, K., Slaghuis, J., Fuchs, T.M., and Goebel, W. (2006). Identification of *Listeria monocytogenes* genes contributing to intracellular replication by expression profiling and mutant screening. J. Bacteriol. *188*, 556–568.

Jules, M., and Buchrieser, C. (2007). *Legionella pneumophila* adaptation to intracellular life and the host response: Clues from genomics and transcriptomics. FEBS. Letters. *581*:2829–2838.

Kerscher, O., Felberbaum, R., and Hochstrasser, M. (2006). Modification of proteins by ubiquitin and ubiquitin-like proteins. Annu. Rev. Cell. Dev. Biol. *22*, 159–180.

Loftus, B., Anderson, I., Davies, R., Alsmark, U.C., Samuelson, J., Amedeo, P., Roncaglia, P., Berriman, M., Hirt, R.P., Mann, B.J., Nozaki, T., Suh, B., Pop, M., Duchene, M., Ackers, J., Tannich, E., Leippe, M., Hofer, M., Bruchhaus, I., Willhoeft, U., Bhattacharya, A., Chillingworth, T., Churcher, C., Hance, Z., Harris, B., Harris, D., Jagels, K., Moule, S., Mungall, K., Ormond, D., Squares, R., Whitehead, S., Quail, M.A., Rabbinowitsch, E., Norbertczak, H., Price, C., Wang, Z., Guillén, N., Gilchrist, C., Stroup, S.E., Bhattacharya, S., Lohia, A., Foster, P.G., Sicheritz-Ponten, T., Weber, C., Singh, U., Mukherjee, C., El-Sayed, N.M., Petri, W.A., Clark, C.G., Embley, T.M., Barrell, B., Fraser, C.M., and Hall, N. (2005). The genome of the protist parasite *Entamoeba histolytica*. Nature. *433*, 865–868.

Luneberg, E., Mayer, B., Daryab, N., Kooistra, O., Zahringer, U., Rohde, M., Swanson, J., and Frosch, M. (2001). Chromosomal insertion and excision of a 30 kb unstable genetic element is responsible for phase variation of lipopolysaccharide and other virulence determinants in *Legionella pneumophila*. Mol. Microbiol. *39*, 1259–1271.

Luo, Z.Q., and Isberg, R.R. (2004). Multiple substrates of the *Legionella pneumophila* Dot/Icm system identified by interbacterial protein transfer. Proc. Natl. Acad. Sci. USA. *101*, 841–846.

Machner, M.P., and Isberg, R.R. (2006). Targeting of host Rab GTPase function by the intravacuolar pathogen *Legionella pneumophila*. Dev. Cell. *11*, 47–56.

Marmiesse, M., Brodin, P., Buchrieser, C., Gutierrez, C., Simoes, N., Vincent, V., Glaser, P., Cole, S.T., and Brosch, R. (2004). Macro-array and bioinformatic analyses reveal mycobacterial 'core' genes, variation in the ESAT-6 gene family and new phylogenetic markers for the *Mycobacterium tuberculosis* complex. Microbiology. *150*, 483–496.

Marra, A., Blander, S.J., Horwitz, M.A., and Shuman, H.A. (1992). Identification of a *Legionella pneumophila* locus required for intracellular multiplication in human macrophages. Proc. Natl. Acad. Sci. USA. *89*, 9607–9611.

Matsubara, M., and Mizuno, T. (2000). The SixA phospho-histidine phosphatase modulates the ArcB phosphorelay signal transduction in *Escherichia coli*. FEBS. Lett. *470*, 118–124.

McDade, J.E., Shepard, C.C., Fraser, D.W., Tsai, T.R., Redus, M.A., and Dowdle, W.R. (1977). Legionnaires' disease: isolation of a bacterium and demonstration of its role in other respiratory disease. N. Engl. J. Med. *297*, 1197–1203.

Mika, F., and Hengge, R. (2005). A two-component phosphotransfer network involving ArcB, ArcA, and RssB coordinates synthesis and proteolysis of sigmaS (RpoS) in *E. coli*. Genes. Dev. *19*, 2770–2781.

Miyamoto, H., Yoshida, S., Taniguchi, H., Qin, M.H., Fujio, H., and Mizuguchi, Y. (1993). Protein profiles of *Legionella pneumophila* Philadelphia-1 grown in macrophages and characterization of a gene encoding a novel 24 kDa *Legionella* protein. Microb. Pathog. *15*, 469–484.

Molofsky, A.B., and Swanson, M.S. (2004). Differentiate to thrive: lessons from the *Legionella pneumophila* life cycle. Mol. Microbiol. *53*, 29–40.

Molofsky, A.B., Shetron-Rama, L.M., Swanson, M.S. (2005). Components of the *Legionella pneumophila* flagellar regulon contribute to multiple virulence traits, including lysosome avoidance and macrophage death. Infect. Immun. *73*:5720–5734.

Monecke, S., Berger-Bachi, B., Coombs, G., Holmes, A., Kay, I., Kearns, A., Linde, H.J., O'Brien, F., Slickers, P., and Ehricht, R. (2007). Comparative genomics and DNA array-based genotyping of pandemic *Staphylococcus aureus* strains encoding Panton-Valentine leukocidin. Clin. Microbiol. Infect. *13*, 236–249.

Morozova, I., Qu, X., Shi, S., Asamani, G., Greenberg, J.E., Shuman, H.A., and Russo, J.J. (2004). Comparative sequence analysis of the *icm/dot* genes in *Legionella*. Plasmid. *51*, 127–147.

Mosavi, L.K., Minor, D.L., Jr., and Peng, Z.Y. (2002). Consensus-derived structural determinants of the ankyrin repeat motif. Proc. Natl. Acad. Sci. USA. *99*, 16029–16034.

Munro, P., Flatau, G., and Lemichez, E. (2007). Bacteria and the ubiquitin pathway. Curr. Opin. Microbiol. *10*, 39–46.

Murata, T., Delprato, A., Ingmundson, A., Toomre, D.K., Lambright, D.G., and Roy, C.R. (2006). The *Legionella pneumophila* effector protein DrrA is a Rab1 guanine nucleotide-exchange factor. Nat. Cell. Biol. *8*, 971–977.

Nagai, H., Cambronne, E.D., Kagan, J.C., Amor, J.C., Kahn, R.A., and Roy, C.R. (2005). A C-terminal translocation signal required for Dot/Icm-dependent delivery of the *Legionella* RalF protein to host cells. Proc. Natl. Acad. Sci.USA. *102*, 826–831.

Nagai, H., Kagan, J.C., Zhu, X., Kahn, R.A., and Roy, C.R. (2002). A bacterial guanine nucleotide exchange factor activates ARF on *Legionella* phagosomes. Science. *295*, 679–682.

Neu, J.M., MacMillan, S.V., Nodwell, J.R., and Wright, G.D. (2002). StoPK-1, a serine/threonine protein kinase from the glycopeptide antibiotic producer *Streptomyces toyocaensis* NRRL 15009, affects oxidative stress response. Mol. Microbiol. *44*, 417–430.

Ninio, S., Zuckman-Cholon, D.M., Cambronne, E.D., and Roy, C.R. (2005). The *Legionella* IcmS-IcmW protein complex is important for Dot/Icm-mediated protein translocation. Mol. Microbiol. 55, 912–926.

Ogata, H., La Scola, B., Audic, S., Renesto, P., Blanc, G., Robert, C., Fournier, P.E., Claverie, J.M., and Raoult, D. (2006). Genome sequence of *Rickettsia bellii* illuminates the role of amoebae in gene exchanges between intracellular pathogens. PLoS. Genet. *2*, e:76.

Ogata, H., Renesto, P., Audic, S., Robert, C., Blanc, G., Fournier, P.E., Parinello, H., Claverie, J.M., Raoult, D., Galvao, M.A., *et al.* (2005). The genome sequence of *Rickettsia felis* identifies the first putative conjugative plasmid in an obligate intracellular parasite. PLoS. Biol. 3, e248.

Peekhaus, N., and Conway, T. (1998). What's for dinner?: Entner-Doudoroff metabolism in *Escherichia coli*. J. Bacteriol. *180*, 3495–3502.

Polesky, A.H., Ross, J.T., Falkow, S., Tompkins, L.S. (2001). Identification of *Legionella pneumophila* genes important for infection of amoebas by signature-tagged mutagenesis. Infect. Immun. 69:977–987.

Rossier, O., and Cianciotto, N.P. (2001). Type II protein secretion is a subset of the PilD-dependent processes that facilitate intracellular infection by *Legionella pneumophila*. Infect. Immun. *69*, 2092–2098.

Rossier, O., and Cianciotto, N.P. (2005). The *Legionella pneumophila tatB* gene facilitates secretion of phospholipase C, growth under iron-limiting conditions, and intracellular infection. Infect. Immun. *73*, 2020–2032.

Rowbotham, T.J. (1980). Preliminary report on the pathogenicity of *Legionella pneumophila* for freshwater and soil amoebae. J Clin Pathol 33, 1179–1183.

Salama, N., Guillemin, K., McDaniel, T.K., Sherlock, G., Tompkins, L., and Falkow, S. (2000). A whole-genome microarray reveals genetic diversity among *Helicobacter pylori* strains. Procv Natl. Acad. Sci. USA. *26*, 14668–14673.

Sansom, F.M., Newton, H. J., Crikis, S., Cianciotto, N.P., Cowan, P.J., d'Apice, A.J., and Hartland, E.L. (2007). A bacterial ecto-triphosphate diphosphohydrolase similar to human CD39 is essential for intracellular multiplication of *Legionella pneumophila*. Cell Microbiol doi:10.1111/j.1462–5822.2007.00924.x.

Sauer, J.D., Bachman, M.A., and Swanson, M.S. (2005). The phagosomal transporter A couples threonine acquisition to differentiation and replication of *Legionella pneumophila* in macrophages. Proc. Natl. Acad. Sci. USA. *102*, 9924–9929.

Schnappinger, D., Ehrt, S., Voskuil, M.I., Liu, Y., Mangan, J.A., Monahan, I.M., Dolganov, G., Efron, B., Butcher, P.D., Nathan, C., and Schoolnik, G K. (2003). Transcriptional Adaptation of *Mycobacterium tuberculosis* within Macrophages: Insights into the Phagosomal Environment. J. Exp. Med. *198*, 693–704.

Segal, G., Feldman, M., and Zusman, T. (2005). The Icm/Dot type-IV secretion systems of *Legionella pneumophila* and *Coxiella burnetii*. FEMS. Microbiol. Rev. *29*, 65–81.

Sexton, J.A., and Vogel, J.P. (2002). Type IVB secretion by intracellular pathogens. Traffic. *3*, 178–185.

Shohdy, N., Efe, J.A., Emr, S.D., and Shuman, H.A. (2005). Pathogen effector protein screening in yeast identifies *Legionella* factors that interfere with membrane trafficking. Proc. Natl. Acad. Sci. USA. *102*, 4866–4871.

Simm, R., Morr, M., Kader, A., Nimtz, M., and Romling, U. (2004). GGDEF and EAL domains inversely regulate cyclic di-GMP levels and transition from sessility to motility. Mol. Microbiol. *53*, 1123–1134.

Snyder, J.A., Haugen, B.J., Buckles, E.L., Lockatell, C.V., Johnson, D.E., Donnenberg, M.S., Welch, R.A., and Mobley, H.L. (2004). Transcriptome of uropathogenic *Escherichia coli* during urinary tract infection. Infect. Immun. *72*, 6373–6381.

Soderberg, M.A., Rossier, O., and Cianciotto, N.P. (2004). The type II protein secretion system of *Legionella pneumophila* promotes growth at low temperatures. J. Bacteriol. *186*, 3712–3720.

Steinert, M., Heuner, K., Buchrieser, C., Albert-Weissenberger, C., and Glöckner, G. (2007). *Legionella* pathogenicity: genome structure, regulatory networks and the host cell response. Int. J. Med. Microbiol. doi:10.1016/j.ijmm.2007.03.009

Stone, B.J., and Abu Kwaik, Y. (1998). Expression of multiple pili by *Legionella pneumophila*: identification and characterization of a type IV pilin gene and its role in adherence to mammalian and protozoan cells. Infect. Immun. *66*, 1768–1775.

Stover, C.K., Pham, X.Q., Erwin, A.L., Mizoguchi, S.D., and Warrener, P. (2000). Complete genome sequence of *Pseudomonas aeruginosa* PA01, an opportunistic pathogen. Nature. *406*, 959–964.

Tal, R., Wong, H.C., Calhoon, R., Gelfand, D., Fear, A.L., Volman, G., Mayer, R., Ross, P., Amikam, D., Weinhouse, H., Cohen, A., Sapir, S., Ohana, P., and Benziman, M. (1998). Three *cdg* operons control cellular turnover of cyclic di-GMP in *Acetobacter xylinum*: genetic organization and occurrence of conserved domains in isoenzymes. J. Bacteriol. *180*, 4416–4425.

Taya, Y., Tanaka, Y., and Nishimura, S. (1978). 5′-AMP is a direct precursor of cytokinin in *Dictyostelium discoideum*. Nature *271*, 545–547.

Tesh, M.J., and Miller, R.D. (1981). Amino acid requirements for *Legionella pneumophila* growth. J. Clin. Microbiol. *13*, 865–869.

Trad, S., Allignet, J., Frangeul, L., Davi, M., Vergassola, M., Couve, E., Morvan, A., Kechrid, A., Buchrieser, C., Glaser, P., and El-Solh, N. (2004). DNA macroarray for identification and typing of *Staphylococcus aureus* isolates. J. Clin. Microbiol. *42*, 2054–2064.

Umeyama, T., Lee, P.C., and Horinouchi, S. (2002). Protein serine/threonine kinases in signal transduction for secondary metabolism and morphogenesis in *Streptomyces*. Appl. Microbiol. Biotechnol. *59*, 419–425.

Veiga, E., and Cossart, P. (2005). Ubiquitination of intracellular bacteria: a new bacteria-sensing system? Trends. Cell. Biol. *15*, 2–5.

Vergunst, A.C., Schrammeijer, B., den Dulk-Ras, A., de Vlaam, C.M., Regensburg-Tuink, T.J., and Hooykaas, P.J. (2000). VirB/D4-dependent protein translocation from *Agrobacterium* into plant cells. Science. *290*, 979–982.

Vogel, J.P., Andrews, H.L., Wong, S.K., and Isberg, R.R. (1998). Conjugative transfer by the virulence system of *Legionella pneumophila*. Science. 279:873–876.

Wang, N., Soderbom, F., Anjard, C., Shaulsky, G., and Loomis, W.F. (1999). SDF-2 induction of terminal differentiation in *Dictyostelium discoideum* is mediated by the membrane-spanning sensor kinase DhkA. Mol. Cell. Biol. *19*, 4750–4756.

Weber, S.S., Ragaz, C., Reus, K., Nyfeler, Y., and Hilbi, H. (2006). *Legionella pneumophila* exploits PI(4)P to anchor secreted effector proteins to the replicative vacuole. PLoS. Pathog. *2*, e46.

Welchman, R.L., Gordon, C., and Mayer, R.J. (2005). Ubiquitin and ubiquitin-like proteins as multifunctional signals. Nat. Rev. Mol. Cell. Biol. *6*, 599–609.

Wu, M., Sun, L.V., Vamathevan, J., Riegler, M., Deboy, R., Brownlie, J.C., McGraw, E.A., Martin, W., Esser, C., Ahmadinejad, N., Wiegand, C., Madupu, R., Beanan, M.J., Brinkac, L.M., Daugherty, S.C., Durkin, A.S., Kolonay, J.F., Nelson, W.C., Mohamoud, Y., Lee, P., Berry, K., Young, M.B., Utterback, T., Weidman, J., Nierman, W.C., Paulsen, I.T., Nelson, K.E., Tettelin, H., O'Neill, S.L., and Eisen, J.A. (2004). Phylogenomics of the Reproductive Parasite *Wolbachia pipientis* wMel: A Streamlined Genome Overrun by Mobile Genetic Elements. PLoS. Biol. *2*, e69.

Yu, V.L., Plouffe, J.F., Pastoris, M.C., Stout, J.E., Schousboe, M., Widmer, A., Summersgill, J., File, T., Heath, C.M., Paterson, D.L., and Chereshsky, A. (2002). Distribution of *Legionella* species and serogroups isolated by culture in patients with sporadic community-acquired legionellosis: an international collaborative survey. J. Infect. Dis. *186*, 127–128.

Zusman, T., Aloni, G., Halperin, E., Kotzer, H., Degtyar, E., Feldman, M., and Segal, G. (2007). The response regulator PmrA is a major regulator of the icm/dot type IV secretion system in *Legionella pneumophila* and *Coxiella burnetii*. Mol. Microbiol. *63*, 1508–1523.

Secretion and Export in *Legionella*

8

Nicholas P. Cianciotto

Abstract

L. pneumophila secretes many factors that promote its growth and persistence within the environment, various types of host cells, and the mammalian lung. These factors include both proteins and non-proteinaceous molecules. This chapter will focus on two topics. First will be the type II protein secretion system that elaborates a large number and wide variety of enzymes, some of which have recently been implicated as factors that promote infection. Second will be the export of siderophores, non-protein, high-affinity iron chelators that stimulate growth in low-iron conditions.

Introduction

L. pneumophila secretes many different factors that promote environmental survival and growth with the mammalian host. These factors include both proteins and non-proteinaceous molecules. Among the exoproteins, there are a variety of degradative enzymes, including ones with pathogenic function. In *L. pneumophila,* as in other Gram-negative bacteria, the secretion of exoproteins is a complex process that requires transport across the inner membrane, the periplasmic space, and finally the outer membrane. Gram-negative organisms can have as many as six secretion systems that permit the export of proteins from within the bacterial cell to the extracellular milieu and/or into target host cells; i.e., type I, II, III, IV, V, or VI secretion (Kostakioti *et al.*, 2005; Economou *et al.*, 2006). At least two of these systems are in operation in *L. pneumophila* (Bruggemann *et al.*, 2006a). A great deal of work has shown that both type II and type IV systems are functional and critical for *L. pneumophila*. The type II system will be the primary focus of this chapter, whereas the type IV systems will be covered in Chapter 9. Recent genomic sequencing suggests that type I and type V secretion systems may also exist in strains of *L. pneumophila* (Jacobi and Heuner, 2003; Cazalet *et al.*, 2004). However, whether these systems are functional remains to be determined (Lammertyn and Anne, 2004; Bruggemann *et al.*, 2006a). That *L. pneumophila* has a flagellar system that is related to type III secretion was discussed in Chapter 6. Turning to the export of non-protein factors, a novel siderophore is implicated in *L. pneumophila* iron acquisition. This iron chelator, named legiobactin, will be discussed later in this chapter.

Type II protein secretion

The mechanism of Gram-negative type II secretion

Type II protein secretion (T2S) is a two-step process (Fig. 8.1) (Johnson *et al.*, 2006). In the first step, proteins destined for secretion are translocated across the inner membrane by the Sec pathway or, in fewer instances, the Tat pathway, and then N-terminally cleaved by signal peptidase. In the second step, the proteins are transported from the periplasm to the cell's exterior by a complex of proteins specifically dedicated to T2S, ultimately involving passage of the proteins through an outer membrane pore. The mechanism by which proteins are rec-

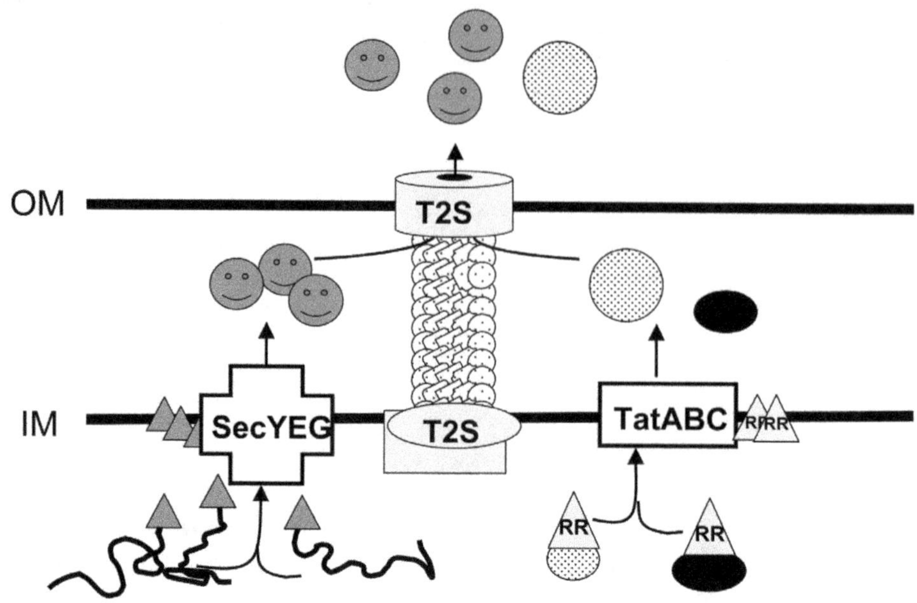

Figure 8.1 Type II secretion (T2S) in *L. pneumophila*.

ognized for entry into the T2S apparatus is not understood but likely involves tertiary structural determinants.

Evolutionarily speaking, the T2S apparatus is related to the type 4 pilus apparatus (Peabody *et al.*, 2003). The T2S apparatus consists of 12 core proteins; i.e., a cytoplasmic ATPase (T2S E), three inner membrane proteins that create a platform and binding site for the ATPase (T2S F, L, M), major and minor 'pseudopilins' which form a pilus-like structure that can span the periplasm (T2S G, H, I, J, K), an inner membrane peptidase that cleaves pre-pseudopilins and then methylates the pseudopilins prior to their integration into the apparatus (T2S O, which is also commonly known as PilD), an outer membrane 'secretin' that oligomerizes to form the exit pore (T2S D), and a protein that may be involved in substrate recognition and/or secretin interactions (T2S C) (Peabody *et al.*, 2003; Filloux, 2004; Johnson *et al.*, 2006). In summary, proteins that are destined for T2S are recognized in the periplasm and then delivered to the T2S apparatus, whereupon, using energy generated at the inner membrane, a pilus-like structure behaves like a piston to push the proteins out through an outer membrane pore (Fig. 8.1). Analysis of sequenced genomes reveals that T2S genes are common but not universal within Gram-negatives (Cianciotto, 2005). Functional studies indicate that T2S promotes the physiology of various environmental bacteria as well as the virulence of numerous human, animal, and plant pathogens (Cianciotto, 2005). In many ways, though, work in *L. pneumophila* has given us the broadest appreciation of the biological significance of T2S.

The discovery of type II secretion genes in *L. pneumophila*

The first indication that *L. pneumophila* employs T2S was the discovery of a *pilBCD* locus, encoding the pseudopilin peptidase PilD (T2S O), along with PilB and PilC, which are proteins that are generally only associated with type 4 pilus biogenesis (Liles *et al.*, 1998). Next, it was shown that mutation of *pilD* in serogroup-1 strain 130b abolished secretion, as evidenced by the loss of proteins in mutant culture supernatants visualized by 1D-SDS-PAGE (Liles *et al.*, 1999). Although *pilD* transcripts are produced at 37°C, Northern blot and *lacZ* gene fusion analyses demonstrated that the gene is more highly expressed at lower temperatures such as 30°C (Liles *et al.*, 1998; Söderberg *et al.*, 2004). A study of strain Philadelphia-1 then revealed the presence of *lspFGHIJK*, predicted to encode the T2S FGHIJK proteins (Hales and Shuman, 1999). As was seen with mutation of *pilD*, inactivation of *lspGH* resulted in loss of protein secretion. In two further studies of strain 130b, genes encoding homologues of T2S DE (*lspDE*), C (*lspC*), and LM (*lspLM*) were reported, with mutational analysis of *lspDE* confirming their role in secretion activity (Rossier and Cianciotto, 2001; Rossier *et al.*, 2004). Taken together, these studies indicated that *L. pneumophila* has a complete and functional T2S system. This was confirmed when the complete genomes of serogroup-1 strains Philadelphia 1, Paris, and Lens were published (Cazalet *et al.*, 2004; Chien *et al.*, 2004; Cianciotto, 2005). Southern hybridizations have shown that T2S genes are also in a variety of other serogroups of *L. pneumophila* as well as other species of *Legionella* (Rossier *et al.*, 2004). A variety of T2S-specific mutants of *L. pneumophila* have been constructed and characterized; in all cases, the mutants replicate normally in standard buffered yeast extract (BYE) broth at 37°C (Rossier and Cianciotto, 2001; Rossier *et al.*, 2004). T2S-specific mutants also have a typical efficiency of plating on buffered charcoal yeast extract (BCYE) agar at 37°C, but they do show slightly altered colony morphology (Rossier and Cianciotto, 2001; Rossier *et al.*, 2004).

Sec translocation in *L. pneumophila*

The presence and functioning of T2S-specific genes in *L. pneumophila* strongly implied that there must also be a mechanism for translocating *Legionella* proteins across the inner membrane prior to their entry into and secretion through the T2S apparatus. There was little doubt, though, that Sec provides this function, because of its seemingly universal conservation within bacteria (van der Sluis and Driessen, 2006). Indeed, subsequent genome sequencing confirmed that *L. pneumophila* encodes a complete Sec system for translocation across the inner membrane (Cazalet *et al.*, 2004; Chien *et al.*, 2004; Lammertyn and Anne, 2004).

Like other Gram-negative proteobacteria, *L. pneumophila* contains the machinery for both co-translational and post-translational translocation. In co-translational translocation, nascent polypeptides are delivered by the signal recognition particle, with the aid of YidC, to an inner membrane channel composed of the trimeric SecYEG complex (Fig. 8.1) (Lee and Schneewind, 2001; Robson and Collinson, 2006). In post-translational translocation, the cytoplasmic SecA ATPase and SecB chaperone deliver newly made protein to SecYEG in a process that also requires SecDF and YajC (Lee and Schneewind, 2001; Robson and Collinson, 2006). SecM, a regulator of SecA expression in *E. coli* and other Enterobacteriales (van der Sluis and Driessen, 2006), is absent from *L. pneumophila*. As expected, *L. pneumophila* encodes the inner membrane type I signal peptidase (LepB) that cleaves the N-terminus from proteins as they emerge, via Sec, on the periplasmic face of the inner membrane (Lammertyn *et al.*, 2004). The type II signal peptidase that cleaves translocated lipoproteins has also been characterized in *L. pneumophila* (Geukens *et al.*, 2006). Given the overwhelming evidence obtained from the study of other bacteria, there is no doubt that the Sec pathway is essential for *L. pneumophila* growth, providing the main conduit for many proteins to localize to the periplasm, outer membrane, and/or extracellular milieu. Indeed, recent *in silico* analysis, using the SignalP algorithm, has determined that there are as many as 576 proteins that contain a typical signal peptide and are potential substrates for Sec translocation (DebRoy *et al.*, 2006b). Additional analyses using PSORTb and PA-SUB, which are designed to further predict cellular localization, suggested that there might be 60 and perhaps even more proteins that are fully secreted into the extracellular space, with presumably a significant portion of those being handled by the T2S system (DebRoy *et al.*, 2006b).

Twin arginine translocation in *L. pneumophila*

Genome sequencing also confirmed that *L. pneumophila* possesses the genes encoding twin-arginine translocation (Tat), another inner membrane translocation pathway that is conserved in many organisms (Cazalet *et al.*, 2004; Chien *et al.*, 2004; De Buck *et al.*, 2004; Rossier and Cianciotto, 2005). Although, for years, Tat had been viewed as a system for localizing periplasmic enzymes, the study of phospholipase C enzymes in *Pseudomonas aeruginosa* revealed that Tat substrates can also feed into the T2S system for extracellular secretion (Ochsner *et al.*, 2002). Thus, the presence of Tat in *L. pneumophila* has potential significance for *Legionella* T2S.

Unlike the Sec system, the Tat machinery promotes the transport of fully folded proteins across membranes (Fig. 8.1) (Lee *et al.*, 2006; Sargent *et al.*, 2006). Tat is mediated by three integral membrane proteins, named TatA, TatB, and TatC. Although the signal peptides of Tat-dependently secreted proteins resemble Sec-dependent signal peptides in their overall structure, they differ in several ways from their Sec counterparts. First, they have a twin arginine consensus motif (RRXΦΦ, where Φ is a hydrophobic residue), with that motif being followed by a region that is less hydrophobic than that of Sec signal peptides. Second, basic residues, which could serve as a Sec-avoidance signal, are commonly found just before the signal peptidase cleavage site. Finally, Tat signal peptides can be much longer than the Sec signal sequences. In the current *E. coli* model, Tat is viewed as a three-step event (Fig. 8.1) (Sargent *et al.*, 2006). First, the twin-arginine signal peptides are bound by an inner membrane complex consisting of TatB and TatC, with TatC probably being the main factor in peptide recognition. Then, a complex consisting of TatA is recruited to the site in a process that is dependent upon the membrane's electrochemical gradient. Finally, the folded exoprotein passes through a channel in the TatA complex.

Once the *tatA*, *tatB*, and *tatC* genes were found in *L. pneumophila*, RT-PCR analysis revealed that they are expressed during both extra- and intracellular growth (De Buck *et al.*, 2004). Since *L. pneumophila tatB* mutants grow normally in standard BYE, Tat is not required for general extracellular growth (Rossier and Cianciotto, 2005). However, *L. pneumophila tat* mutants are defective for growth in iron-limiting conditions and during intracellular infection of macrophages and amoebae, indicating that Tat substrates are critical for *Legionella* survival under specialized conditions, including those that exist during infection (De Buck *et al.*, 2005; Rossier and Cianciotto, 2005). As regards more specific effects of Tat, it was demonstrated that *L. pneumophila* Tat is required for cytochrome *c*-dependent respiration and the associated processing of the cytochrome *c* reductase PetA (Rossier and Cianciotto, 2005). *In silico* analysis that looked for the presence of twin arginine residues in signal peptides revealed ca. 35 more putative Tat substrates, which are predicted to be involved in a variety of other cellular processes (De Buck *et al.*, 2004; Rossier and Cianciotto, 2005).

Factors secreted by the type II system of *L. pneumophila*

With the knowledge that *L. pneumophila* possesses the full set of T2S-specific genes, as well as the Sec and Tat machinery for delivering proteins into the periplasm in order that they might then enter the secretion apparatus, we can now focus on the factors that are secreted by *Legionella* T2S. By 2005, at least eleven enzymes were determined to be secreted via the *L. pneumophila* T2S system (Rossier *et al.*, 2004; Banerji *et al.*, 2005; Cianciotto, 2005). This conclusion was based upon the loss of enzymatic activities from the culture supernatants of *lspDE*, *lspF*, *lspG*, *lspGH*, or *pilD* mutants grown in standard BYE broth at 37°C (Hales and Shuman, 1999; Liles *et al.*, 1999; Aragon *et al.*, 2000; Rossier and Cianciotto, 2001; Rossier *et al.*, 2004). The T2S-dependent activities include tartrate-sensitive and tartrate-resistant acid phosphatases, phospholipase A, phospholipase(s) C, lysophospholipase A, glycerophospholipid:cholesterol acyltransferase (GCAT), mono-, di-, and triacylglycerol lipases, ribonuclease, and protease (Hales and Shuman, 1999; Liles *et al.*, 1999; Aragon *et al.*, 2000; Aragon *et al.*, 2001; Flieger *et al.*, 2001b; Rossier and Cianciotto, 2001; Aragon *et al.*, 2002; Flieger *et al.*, 2002; Rossier *et al.*, 2004; Banerji *et al.*, 2005). Because mutants specifically lacking the type 4 pilus are not defective for these activities, the altered secretion of the *pilD* mutants is due to the loss of T2S (Rossier and Cianciotto, 2001; Rossier *et al.*, 2004). Several of the secreted enzymatic activities have now been identified in other species of *Legionella* (Flieger *et al.*, 2001a; Rossier *et al.*, 2004).

In some cases, the structural genes (proteins) encoding T2S-dependent secreted activities have been identified. These include *map* (Map) for the tartrate-sensitive acid phosphatase (Aragon *et al.*, 2001), *plcA* (PlcA) for phospholipase C activity (Aragon *et al.*, 2002), *plaA* (PlaA) for the lysophospholipase A (Flieger *et al.*, 2002), *plaC* (PlaC) for GCAT (Banerji *et al.*, 2005), *lipA* (LipA) and *lipB* (LipB) for mono- and triacylglycerol lipases (Aragon *et al.*, 2002), and *proA/msp* (ProA/Msp) for the zinc metalloprotease (Hales and Shuman, 1999; Liles *et al.*, 1999). Notably, mutations in any one of these genes do not completely abolish the corresponding activity. Accordingly, *L. pneumophila* appears to have more than one secreted phosphatase, phospholipase C, lysophospholipase A, lipase, and protease/peptidase, and T2S likely mediates the secretion of more than eleven proteins. The analysis of supernatants from *proA* mutants demonstrates that some secreted proteins are subject to cleavage and perhaps activation by the T2S-dependent metalloprotease (Flieger *et al.*, 2002; Banerji *et al.*, 2005). By virtue of a twin-arginine in its signal peptide, PlcA of *L. pneumophila* strain 130b was predicted to be a Tat substrate; indeed, mutations in *tatB* diminish secreted phospholipase C activity in that strain (Rossier and Cianciotto, 2005).

To further define the proteins secreted by *L. pneumophila* T2S, proteins in wild type and *lsp* mutant supernatants were compared by two-dimensional PE, and then mass spectrometry was used to identify the secreted proteins (DebRoy *et al.*, 2006b). Twenty-seven proteins present for wild type but absent or greatly reduced for an *lspF* mutant were identified. Three of these (i.e., ProA, PlaA, and Map) were previously defined as T2S exoproteins, as noted above. A fourth protein was identified as a ribonuclease, most likely representing the T2S-dependent ribonuclease activity that had been reported before, but for which the gene had been unknown. Of the remaining 23 proteins, 13 were also predicted to contain a signal sequence and are quite likely to represent new type II dependent exoproteins. Two of these were annotated in the database as a leucine aminopeptidase and a chitinase, assignments subsequently confirmed by mutational analysis of the cognate genes (DebRoy *et al.*, 2006b; Rossier and Cianciotto, unpublished). Four others were annotated as hypothetical proteins but having similarity to bacterial enzymes; they included an amidase, another aminopeptidase, a cysteine protease, and an endoglucanase/cellulase. Two others were most similar to eukaryotic proteins, i.e. one had collagen-like repeats, and the other homology to a eukaryotic zinc proteinase. The remaining five proteins that were clearly T2S dependent were annotated as hypothetical with no similarities to any known protein or domain in the database and thus may represent novel activities. Three more proteins, IcmX, LvrE, and a VirK-like protein, identified in wild-type but not mutant supernatants (Matthews and Roy, 2000; De Buck *et al.*, 2005; Albert *et al.*, 2006; DebRoy *et al.*, 2006b), had signal sequences, but their connection to T2S seems more complicated. Indeed, IcmX and LvrE are known for their links, respectively, to the Dot/Icm and Lvh type IV systems of *L. pneumophila* (Brand *et al.*, 1994; Segal *et al.*, 1999; Matthews and Roy, 2000; Ridenour *et al.*, 2003), and VirK is associated with a type IV secretion system in *Agrobacterium tumefaciens* (Kalogeraki and Winans, 1998; Li *et al.*, 2005). These data suggest a possible connection between type II and IV secretion. Alternately, although the genes encoding IcmX, LvrE, and the VirK-like protein are linked to type IV secretion genes, the secretion of the proteins may in fact occur by T2S.

The results of the recent proteomic analysis combined with the earlier assessments of culture supernatant enzymatic activities indicate that the number of proteins secreted by the *L. pneumophila* T2S system is at least 25. However, the total output of the *L. pneumophila* T2S system is likely greater, for several reasons: i) not every protein observed by 2D-PE was submitted for identification by mass spectroscopy, ii) some T2S effectors might have been missed because of low-level expression or degradation due to the processing of the sample or the action of a secreted protease, iii) the supernatant samples that were examined were from bacteria cultured under only a single growth condition, iv) an *in silico* screen of the *L. pneumophila* genome revealed 60 proteins that are predicted to be extracellular by at least one programme; and 13 meet the criteria of two programmes (DebRoy *et al.*, 2006b). Thus, the most likely scenario is that the *L. pneumophila* T2S processes between 25 and 60 substrates. Even at 25, the experimentally defined catalogue of *L. pneumophila* T2S effectors is the largest known in bacteria (Cianciotto, 2005; DebRoy *et al.*, 2006b).

Clearly, *L. pneumophila* secretes a variety of factors via its T2S system. Several of the types of enzymes observed, including proteolytic enzymes, lipolytic enzymes, chitinases, and phosphatases, and potential cellulases, are also secreted by other bacterial T2S systems (Cianciotto, 2005). In some cases, the T2S-dependent effectors of *L. pneumophila* are actually closely related to the secreted proteins of other bacteria; e.g. ProA/Msp is homologous with well characterized proteases from a variety of organisms, and PlcA is highly related to a new type of phospholipase C found in *Pseudomonas* (Aragon *et al.*, 2002). But, the type II secretome of *L. pneumophila* also includes proteins that show their greatest similarity to eukaryotic proteins; e.g. the Map acid phosphatase, a collagen-like protein, and a eukaryotic-like proteinase (Aragon *et al.*, 2001; DebRoy *et al.*, 2006b). *L. pneumophila* is also unique in its secretion of an RNAse and perhaps an amidase. Perhaps most significantly, a number of the *Legionella* exoproteins do not bear any similarity to known proteins, raising the possibility of there being novel effectors secreted by T2S. From proteomic data and *in silico* predictions, the type of proteins that appear to be most represented in the T2S repertoire are proteases/peptidases, a finding consistent with the fact that amino acids are the main source of carbon and energy for broth-grown *L. pneumophila* (George *et al.*, 1980). The legionellae can also catabolize carbohydrate-derivatives (Weiss *et al.*, 1980; Tesh *et al.*, 1983; Bruggemann *et al.*, 2006b), and the putative Lsp exoproteins do also include proteins with predicted glycosidase activity such as a eukaryotic-like glucoamylase. Based on both the number of proteins uncovered and the types of factors detected, the analysis of *L. pneumophila* highlights more than ever the impact that T2S has on bacterial function.

A connection between Mip and type II secretion

The recent study of *L. pneumophila* has given new insight into the mechanisms of secretion, i.e. the involvement of Mip, a surface associated peptidylproline *cis–trans* isomerase (PPIase), in the elaboration of secreted activity (DebRoy *et al.*, 2006a). Among the activities that are diminished in *lsp* mutant supernatants is the ability to hydrolyse *p*-nitrophenol phosphorylcholine (*p*-NPPC), and in many bacteria, *p*-NPPC hydrolysis is attributed to phospholipase C enzymes (Merino *et al.*, 1999; Terada *et al.*, 1999). However, since mutation of *plcA* only partly diminished secreted *p*-NPPC hydrolase activity (see above), it had been hypothesized that *L. pneumophila* secretes another phospholipase C (Aragon *et al.*, 2002; Rossier and Cianciotto, 2005). When a screen was done to identify other factors that promote secreted hydrolase activity, a *mip* mutant was isolated (DebRoy *et al.*, 2006a).

Although Mip, originally identified for its role in promoting intracellular infection and virulence, is one of the most heavily studied proteins in *Legionella* (Cianciotto *et al.*, 1989; Cianciotto *et al.*, 1990; Cianciotto and Fields, 1992; Fischer *et al.*, 1992; Ludwig *et al.*, 1994; Cianciotto *et al.*, 1995; Wintermeyer *et al.*, 1995; Helbig *et al.*, 2001; Riboldi-Tunnicliffe *et al.*, 2001; Kohler *et al.*, 2003), its molecular target(s) had remained obscure. Five independently derived *mip* mutants showed a 40–70% reduction in secreted activity, and the defects were complemented by providing *mip* in trans. Thus, Mip promotes the presence of a *p*-NPPC hydrolase activity in culture supernatants. When supernatants were examined by chromatography, the *p*-NPPC hydrolase activity associated with Mip proved to be T2S-dependent but distinct from PlcA, indicating that Mip promotes the elaboration of a new T2S-exoprotein.

Two hypotheses are proposed to account for these observations (DebRoy *et al.*, 2006a). In the first scenario, Mip is involved in the extracellular release of an active enzyme that has *p*-NPPC hydrolase activity; Mip could act directly on the exoprotein, or it might promote the maturation of proteins that form the secretion pathway. In the second scenario, Mip interacts with a newly secreted protein and causes changes that convert the protein from being enzymatically inactive to active. These data are the first evidence for a target for Mip and the first case of a surface PPIase being linked to the secretion or activation of proteins beyond the outer membrane (DebRoy *et al.*, 2006a).

The role of type II secretion in *L. pneumophila* pathogenesis

L. pneumophila T2S mutants (i.e., derivatives containing an insertion mutation in *lspDE, lspF, lspG, lspK*, or *pilD*) show a reduced ability to infect human macrophages, including the U937 cell line and peripheral blood mononuclear cell-derived macrophages (Liles *et al.*, 1999; Polesky *et al.*, 2001; Rossier and Cianciotto, 2001; Rossier *et al.*, 2004). In the U937 cells, the mutants displayed a 10-fold reduced yield at 48 hours post inoculation (Polesky *et al.*, 2001; Rossier and Cianciotto, 2001; Rossier *et al.*, 2004). In contrast, an *lsp* mutant did not grow in macrophages derived from peripheral blood cells, suggesting that the magnitude of the mutant phenotype is influenced by the nature of the host cell. Indeed, when tested in HL60 cells, an *lspGH* mutant behaved like wild type (Hales and Shuman, 1999). The reduced infectivity of mutants was complemented by reintroduction of the corresponding *pilD* or *lsp* genes (Rossier *et al.*, 2004). Since mutants specifically lacking type 4 pili are not defective for macrophage infection, the infectivity defect of the *pilD* mutant is due to the loss of T2S (Rossier *et al.*, 2004). Beyond their apparent defect in intracellular replication, Lsp mutants, under some growth conditions, exhibit a defect in invasion (Polesky *et al.*, 2001). The T2S mutants are not impaired for the induction of apoptosis or a pore-forming activity linked to egress from host cells (Molmeret *et al.*, 2002; Zink *et al.*, 2002). Together, these data indicate that T2S promotes *L. pneumophila* infection of macrophages.

Recently, it has been shown that *lsp* mutants are defective in an animal model of Legionnaires' disease (Rossier *et al.*, 2004). Upon intratracheal inoculation into A/J mice, *lspDE, lspF* and *pilD* mutants, but not pilus mutants, display a reduced ability to grow in the lung, as measured by competition assays. The *lspF* was also defective in an *in vivo* growth kinetics assay; i.e., whereas wild type increased at least 10-fold in the lungs, the T2S mutant gave no evidence of replication. The examination of the mouse sera revealed that type II-secreted proteins are expressed *in vivo*. Therefore, *L. pneumophila* T2S is an important virulence factor. The reduced survival of the *lsp* mutants in the lung is likely due, at least in part, to reduced growth in alveolar macrophages. However, that mutant numbers do not increase in the lungs, whereas *in vitro* they do, although not optimally suggests that the *lsp* mutants are also defective for extracellular processes that are operative in the lungs, such as combating other aspects of the host response. Together, these studies are the first documentation of a role for T2S in intracellular infection and in an animal model of disease.

Mutants that lack particular T2S effectors have also been tested for alterations in infection. The *map, plcA, plaA, plaC, lipA, lipB, proA/msp* and *chiA* mutants all grow normally in macrophages *in vitro*, indicating that the Map tartrate-sensitive acid phosphatase, PlcA phospholipase C, PlaA lysophospholipase A, PlaC GCAT, LipA and LipB lipases, ProA zinc metalloprotease, and ChiA chitinase are not required for macrophage infection (Aragon *et al.*, 2001; Aragon *et al.*, 2002; Flieger *et al.*, 2002; Banerji *et al.*, 2005; DebRoy *et al.*, 2006b). These results imply that the T2S pathway secretes a yet-to-be-defined factor that is necessary for macrophage infection. On the other hand, there may be redundancy in the effectors, such that one factor can compensate for the loss of another. Interestingly, the ChiA chitinase mutant proved to be defective when tested in the murine model of Legionnaires' disease (DebRoy *et al.*, 2006b). Following introduction into the lung, the mutant had a ca. 4-fold deficiency compared to wild type. That independently derived mutants behaved similarly indicated that the defects were due to loss of monocistronic *chiA* and not spontaneous second-site mutation. Western blot analysis using antisera from animals intratracheally inoculated with wild-type bacteria showed that ChiA is expressed *in vivo*.

Taken together, these data indicate that ChiA is, directly or indirectly, required for optimal survival of *L. pneumophila* in the lung. ChiA is the first T2S effector to be implicated in *Legionella* virulence. Other effector mutants tested, including strains lacking acid phosphatase, lipases A and B, lysophospholipase A, metalloprotease, or phospholipase C, displayed no defect in the mice (DebRoy *et al.*, 2006b).

Given that the *chiA* mutant grew normally in macrophages *in vitro*, ChiA would seem to be critical for more than intracellular infection. That the reduced survival of the mutant in the lung was manifest in the later stages of infection implies that ChiA promotes persistence vs. replication. Since mammals do not have chitin, these observations lead to the hypothesis that there is a chitin-like substance in the lung whose degradation aids bacterial persistence. Alternatively, ChiA might be a bi-functional enzyme that has another, unidentified substrate, whose loss enhances bacterial survival. Clearly, more work is needed to determine whether these or other hypotheses are correct. Yet, already, these results are the first documentation of a protein having chitinase activity also promoting the survival of a pathogen in a mammalian host (DebRoy *et al.*, 2006b). Thus, factors, such as chitinases, that are traditionally viewed as only being relevant in the environment may actually have great importance in infection.

The role of type II secretion in environmental survival

L. pneumophila lsp and *pilD* mutants are very defective for intracellular growth in amoebae, including *Hartmannella vermiformis* and *Acanthamoeba castellanii* (Hales and Shuman, 1999; Liles *et al.*, 1999; Polesky *et al.*, 2001; Rossier and Cianciotto, 2001; Rossier *et al.*, 2004). Indeed, the mutants show little, if any, evidence of growth in the protozoa, and their reduced infectivity is complementable. Thus, *Legionella* T2S is required for amoebal infection. As was the case for macrophage infection, studies using mutants lacking particular enzymes have not identified a T2S effector that is key to protozoan infection (DebRoy *et al.*, 2006b). However, given the importance of protozoa for *L. pneumophila* survival in water, these data establish T2S as a significant promoter of *Legionella* persistence in the environment. Since infected protozoa may be part of the infective dose (Brieland *et al.*, 1996; Cirillo *et al.*, 1999), the fact that T2S promotes interactions with protozoa also increases its relevance for pathogenesis.

Following upon the finding that *L. pneumophila pilD* transcripts are more abundant at 30 vs. 37°C (Liles *et al.*, 1998), it was found that *pilD* mutants have a reduced ability to grow in BYE broth and to form colonies on BCYE agar at lower temperatures (Söderberg *et al.*, 2004). Mutants specifically impaired in T2S are also greatly impaired for colony formation at 25, 17, and 12°C (Söderberg *et al.*, 2004). These growth defects were complemented by reintroduction of the corresponding *pilD* or *lsp* gene. In experiments that more closely mimic aquatic habitats, *lsp* mutants show reduced survival in tap water incubated at 25, 17, 12, and 4°C (M. A. Söderberg and N. P. Cianciotto, unpublished). The mutants display improved growth at 25°C when plated next to a streak of wild-type but not mutant bacteria, suggesting that a secreted, diffusible factor promotes low-temperature growth (Söderberg *et al.*, 2004). Indeed, growth stimulation could be seen using supernatants from wild type. Mutants lacking the known acid phosphatase, lipases, phospholipase C, lysophospholipase A, or protease grow normally at 25°C, suggesting the existence of a critical, yet-to-be-defined exoproteins(s) (Söderberg *et al.*, 2004). In sum, these data show that *L. pneumophila* replicates at temperatures below 20°C and that a T2S system can stimulate growth at low temperatures.

Concluding comments

Studies with macrophages, animals, protozoa, and low-temperature conditions illustrate how T2S promotes *L. pneumophila* growth in a mammalian host and in environmental niches. Much more work is needed to ascertain which T2S effectors promote growth in the disparate niches and how they function. Nonetheless, *L. pneumophila* already provides the best-characterized T2S system in terms of showing its overall significance for bacterial physiology, pathogenesis, and ecology (Cianciotto, 2005; DebRoy *et al.*, 2006b).

The export of siderophores

L. pneumophila and iron

Iron is a key factor in *L. pneumophila* growth (Reeves *et al.*, 1981). This iron requirement was first thought to be 3–13 μM for minimal growth and >20 μM for optimal growth (Reeves *et al.*, 1983; Johnson *et al.*, 1991; Mengaud and Horwitz, 1993). But, recent work shows it to be <1 μM (James *et al.*, 1995; Liles *et al.*, 2000). Upon incubation with $^{55}FeCl_3$, *L. pneumophila* takes up iron in an energy-dependent, protease-resistant process (Johnson *et al.*, 1991). As in other bacteria, iron serves as a cofactor in a variety of enzymes (Mengaud and Horwitz, 1993). The ability of *L. pneumophila* to replicate in the mammalian host is also highly dependent on iron. For example, human macrophages treated with iron chelators do not support bacterial growth, and interferon-γ inhibits intracellular growth by limiting iron (Byrd and Horwitz, 2000; Viswanathan *et al.*, 2000). Additionally, some murine macrophages become permissive for *Legionella* infection only after the addition of iron (Gebran *et al.*, 1994). Furthermore, iron supplementation increases the susceptibility of animals to infection, and legionellae grown under iron-limiting conditions exhibit reduced virulence (James *et al.*, 1995).

The first genetic data on the role of iron in *Legionella* was the identification of the gene for the transcriptional repressor Fur (Hickey and Cianciotto, 1994). Its repressive activity is, not surprisingly, highest in legionellae grown in iron-rich media (Hickey and Cianciotto, 1994, 1997). Since *L. pneumophila fur* could not be insertionally inactivated, Fur is likely essential for *Legionella* aerobic growth. The importance of *L. pneumophila fur* is evident from the identification of multiple iron- and Fur-regulated genes (Hickey and Cianciotto, 1997). Although the importance of iron for *Legionella* has always been clear, the mechanisms used by the bacterium to acquire and transport iron have been rather elusive. Traditional biochemical methods nicely determined that *L. pneumophila* encodes a cytoplasmic and a periplasmic ferric reductase (James *et al.*, 1997; Johnson *et al.*, 1991; Poch and Johnson, 1993). Nevertheless, as will be described momentarily, proof of the existence of a secreted siderophore was nearly 25 years in waiting.

The discovery of a *Legionella* siderophore

In 1983, it was reported that *L. pneumophila* does not make siderophores (Reeves *et al.*, 1983). This conclusion was based largely on results from assays that detect typical catecholate and hydroxamate structures. Eight years later, the question of *Legionella* siderophores was revisited using the CAS assay, which detects iron chelators regardless of their structure (Goldoni *et al.*, 1991). CAS reactivity was detected in statically grown cultures, suggesting the existence of a non-catecholate, -hydroxamate siderophore. Yet, a later study determined that the CAS reactivity was due to cysteine in the medium (Liles and Cianciotto, 1996). When supernatants where obtained from cultures made with cysteine-free media, no CAS reactivity was seen (Liles and Cianciotto, 1996; James *et al.*, 1997). In 2000, however, it was shown that *L. pneumophila* could excrete a high-affinity iron-chelator (legiobactin) when grown at 37°C in a low-iron, chemically defined medium (Liles *et al.*, 2000). Importantly, the CAS reactivity was only observed when cultures are inoculated with bacteria that had been grown to log or early stationary phase. Inocula derived from late-stationary phase cultures, despite growing in the CDM, do not display the siderophore-like activity. CAS reactivity was observed for multiple serogroup 1 strains as well as isolates representing all nine of the other serogroups tested (Liles *et al.*, 2000). The CAS reactivity has been studied further in *L. pneumophila* serogroup-1 strain 130b, and all data signal that it is a true siderophore (Liles *et al.*, 2000). For example, CAS reactivity correlates with enhanced aerobic growth in an iron-deplete defined medium, and the chelating activity is subject to iron-repression. Moreover, the CAS reactivity is less than 1 kDa in size, is resistant to heat and proteases, and appears rapidly and intensely. Finally, CAS-positive supernatants stimulate the growth of wild-type in BCYE agar containing otherwise inhibitory concentrations of iron chelator (Allard *et al* 2006). Most other *Legionella* species tested also secrete CAS reactivity, although it cannot be concluded yet that they are producing legiobactin (Liles *et al.*, 2000; Starkenburg *et al.*, 2004).

Genes involved in siderophore production

Two linked genes, *lbtA* and *lbtB*, have recently been identified as being involved in the expression of legiobactin (Allard *et al.*, 2006). The *lbtA* gene is predicted to encode a protein with homology to siderophore synthetases, including enzymes involved in the biosynthesis of aerobactin in *E. coli*, alcaligin in *Bordetella* species, and rhizobactin 1021 in *Sinorhizobium meliloti* (Allard *et al.*, 2006). Very recent studies have also uncovered an LbtA-like gene in *Francisella tularensis* (Deng *et al.*, 2006; Sullivan *et al.*, 2006). LbtA is also related to FrgA, a *L. pneumophila* protein that had previously been shown to be iron-regulated but not required for siderophore expression (Hickey and Cianciotto, 1997; Liles *et al.*, 2000; Allard *et al.*, 2006). The *lbtB* gene encodes a protein that is homologous with members of the major facilitator superfamily of multidrug efflux pumps. Significantly, members of this superfamily have recently been implicated in siderophore export, including proteins required for the export of enterobactin in *E. coli*, protochelin in *Azotobacter vinelandii*, achromabactin in *E. chrysanthemi*, and alcaligin in *Bordetella* species (Furrer *et al.*, 2002; Brickman and Armstrong, 2005; Franza *et al.*, 2005). Thus, LbtA is predicted to be involved in the biosynthesis of legiobactin, and LbtB in its export. Mutants inactivated for *lbtA* or *lbtB* are impaired in legiobactin expression, producing 40–70% less CAS reactivity in iron-deplete CDM (Allard *et al.*, 2006). In bioassays, mutant supernatants, unlike those of wild type, do not support the growth of iron-limited wild-type bacteria or a ferrous iron transport mutant. The *lbtA* mutant was also defective for growth in media containing the iron chelator citrate, indicating that legiobactin is needed in conditions of severe iron limitation. The iron (III) K_s for legiobactin has been estimated to be between 10^{11} and 10^{31} (Allard *et al.*, 2006). *Trans*-complementation of the *lbt* mutants restored both legiobactin expression as measured by the CAS assay and bioassays and bacterial growth in citrate-containing media (Allard *et al.*, 2006). Fur boxes precede *lbtAB*, suggesting that Fur mediates the iron-regulation of legiobactin. *lbtA* is widely distributed among *Legionella* strains (Allard *et al.*, 2006).

Bacteria usually have large siderophore operons encoding a number of biosynthetic enzymes and a ferrisiderophore receptor; e.g. there appear to be six biosynthetic genes for rhizobactin, and three to five for alcaligin (Challis, 2005). The *L. pneumophila* system seems different. There is a gene (*lbtC*) downstream of *lbtAB* that is predicted to be another MFS member, but mutations in it do not alter siderophore expression (Allard *et al.*, 2006). Furthermore, examination of the genome reveals no other obvious candidate legiobactin genes unlinked to *lbtABC*. Hence, the biosynthesis of legiobactin may be simple, potentially involving LbtA and one or two precursors. On the other hand, more legiobactin genes may exist, but they would appear to be unique in content and location. In a similar vein, the uptake of legiobactin would appear to be mediated by a novel process since the *L. pneumophila* genome does not contain TonB, the membrane spanning protein that has been typically implicated in siderophore import (Koebnik, 2005).

The role of siderophores in infection and ecology

To begin to assess the role of siderophores in *Legionella* infection and environmental persistence, *lbtAB* mutants were tested for their ability to infect macrophages and *Hartmannella* amoebae (Allard *et al.*, 2006). Although the *lbtA* gene was shown to be expressed intracellularly, the mutants were not defective for macrophage or amoeba infection. The mutant was also not impaired for infection of the mouse lung, suggesting that legiobactin is not required for infection (Allard *et al.*, 2006). These data do not, however, indicate that siderophores are not relevant for *L. pneumophila*: a second siderophore could compensate for the loss of legiobactin, as is very often the case in other bacteria. Interestingly, FrgA mutants are defective in macrophages (Hickey and Cianciotto, 1997), supporting the notion that *L. pneumophila* has a second siderophore that promotes intracellular infection. The fact that mutations in *lbtAB* do not completely abolish CAS reactivity further suggests that additional iron chelators are secreted by *L. pneumophila*.

Concluding comments

With the discovery of legiobactin, we now have, after years of disbelief, evidence of a *Legionella* siderophore. It is likely that the novel influence of the bacterial inoculum on siderophore expression is the primary reason that legiobactin was not found earlier. It is reasonable that *L. pneumophila* produces a siderophore(s), since many other aquatic bacteria produce these iron scavengers.

Given the phenotype of *lbtAB* mutants, legiobactin is likely to be most critical for extracellular, rather than intracellular, growth; e.g. it might be vital for growth in aquatic environments such as biofilms. Clearly, further work is needed in this area as well as to address the question of whether *L. pneumophila* and the other legionellae export siderophores besides legiobactin.

Acknowledgements

The author wishes to thank both past and present lab members for all of their work and dedication. Our studies are supported by NIH grants AI34937 and AI43987.

References

Albert, C., Jacobi, S., De Buck, E., Lammertyn, E., and Heuner, K. (2006). Identification of target proteins of the Lss secretion system of *Legionella pneumophila* Corby. In *Legionella*: State of the art 30 years after its recognition, N.P. Cianciotto, Y. Abu Kwaik, P.H. Edelstein, B.S. Fields, D.F. Geary, T.G. Harrison, C.A. Joseph, R.M. Ratcliff, J.E. Stout, and M.S. Swanson, eds. (Washington, DC, ASM Press), pp. 221–223.

Allard, K.A., Viswanathan, V.K., and Cianciotto, N.P. (2006). *lbtA* and *lbtB* are required for production of the *Legionella pneumophila* siderophore legiobactin. J. Bacteriol. *188*, 1351–1363.

Aragon, V., Kurtz, S., and Cianciotto, N.P. (2001). *Legionella pneumophila* major acid phosphatase and its role in intracellular infection. Infect. Immun. *69*, 177–185.

Aragon, V., Kurtz, S., Flieger, A., Neumeister, B., and Cianciotto, N.P. (2000). Secreted enzymatic activities of wild-type and *pilD*-deficient *Legionella pneumophila*. Infect. Immun. *68*, 1855–1863.

Aragon, V., Rossier, O., and Cianciotto, N.P. (2002). *Legionella pneumophila* genes that encode lipase and phospholipase C activities. Microbiol. *148*, 2223–2231.

Banerji, S., Bewersdorff, M., Hermes, B., Cianciotto, N.P., and Flieger, A. (2005). Characterization of the major secreted zinc metalloprotease- dependent glycerophospholipid:cholesterol acyltransferase, PlaC, of *Legionella pneumophila*. Infect. Immun. *73*, 2899–2909.

Brand, B.C., Sadosky, A.B., and Shuman, H.A. (1994). The *Legionella pneumophila icm* locus: a set of genes required for intracellular multiplication in human macrophages. Mol. Microbiol. *14*, 797–808.

Brickman, T.J., and Armstrong, S.K. (2005). *Bordetella* AlcS transporter functions in alcaligin siderophore export and is central to inducer sensing in positive regulation of alcaligin system gene expression. J. Bacteriol. *187*, 3650–3661.

Brieland, J., McClain, M., Heath, L., Chrisp, C., Huffnagle, G., LeGendre, M., Hurley, M., Fantone, J., and Engleberg, C. (1996). Coinoculation with *Hartmannella vermiformis* enhances replicative *Legionella pneumophila* lung infection in a murine model of Legionnaires' disease. Infect. Immun. *64*, 2449–2456.

Bruggemann, H., Cazalet, C., and Buchrieser, C. (2006a). Adaptation of *Legionella pneumophila* to the host environment: role of protein secretion, effectors and eukaryotic-like proteins. Curr. Opin. Microbiol. *9*, 86–94.

Bruggemann, H., Hagman, A., Jules, M., Sismeiro, O., Dillies, M.A., Gouyette, C., Kunst, F., Steinert, M., Heuner, K., Coppee, J.Y., and Buchrieser, C. (2006b). Virulence strategies for infecting phagocytes deduced from the *in vivo* transcriptional program of *Legionella pneumophila*. Cell. Microbiol. *8*, 1228–1240.

Byrd, T.F., and Horwitz, M.A. (2000). Aberrantly low transferrin receptor expression on human monocytes is associated with nonpermissiveness for *Legionella pneumophila* growth. J. Infect. Dis. *181*, 1394–1400.

Cazalet, C., Rusniok, C., Bruggemann, H., Zidane, N., Magnier, A., Ma, L., Tichit, M., Jarraud, S., Bouchier, C., Vandenesch, F., Kunst, F., Etienne, J., Glaser, P., and Buchrieser, C. (2004). Evidence in the *Legionella pneumophila* genome for exploitation of host cell functions and high genome plasticity. Nat. Genet. *36*, 1165–1173.

Challis, G.L. (2005). A widely distributed bacterial pathway for siderophore biosynthesis independent of nonribosomal peptide synthetases. Chembiochem. *6*, 601–611.

Chien, M., Morozova, I., Shi, S., Sheng, H., Chen, J., Gomez, S.M., Asamani, G., Hill, K., Nuara, J., Feder, M., Rineer, J., Greenberg, J.J., Steshenko, V., Park, S.H., Zhao, B., Teplitskaya, E., Edwards, J.R., Pampou, S., Georghiou, A., Chou, I.C., Iannuccilli, W., Ulz, M.E., Kim, D.H., Geringer-Sameth, A., Goldsberry, C., Morozov, P., Fischer, S.G., Segal, G., Qu, X., Rzhetsky, A., Zhang, P., Cayanis, E., De Jong, P.J., Ju, J., Kalachikov, S., Shuman, H.A., and Russo, J.J. (2004). The genomic sequence of the accidental pathogen *Legionella pneumophila*. Science *305*, 1966–1968.

Cianciotto, N.P. (2005). Type II secretion: a protein secretion system for all seasons. Trends Microbiol. *13*, 581–588.

Cianciotto, N.P., Eisenstein, B.I., Mody, C.H., and Engleberg, N.C. (1990). A mutation in the *mip* gene results in an attenuation of *Legionella pneumophila* virulence. J. Infect. Dis. *162*, 121–126.

Cianciotto, N.P., Eisenstein, B.I., Mody, C.H., Toews, G.B., and Engleberg, N.C. (1989). A *Legionella pneumophila* gene encoding a species-specific surface protein potentiates initiation of intracellular infection. Infect. Immun. *57*, 1255–1262.

Cianciotto, N.P., and Fields, B.S. (1992). *Legionella pneumophila mip* gene potentiates intracellular infection of protozoa and human macrophages. Proc. Natl. Acad. Sci. USA *89*, 5188–5191.

Cianciotto, N.P., Kim Stamos, J., and Kamp, D.W. (1995). Infectivity of *Legionella pneumophila mip* mutant for alveolar epithelial cells. Curr. Microbiol. *30*, 247–250.

Cirillo, J.D., Cirillo, S.L.G., Yan, L., Bermudez, L.E., Falkow, S., and Tompkins, L.S. (1999). Intracellular growth in *Acanthamoeba castellani* affects monocyte entry mechansism and enhances virulence of *Legionella pneumophila*. Infect. Immun. *67*, 4427–4434.

De Buck, E., Lebeau, I., Maes, L., Geukens, N., Meyen, E., Van Mellaert, L., Anne, J., and Lammertyn, E. (2004). A putative twin-arginine translocation pathway in *Legionella pneumophila*. Biochem. Biophys. Res. Commun. *317*, 654–661.

De Buck, E., Maes, L., Meyen, E., Van Mellaert, L., Geukens, N., Anne, J., and Lammertyn, E. (2005). *Legionella pneumophila* Philadelphia-1 *tatB* and *tatC* affect intracellular replication and biofilm formation. Biochem. Biophys. Res. Commun. *331*, 1413–1420.

DebRoy, S., Aragon, V., Kurtz, S., and Cianciotto, N.P. (2006a). *Legionella pneumophila* Mip, a surface-exposed peptidylproline *cis-trans*-isomerase, promotes the presence of phospholipase C-like activity in culture supernatants. Infect. Immun. *74*, 5152–5260.

DebRoy, S., Dao, J., Soderberg, M., Rossier, O., and Cianciotto, N.P. (2006b). *Legionella pneumophila* type II secretome reveals unique exoproteins and a chitinase that promotes bacterial persistence in the lung. Proc. Natl. Acad. Sci. USA *103*, 19146–19151.

Deng, K., Blick, R.J., Liu, W., and Hansen, E.J. (2006). Identification of *Francisella tularensis* genes affected by iron limitation. Infect. Immun. *74*, 4224–4236.

Economou, A., Christie, P.J., Fernandez, R.C., Palmer, T., Plano, G.V., and Pugsley, A.P. (2006). Secretion by numbers: Protein traffic in prokaryotes. Mol. Microbiol. *62*, 308–319.

Filloux, A. (2004). The underlying mechanisms of type II protein secretion. Biochim. Biophys. Acta-Mol. Cell. Res. *1694*, 163–179.

Fischer, G., Bang, H., Ludwig, B., Mann, K., and Hacker, J. (1992). Mip protein of *Legionella pneumophila* exhibits peptidyl-prolyl-cis/trans isomerase (PPlase) activity. Mol. Microbiol. *6*, 1375–1383.

Flieger, A., Gong, S., Faigle, M., Northoff, H., and Neumeister, B. (2001a). In vitro secretion kinetics of proteins from *Legionella pneumophila* in comparison to proteins from non-*pneumophila* species. Microbiology *147*, 3127–3134.

Flieger, A., Gong, S., Faigle, M., Stevanovic, S., Cianciotto, N.P., and Neumeister, B. (2001b). Novel lysophospholipase A secreted by *Legionella pneumophila*. J. Bacteriol. *183*, 2121–2124.

Flieger, A., Neumeister, B., and Cianciotto, N.P. (2002). Characterization of the gene encoding the major secreted lysophospholipase A of *Legionella pneumophila*and its role in detoxification of lysophosphatidylcholine. Infect. Immun. *70*, 6094–6106.

Franza, T., Mahe, B., and Expert, D. (2005). *Erwinia chrysanthemi* requires a second iron transport route dependent of the siderophore achromobactin for extracellular growth and plant infection. Mol. Microbiol. *55*, 261–275.

Furrer, J.L., Sanders, D.N., Hook-Barnard, I.G., and McIntosh, M.A. (2002). Export of the siderophore enterobactin in *Escherichia coli*: involvement of a 43 kDa membrane exporter. Mol. Microbiol. *44*, 1225–1234.

Gebran, S.J., Newton, C., Yamamoto, Y., Widen, R., Klein, T.W., and Friedman, H. (1994). Macrophage permissiveness for *Legionella pneumophila* growth modulated by iron. Infect. Immun. *62*, 564–568.

George, J.R., Pine, L., Reeves, M.W., and Harrell, W.K. (1980). Amino acid requirements of *Legionella pneumophila*. J. Clin. Microbiol. *11*, 286–291.

Geukens, N., De Buck, E., Meyen, E., Maes, L., Vranckx, L., Van Mellaert, L., Anne, J., and Lammertyn, E. (2006). The type II signal peptidase of *Legionella pneumophila*. Res. Microbiol. *157*, 836–841.

Goldoni, P., Visca, P., Pastoris, M.C., Valenti, P., and Orsi, N. (1991). Growth of *Legionella* spp. under conditions of iron restriction. J. Med. Microbiol. *34*, 113–118.

Hales, L.M., and Shuman, H.A. (1999). *Legionella pneumophila* contains a type II general secretion pathway required for growth in amoebae as well as for secretion of the Msp protease. Infect. Immun. *67*, 3662–3666.

Helbig, J.H., Luck, P.C., Steinert, M., Jacobs, E., and Witt, M. (2001). Immunolocalization of the Mip protein of intracellularly and extracellularly grown *Legionella pneumophila*. Lett. Appl. Microbiol. *32*, 83–88.

Hickey, E.K., and Cianciotto, N.P. (1994). Cloning and sequencing of the *Legionella pneumophila fur* gene. Gene *143*, 117–121.

Hickey, E.K., and Cianciotto, N.P. (1997). An iron- and fur-repressed *Legionella pneumophila* gene that promotes intracellular infection and encodes a protein with similarity to the *Escherichia coli* aerobactin synthetases. Infect. Immun. *65*, 133–143.

Jacobi, S., and Heuner, K. (2003). Description of a putative type I secretion system in *Legionella pneumophila*. Int. J. Med. Microbiol. *293*, 349–358.

James, B.W., Mauchline, W.S., Dennis, P.J., and Keevil, C.W. (1997). A study of iron acquisition mechanisms of *Legionella pneumophila* grown in chemostat culture. Curr. Microbiol. *34*, 238–243.

James, B.W., Mauchline, W.S., Fitzgeorge, R.B., Dennis, P.J., and Keevil, C.W. (1995). Influence of iron-limited continuous culture on physiology and virulence of *Legionella pneumophila*. Infect. Immun. *63*, 4224–4230.

Johnson, T.L., Abendroth, J., Hol, W.G., and Sandkvist, M. (2006). Type II secretion: from structure to function. FEMS Microbiol. Lett. *255*, 175–186.

Johnson, W., Varner, L., and Poch, M. (1991). Acquisition of iron by *Legionella pneumophila*: role of iron reductase. Infect. Immun. *59*, 2376–2381.

Kalogeraki, V.S., and Winans, S.C. (1998). Wound-released chemical signals may elicit multiple responses from an *Agrobacterium tumefaciens* strain containing an octopine-type Ti plasmid. J. Bacteriol. *180*, 5660–5667.

Koebnik, R. (2005). TonB-dependent trans-envelope signalling: the exception or the rule? Trends Microbiol. *13*, 343–347.

Kohler, R., Fanghanel, J., Konig, B., Luneberg, E., Frosch, M., Rahfeld, J.U., Hilgenfeld, R., Fischer, G., Hacker, J., and Steinert, M. (2003). Biochemical and functional analyses of the Mip protein: influence of the N-terminal half and of peptidylprolyl isomerase activity on the virulence of *Legionella pneumophila*. Infect. Immun. *71*, 4389–4397.

Kostakioti, M., Newman, C.L., Thanassi, D.G., and Stathopoulos, C. (2005). Mechanisms of protein export across the bacterial outer membrane. J. Bacteriol. *187*, 4306–4314.

Lammertyn, E., and Anne, J. (2004). Protein secretion in *Legionella pneumophila* and its relation to virulence. FEMS Microbiol. Lett. *238*, 273–279.

Lammertyn, E., Van Mellaert, L., Meyen, E., Lebeau, I., De Buck, E., Anne, J., and Geukens, N. (2004). Molecular and functional characterization of type I signal peptidase from *Legionella pneumophila*. Microbiol. *150*, 1475–1483.

Lee, P.A., Tullman-Ercek, D., and Georgiou, G. (2006). The bacterial twin-arginine translocation pathway. Annu. Rev. Microbiol. *60*, 373–395.

Lee, V.T., and Schneewind, O. (2001). Protein secretion and the pathogenesis of bacterial infections. Genes Dev *15*, 1725–1752.

Li, J., Wolf, S.G., Elbaum, M., and Tzfira, T. (2005). Exploring cargo transport mechanics in the type IV secretion systems. Trends Microbiol. *13*, 295–298.

Liles, M.R., Aber Scheel, T., and Cianciotto, N.P. (2000). Discovery of a nonclassical siderophore, legiobactin, produced by strains of *Legionella pneumophila*. J. Bacteriol. *182*, 749–757.

Liles, M.R., and Cianciotto, N.P. (1996). Absence of siderophore-like activity in *Legionella pneumophila* supernatants. Infect. Immun. *64*, 1873–1875.

Liles, M.R., Edelstein, P.H., and Cianciotto, N.P. (1999). The prepilin peptidase is required for protein secretion by and the virulence of the intracellular pathogen *Legionella pneumophila*. Mol. Microbiol. *31*, 959–970.

Liles, M.R., Viswanathan, V.K., and Cianciotto, N.P. (1998). Identification and temperature regulation of *Legionella pneumophila* genes involved in type IV pilus biogenesis and type II protein secretion. Infect. Immun. *66*, 1776–1782.

Ludwig, B., Rahfeld, J., Schmidt, B., Mann, K., Wintermeyer, E., Fischer, G., and Hacker, J. (1994). Characterization of Mip proteins of *Legionella pneumophila*. FEMS Microbiol. Letts. *118*, 23–30.

Matthews, M., and Roy, C.R. (2000). Identification and subcellular localization of the *Legionella pneumophila* IcmX protein: a factor essential for establishment of a replicative organelle in eukaryotic host cells. Infect. Immun. *68*, 3971–3982.

Mengaud, J.M., and Horwitz, M.A. (1993). The major iron-containing protein of *Legionella pneumophila* is an aconitase homologous with the human iron-responsive element-binding protein. J. Bacteriol. *175*, 5666–5676.

Merino, S., Aguilar, A., Nogueras, M.M., Regue, M., Swift, S., and Tomas, J.M. (1999). Cloning, sequencing, and role in virulence of two phospholipases (A1 and C) from mesophilic *Aeromonas* sp. serogroup O:34. Infect. Immun. *67*, 4008–4013.

.Molmeret, M., Alli, O.A., Zink, S., Flieger, A., Cianciotto, N.P., and Kwaik, Y.A. (2002). icmT is essential for pore formation-mediated egress of *Legionella pneumophila* from mammalian and protozoan cells. Infect. Immun. *70*, 69–78.

Ochsner, U.A., Snyder, A., Vasil, A.I., and Vasil, M.L. (2002). Effects of the twin-arginine translocase on secretion of virulence factors, stress response, and pathogenesis. Proc. Natl. Acad. Sci. USA *99*, 8312–8317.

Peabody, C.R., Chung, Y.J., Yen, M.R., Vidal-Ingigliardi, D., Pugsley, A.P., and Saier, M.H., Jr. (2003). Type II protein secretion and its relationship to bacterial type IV pili and archaeal flagella. Microbiol. *149*, 3051–3072.

Poch, M.T., and Johnson, W. (1993). Ferric reductases of *Legionella pneumophila*. Biometals *6*, 107–114.

Polesky, A.H., Ross, J.T., Falkow, S., and Tompkins, L.S. (2001). Identification of *Legionella pneumophila* genes important for infection of amoebas by signature-tagged mutagenesis. Infect. Immun. *69*, 977–987.

Reeves, M.W., Pine, L., Hutner, S.H., George, J.R., and Harrell, W.K. (1981). Metal requirements of *Legionella pneumophila*. J. Clin. Microbiol. *13*, 688–695.

Reeves, M.W., Pine, L., Neilands, J.B., and Balows, A. (1983). Absence of siderophore activity in *Legionella* species grown in iron-deficient media. J. Bacteriol. *154*, 324–329.

Riboldi-Tunnicliffe, A., Konig, B., Jessen, S., Weiss, M.S., Rahfeld, J., Hacker, J., Fischer, G., and Hilgenfeld, R. (2001). Crystal structure of Mip, a prolylisomerase from *Legionella pneumophila*. Nature Struct. Biol. *8*, 779–783.

Ridenour, D.A., Cirillo, S.L., Feng, S., Samrakandi, M.M., and Cirillo, J.D. (2003). Identification of a gene that affects the efficiency of host cell infection by *Legionella pneumophila* in a temperature-dependent fashion. Infect. Immun. *71*, 6256–6263.

Robson, A., and Collinson, I. (2006). The structure of the Sec complex and the problem of protein translocation. EMBO Rep. *7*, 1099–1103.

Rossier, O., and Cianciotto, N.P. (2001). Type II protein secretion is a subset of the PilD-dependent processes that facilitate intracellular infection by *Legionella pneumophila*. Infect. Immun. *69*, 2092–2098.

Rossier, O., and Cianciotto, N.P. (2005). The *Legionella pneumophila tatB* gene facilitates secretion of phospholipase C, growth under iron-limiting conditions, and intracellular infection. Infect. Immun. *73*, 2020–2032.

Rossier, O., Starkenburg, S., and Cianciotto, N.P. (2004). *Legionella pneumophila* type II protein secretion promotes virulence in the A/J mouse model of Legionnaires' disease pneumonia. Infect. Immun. *72*, 310–321.

Sargent, F., Berks, B.C., and Palmer, T. (2006). Pathfinders and trailblazers: a prokaryotic targeting system for transport of folded proteins. FEMS Microbiol. Lett. *254*, 198–207.

Segal, G., Russo, J.J., and Shuman, H.A. (1999). Relationships between a new type IV secretion system and the *icm/dot* virulence system of *Legionella pneumophila*. Mol. Microbiol. *34*, 799–809.

Söderberg, M.A., Rossier, O., and Cianciotto, N.P. (2004). The type II protein secretion system of *Legionella pneumophila* promotes growth at low temperatures. J. Bacteriol. *186*, 3712–3720.

Starkenburg, S.R., Casey, J.M., and Cianciotto, N.P. (2004). Siderophore activity among members of the *Legionella* genus. Curr. Microbiol. *49*, 203–207.

Sullivan, J.T., Jeffery, E.F., Shannon, J.D., and Ramakrishnan, G. (2006). Characterization of the siderophore of Francisella tularensis and role of *fslA* in siderophore production. J. Bacteriol. *188*, 3785–8795.

Terada, L.S., Johansen, K.A., Nowbar, S., Vasil, A.I., and Vasil, M.L. (1999). *Pseudomonas aeruginosa* hemolytic phospholipase C suppresses neutrophil respiratory burst activity. Infect. Immun. *67*, 2371–2376.

Tesh, M.J., Morse, S.A., and Miller, R.D. (1983). Intermediary metabolism in *Legionella pneumophila*: utilization of amino acids and other compounds as energy sources. J. Bacteriol. *154*, 1104–1109.

van der Sluis, E.O., and Driessen, A.J. (2006). Stepwise evolution of the Sec machinery in Proteobacteria. Trends Microbiol. *14*, 105–108.

Viswanathan, V.K., Edelstein, P.H., Pope, C.D., and Cianciotto, N.P. (2000). The *Legionella pneumophila iraAB* locus is required for iron assimilation, intracellular infection, and virulence. Infect. Immun. *68*, 1069–1079.

Weiss, E., Peacock, M.G., and Williams, J.C. (1980). Glucose and Glutamate Metabolism of *Legionella pneumophila*. Curr. Microbiol. *4*, 1–6.

Wintermeyer, E., Ludwig, B., Steinert, M., Schmidt, B., Fischer, G., and Hacker, J. (1995). Influence of site specifically altered Mip proteins on intracellular survival of *Legionella pneumophila* in eukaryotic cells. Infect. Immun. *63*, 4576–4583.

Zink, S.D., Pedersen, L., Cianciotto, N.P., and Abu Kwaik, Y. (2002). The Dot/Icm type IV secretion system of *Legionella pneumophila* is essential for the induction of apoptosis in human macrophages. Infect. Immun. *70*, 1657–1663.

The Dot/Icm Type IVB Secretion System of *Legionella*

9

Carr D. Vincent and Joseph P. Vogel*

Abstract

The major virulence system of *Legionella pneumophila* is a specialized secretion system encoded by 26 *dot/icm* genes. This secretion machine is classified as a type IVB secretion system due to its homology to the transfer apparatus of IncI conjugative plasmids. In *Legionella*, the Dot/Icm secretion system is used to export a large number of protein substrates into the cytoplasm of phagocytic cells, thereby allowing *Legionella* to survive and replicate within these normally bactericidal cells.

Legionella introduction

Legionella pneumophila is a Gram-negative bacterium that is found in the environment as a parasite of freshwater amoebae (Fields *et al.*, 2002). Unlike most bacteria, which are rapidly killed after being phagocytosed by an amoeba, *L. pneumophila* cells are able to survive in these eukaryotic hosts and replicate, eventually killing the host cell. When humans inhale aerosolized water droplets containing these bacteria or infected amoebae, the bacteria can cause either a mild disease with flu-like symptoms known as Pontiac fever or a severe form of pneumonia known as Legionnaires' disease.

L. pneumophila is able to cause disease because it can prevent being killed by human alveolar macrophages, immune cells that are normally highly bacteriocidal (Horwitz and Silverstein, 1980). The life cycle of *L. pneumophila* in macrophages appears to be similar to its life cycle in amoebae. Immediately upon uptake by a macrophage, the *Legionella* containing vacuole (LCV) avoids fusion with endocytic vesicles and lysosomes (Horwitz, 1983b). Instead, the LCV recruits endoplasmic reticulum (ER)-derived vesicles, acquiring the ER protein markers Rab1, Sec22b, Arf1, and calnexin (Kagan and Roy, 2002; Derre and Isberg, 2004; Kagan *et al.*, 2004). After transiently associating with mitochondria, the LCV becomes surrounded by rough ER, forming a compartment known as the replicative phagosome (Horwitz, 1983a; Swanson and Isberg, 1995). It is in this compartment that the bacteria multiply. During the replication cycle in A/J mouse bone marrow-derived macrophages, the LCV becomes acidified and acquires lysosomal markers (Sturgill-Koszycki and Swanson, 2000). Thereafter, the bacteria lyse the macrophage and spread to uninfected host cells.

Several genetic screens have been performed to identify virulence factors required for *L. pneumophila* to grow in macrophages (Berger and Isberg, 1993; Sadosky *et al.*, 1993; Vogel *et al.*, 1996; Andrews *et al.*, 1998; Segal *et al.*, 1998; Vogel *et al.*, 1998). These screens primarily identified a set of genes necessary for avoidance of the endocytic pathway that were named *dot* (defect in organelle trafficking) or *icm* (intracellular multiplication). These genes are linked in two pathogenicity islands on the *L. pneumophila* chromosome and were subsequently shown to encode a specialized secretion system called a type IV secretion system (T4SS) that is related to the conjugation system of an IncI plasmid

*Corresponding author.

(Segal *et al.*, 1998; Vogel *et al.*, 1998). This specialized secretion system is the focus of this review.

Type IV secretion systems and type IV coupling proteins

T4SSs include both plasmid transfer systems and adapted conjugation systems utilized by pathogens to deliver proteins and/or DNA into eukaryotic host cells. As plasmid transfer systems are better characterized, it is worth reviewing what is known about these systems prior to discussing adapted conjugation systems. Components involved in plasmid transfer systems can be grouped into three classes of proteins; Mpf, Dtr, and T4CP (Schroder and Lanka, 2005). The Mpf (mating pair formation) proteins are involved in assembling the conjugal pilus and the channel structure that spans the bacterial inner and outer membranes. The Dtr (DNA transfer and replication) proteins are involved in plasmid replication and generation of the protein-DNA substrate that is delivered to recipient cells. Finally, T4CPs (type IV coupling proteins) are inner membrane proteins that mediate interactions between the Mpf and Dtr proteins, 'coupling' the substrates to the export apparatus. All T4CPs contain an ATPase domain that is required for their function, suggesting that these proteins provide energy for substrate export.

The T4CPs perform several critical roles in type IV secretion systems. First, they link substrates to the Mpf export apparatus by functioning as the inner membrane receptors for Mpf complexes. Second, T4CPs play a critical role in determining the specificity of the Mpf complex for a particular substrate or class of substrates. Third, they are thought to play an active role in substrate translocation by functioning as molecular motors, providing energy through ATP hydrolysis to drive export of substrates.

T4CPs have been shown to interact with both their cognate Dtr and Mpf proteins. Direct interactions between several plasmid transfer coupling proteins and their substrates (a relaxase that is covalently attached to the single-stranded DNA substrate) have been experimentally demonstrated (Cabezon *et al.*, 1997; Hamilton *et al.*, 2000; Szpirer *et al.*, 2000; Schroder *et al.*, 2002; Atmakuri *et al.*, 2003). Coupling proteins have also been shown to interact with Mpf proteins. In *A. tumefaciens*, the coupling protein VirD4 has been shown to interact with the core subcomplex protein VirB10, as well as with the ATPases VirB4 and VirB11 (Atmakuri *et al.*, 2004; de Paz *et al.*, 2005). Likewise, in the plasmid transfer systems R27 and R388, the coupling proteins TraG and TrwB have been shown to interact with the respective VirB10 homologues of these systems, TrhB and TrwE (Gilmour *et al.*, 2003; Llosa *et al.*, 2003). Thus, there is a growing body of evidence that the coupling proteins are the inner membrane receptors for secreted substrates of T4SS, linking substrates to the secretion apparatus.

Several studies indicate that the T4CP is the primary factor that determines the specificity of the secretion system for a particular substrate (Cabezon *et al.*, 1994; Hamilton *et al.*, 2000; Llosa *et al.*, 2003). This has been demonstrated in studies using chimeric secretion systems, consisting of a T4CP from one system expressed with the Mpf complex from another. For example, Cabezon and colleagues investigated the specificities of the R388 and RP4 plasmid transfer systems (Cabezon *et al.*, 1994). The R388 plasmid transfer system can transfer both the R388 plasmid and the broad host range plasmid RSF1010. Similarly, the RP4 transfer system can transfer both the RP4 plasmid and RSF1010. When the coupling protein from RP4 (TraG) was expressed with the Mpf complex from R388, this hybrid T4SS retained activity, as it was able to transfer the RSF1010 plasmid. However, the hybrid T4SS could no longer transfer the native substrate of the R388 Mpf, the R388 plasmid. Thus, the Mpf complex does not confer substrate specificity, nor does it exhibit specificity for a particular T4CP. However, the individual coupling proteins may interact with particular substrates, demonstrating that the specificity of the transfer system is determined solely by the coupling protein.

The third proposed function of T4CPs in substrate secretion is that they serve as molecular motors, providing mechanical energy through hydrolysis of ATP to 'pump' substrates through the complex. In support of this proposed role, T4CPs exhibit limited sequence similarity to the DNA motor proteins FtsK and SpoIIIE, both of which are involved in chromosome segregation (Llosa *et al.*, 2002). These proteins have been shown to translocate along DNA in the presence of ATP, suggesting a similar mechanism for T4CPs in plasmid translocation. T4CPs contain highly conserved Walker box motifs, indicating that ATP hydrolysis is likely to be essential for their function (Walker *et al.*, 1982). DNA dependent ATP hydrolysis activity was recently demonstrated for the R388 coupling protein TrwB and several T4CPs have been shown to require their Walker box motifs for plasmid transfer (Balzer *et al.*, 1994; Moncalian *et al.*, 1999; Kumar and Das, 2002; Tato *et al.*, 2005). Moreover, the crystal structure of TrwB reveals that the protein forms a hexamer with a quaternary structure similar to F_1-ATPase (Gomis-Ruth *et al.*, 2001). The hexamer has a central channel approximately 22 Angstroms in width that constricts to seven Angstroms at the cytoplasmic face of the protein. It has been proposed that this channel may serve as the conduit through which secreted proteins and DNA are pumped by the T4CP across the inner membrane.

Adapted conjugation systems

In addition to their function as plasmid transfer systems, T4SS contribute to the virulence of a variety of bacterial pathogens, including the plant pathogen *Agrobacterium tumefaciens* and the mammalian pathogens *Bartonella henselae, Bordetella pertussis, Coxiella burnetii, Helicobacter pylori*, and *Legionella pneumophila* (Cascales and Christie, 2003). These T4SS, termed adapted conjugation systems, are used by pathogens to deliver toxins into host cells and are thought to have evolved from plasmid transfer systems, as they have significant sequence homology (Winans *et al.*, 1996; Christie and Vogel, 2000). In the case of the *L. pneumophila* Dot/Icm T4SS, this evolutionary relationship is clear, as the secretion system retains the ability to transfer the broad host-range RSF1010 plasmid to a bacterial recipient (Segal *et al.*, 1998; Vogel *et al.*, 1998).

Adapted conjugation systems can be divided into two groups, type IVA and type IVB, based on sequence homology (Christie and Vogel, 2000). Although both classes are thought to be functionally homologous, they share very little sequence similarity and type IVB systems are typically composed of approximately twice as many proteins. The best-characterized type IV secretion systems are the type IVA systems, which include the extensively characterized *A. tumefaciens* VirB/D4 secretion system and related T4SS. The best-characterized type IVB secretion system is the *Legionella pneumophila* Dot/Icm system.

The *A. tumefaciens* system is composed of twelve proteins, VirB1–11 and VirD4 (Christie *et al.*, 2005; Schroder and Lanka, 2005). Much is known about the assembly and structure of the proteins that make up these secretion machines, including protein interactions between subunits, the biochemical functions of individual components, and even the translocation route of substrates as they pass through the secretion apparatus. This secretion system contains three energy transducing proteins: VirB4, VirB11, and the coupling protein VirD4. Each of these proteins contains a conserved Walker box nucleoside triphosphate binding motif that is required for substrate secretion. The 'core' of the secretion complex is composed of the VirB6, VirB7, VirB8, VirB9, and VirB10 proteins. Three of these proteins, VirB6, VirB8, and VirB10 are inner membrane proteins; VirB9 and the lipoprotein VirB7 are found in the outer membrane linked to each other by a disulfide bond. Conformation of VirB10 has been shown to change in response to ATP hydrolysis by the VirB11 and VirD4 proteins. Accordingly, VirB10 is thought to function as an energy 'sensor' that regulates the formation of a bridge between the inner and outer membrane components (Cascales and Christie, 2004). The secretion system also assembles a pilus, composed of VirB2 (the pilin) and a minor component VirB5. The pilus is thought to be attached to the core complex through interactions with the lipoprotein VirB7.

It should be noted that two adapted conjugation systems that lack a coupling protein have been identified, the *Brucella* spp. Vir T4SS and the *Bordetella pertussis* Ptl T4SS. Little is known about the *Brucella* T4SS due to the intractability of this organism to genetic analysis. It is known, however, that the *B. pertussis* Ptl substrate, pertussis toxin, contains a Sec signal sequence and is transported across the cytoplasmic membrane by the Sec machinery prior to secretion across the outer membrane by the Ptl T4SS (Burns, 2003). Thus, this exceptional case hints that the function of the T4CPs in other systems is similar to the role of the Sec machinery in transporting substrates across the cytoplasmic membrane (Schroder and Lanka, 2005).

In contrast to the well-characterized type IVA system of *A. tumefaciens*, relatively little is known about the type IVB systems. Type IVB systems include the Dot/Icm T4SS of *L. pneumophila* and *Coxiella burnetii*, as well as the plasmid transfer systems R64 and ColIb-P9 (Christie and Vogel, 2000; Sexton and Vogel, 2002). Only three Dot/Icm proteins have detectable homology to the type IVA systems: DotB is homologous to VirB11, the carboxy-terminus of DotG is homologous to VirB10, and DotL has limited homology to the coupling protein VirD4. The remaining Dot/Icm proteins have no homology to the VirB/D4 system, making predictions of protein functions difficult.

Overview of the *Legionella dot/icm* genes

Twenty-six *dot/icm* genes have been identified. Table 9.1 includes a description of the Dot/Icm proteins including their size, predicted localization, experimentally determined localization via fractionation experiments, other relevant information, and key references. Following is a brief description of the Dot/Icm proteins that have been characterized to date.

DotA

DotA was the first Dot protein identified (Berger and Isberg, 1993). Historically, a *dotA* mutant (Lp03) has served as the 'standard' *dot/icm* mutant in many laboratories, as this mutant is completely defective for all of the known *dot/icm* phenotypes including intracellular growth, alteration of the host endocytic pathway, contact dependent cytotoxicity, and plasmid conjugation. DotA is a large (113 kDa) protein with eight transmembrane domains (Roy and Isberg, 1997). Surprisingly, this protein has been shown to be secreted by wild-type *L. pneumophila* during growth in broth (Nagai and Roy, 2001). Secreted DotA forms ring-shaped structures, and it has been speculated that these structures may form channels in host membranes for the delivery of effectors. However, it is unclear how DotA could function as both an inner membrane component of the Dot/Icm secretion apparatus and as a secreted substrate. Moreover, there is no precedent for secretion of a polytopic inner membrane protein by any known bacterial secretion system. Thus, given the present evidence, it is unlikely that DotA is a secreted substrate of the Dot/Icm T4SS. Instead, release of DotA is likely due to shedding of membrane proteins by *L. pneumophila* during growth in liquid media.

DotB

One of the most extensively characterized Dot/Icm proteins is DotB, a distant homologueof the *A. tumefaciens* protein VirB11. In *A. tumefaciens*, VirB11 is thought to provide energy to the secretion machinery and to be involved in assembly of the secretion apparatus, pilus assembly, and substrate export (Sagulenko *et al.*, 2001; Schroder and Lanka, 2005). DotB is one of three Dot/Icm proteins with a Walker box motif, suggesting that DotB is involved in providing energy to the complex through hydrolysis of ATP. DotB was shown to have ATPase activity, form hexameric rings, and to be peripherally associated with the inner membrane, presumably bound to inner membrane components of the Dot/Icm complex (Sexton *et al.*, 2004b). Although DotB interaction partners have not yet been identified, the DotL protein is a likely candidate based on a report demonstrating interactions between VirB11 and VirD4, a DotL homologue(Atmakuri *et al.*, 2004). Three roles have been proposed for DotB (Sexton *et al.*, 2005). The first role is in assembly of the Dot/Icm complex. Although this has not been definitively demonstrated, this role of DotB seems likely based on the function of its homolog, the VirB11 protein in *A. tumefaciens*. The second function of DotB is in protein export. This role has been confirmed, as single amino acid changes in DotB conferred defects in the export of a subset of Dot/Icm substrates (Sexton *et al.*, 2005). Third, DotB may have a role in retraction of the Dot/Icm pilus, as DotB has homology to PilT, a protein involved in retraction of pili in type II secretion systems. This activity has not been confirmed, since the Dot/Icm pilus has not been identified.

Table 9.1 General information about the *L. pneumophila* Dot/Icm proteins

Protein[a]	Size (aa)	#kDa	Predicted localization[b]	Localization (wild type)[c]	Information[d]	Reference[e]
DotA	1048	113	IM[f]	IM	Polytopic membrane protein	(Roy and Isberg, 1997)
DotB	377	42	Cytoplasm	Cytoplasm	ATPase/PilT and VirB11[g] homology	(Sexton *et al.*, 2004b)
DotC	303	34	IM or OM[h]	OM	Putative lipoprotein	(Yerushalmi *et al.*, 2005)
DotD	163	18	IM or OM	OM	Putative lipoprotein	(Yerushalmi *et al.*, 2005)
DotE/IcmC	194	20	IM		Putative pilin	
DotF/IcmG	269	30	IM	OM	Interacts with secreted substrates	(Luo and Isberg, 2004)
DotG/IcmE	1048	108	IM	OM	VirB10[g] homology	
DotH/IcmK	360	39	Periplasm	Periplasm and OM		(Andrews *et al.*, 1998)
DotI/IcmL	212	23	IM	IM		(Andrews *et al.*, 1998)
DotJ/IcmM	94	11	IM			
DotK/IcmN	189	21	IM or OM	OM	Putative lipoprotein	(Yerushalmi *et al.*, 2005)
DotL/IcmO	783	87	IM	IM	VirD4[g] homology/T4CP/ essential gene	(Buscher *et al.*, 2005)
DotM/IcmP	376	43	IM	IM	Essential gene	
DotN/IcmJ	208	24	Cytoplasm or IM	Cytoplasm and IM	Essential gene	
DotO/IcmB	1009	112	Cytoplasm	IM		(Andrews *et al.*, 1998)
DotP/IcmD	132	14	IM	IM	Putative pilin	
DotU	261	30	IM	IM	Accessory factor (interacts w/IcmF)	(Sexton *et al.*, 2004a)
DotV	180	20	IM	IM	Homology to DotE	(Sexton *et al.*)
IcmF	973	111	IM	IM	Accessory factor (interacts w/DotU)	(Sexton *et al.*, 2004a)
IcmQ	191	22	Cytoplasm	Cytoplasm	Pore forming molecule	(Dumenil and Isberg, 2001; Dumenil *et al.*, 2004)
IcmR	120	13	Cytoplasm	Cytoplasm	Chaperone for IcmQ	(Dumenil and Isberg, 2001; Dumenil *et al.*, 2004)
IcmS	114	13	Cytoplasm	Cytoplasm	Adaptor for substrates	(Coers *et al.*, 2000; Ninio *et al.*, 2005)
IcmT	86	9.9	IM			(Bitar *et al.*, 2005)
IcmV	151	18	IM			

Table 9.1 *continued*

Protein[a]	Size (aa)	#kDa	Predicted localization[b]	Localization (wild type)[c]	Information[d]	Reference[e]
IcmW	151	17	Cytoplasm	Cytoplasm	Adaptor for substrates	(Coers *et al*., 2000; Ninio *et al*., 2005)
IcmX	466	51	Periplasm	Periplasm		(Matthews and Roy, 2000)
LvgA	208	27	Cytoplasm	Cytoplasm	Adaptor for substrates	(Vincent and Vogel, 2006)

[a]The protein name is shown as Dot or Icm if only one names exists or as a Dot/Icm if both names exist. For the sake of brevity, this review uses the only the Dot name when appropriate.
[b]Based on analysis of a Kyte–Doolittle hydrophilicity plot.
[c]Experimental determination by ultracentrifugation and sucrose gradient separation.
[d]Homology and/or motifs.
[e]Only most relevant references listed.
[f]Inner membrane
[g]*Agrobacterium tumefaciens* type IV secretion protein.
[h]Outer membrane.

DotL

The DotL protein has two predicted transmembrane domains at its amino-terminus and has been experimentally localized to the inner membrane (Buscher *et al.*, 2005). It also contains a highly conserved Walker box motif, suggesting ATP binding and hydrolysis is important for the function of DotL. DotL has limited homology to type IV coupling proteins (Buscher *et al.*, 2005). For example, DotL is similar to a well characterized T4CP, TrwB, from the plasmid R388, although this low level of homology (BLAST score 7e-2) is not surprising given the generally low sequence homology between members of the T4CP family. As the type IV coupling protein, DotL is thus predicted to perform two critical functions in the secretion apparatus: (1) as the receptor for secreted substrates, and (2) to provide energy to the secretion apparatus to drive substrate translocation.

An unusual property of DotL is that it is not only essential for growth inside macrophages, it is also essential for growth on bacteriological media in the wild-type *L. pneumophila* strain Lp02 (Buscher *et al.*, 2005). This phenotype is not shared by any other known type IV coupling protein. In addition to *dotL*, two other *dot/icm* genes are essential, *dotM* and *dotN*. Although the functions of these proteins are not known, the conservation of this unusual phenotype suggests that DotL, DotM, and DotN may function together as a subcomplex (discussed further below). Interestingly, although the *dotL* gene is essential in a wild-type Lp02 strain background, suppressor mutations of the Δ*dotL* lethality phenotype can easily be obtained. The high frequency of spontaneous suppressor mutations suggests that there are a large number of genes that will suppress the loss of *dotL*, and it has been demonstrated by our laboratory that mutations in 20 of the 23 non-essential *dot/icm* genes can suppress the loss of *dotL* (Buscher *et al.*, 2005). To explain these observations, it was proposed that DotL may regulate the secretion channel and control whether the complex is open or closed. Loss of DotL could result in a constitutively open complex, which would be detrimental to the cell. Suppressor mutations in other *dot/icm* genes that disrupt the secretion pore would rescue the detrimental effects of the loss of the channel 'gatekeeper.' Thus, the surprising phenotype of Δ*dotL* lethality distinguishes it from other coupling proteins, but at the same time supports the model that coupling proteins regulate secretion by the T4SSs.

Interestingly, a *dotL* deletion is not lethal in another wild-type *L. pneumophila* strain, JR32, although the mutant cells become hypersensitive to low salt concentrations on agar medium (Buscher *et al.*, 2005). Because both Lp02 and JR32 were derived from the same clinical isolate, Philadelphia-1, it is currently unknown how this difference between strains arose. One possible explanation is that during passage *in vitro*, Lp02 acquired a mutation that confers Δ*dotL* lethality. Alternatively, explanation JR32 may have acquired a mutation that mediates resistance to Δ*dotL* lethality.

The Δ*dotL* lethality phenotype has proven useful in a number of studies. In the Vogel laboratory, a screen to identify suppressors of Δ*dotL* lethality identified many known *dot/icm* genes as well as a novel *dot/icm* gene, *dotV* (J.A. Sexton and J.P. Vogel, unpublished). In addition, Δ*dotL* lethality was used as a selection for dominant negative alleles of DotB, aiding in the study of the structure and function of this protein (Sexton *et al.*, 2005). We have also used the Δ*dotL* lethality phenotype to identify factors necessary for the assembly and/or activation of the Dot/Icm complex (Vincent *et al.*, 2006a). In summary, Δ*dotL* lethality supports the current model for the role of coupling proteins as regulators of type IV secretion and has proven useful in the identification and characterization of the Dot/Icm proteins.

DotO

DotO has been proposed to be a functional homologueof the *A. tumefaciens* VirB4 protein, an inner membrane ATPase that is required for assembly of the secretion apparatus (Schroder and Lanka, 2005; Segal *et al.*, 2005). Like DotB and DotL, DotO contains a Walker box domain, indicating that this protein is likely to provide energy to the Dot/Icm apparatus through hydrolysis of ATP. Although DotO does not contain any clear transmembrane motifs, it localizes exclusively to the inner membrane (Vincent *et al.*, 2006b). Interestingly, DotO and DotH were observed by immunofluorescence microscopy in a fibrous structure on the surface of *L. pneumophila* after exposure to macrophages or to macrophage-conditioned medium (Watarai *et al.*, 2001). This surface structure was also observed late in the intracellular growth cycle, shortly before the bacteria escape from the host cell. Based on this, it was proposed that this structure might be involved in escape from the host cell or in infection of new host cells. However, an alternate explanation for the exposure of DotO and DotH is that after infection of macrophages, wild-type *L. pneumophila* cells become susceptible to the fixation conditions used for microscopy, and fixation results in the release of intracellular contents. Unfortunately, it is difficult to test this hypothesis and the reason for the surface exposure of DotO and DotH remains unclear.

DotU/IcmF

DotU and IcmF are somewhat misleadingly named, as these proteins are not thought to be part of the Dot/Icm secretion apparatus itself, but rather function as accessory factors that are required for stabilization of other Dot/Icm proteins (Sexton *et al.*, 2004a; VanRheenen *et al.*, 2004). Supporting this idea, the DotF, DotG, and DotH proteins are destabilized in a Δ*dotU* Δ*icmF* mutant. Interestingly, homologues of *dotU* and *icmF* are found in a wide variety of bacterial species, where they are frequently associated with genes encoding the recently discovered type VI secretion systems (IAHP loci) (Sexton *et al.*, 2004a; Mougous *et al.*, 2006). Thus, the function of DotU and IcmF as 'macromolecular chaperones' appears to be conserved; once horizontally acquired by *L. pneumophila* these proteins may have been become adapted to the Dot/Icm type IV secretion system.

IcmQ

IcmQ is found in the bacterial cytoplasm and purified IcmQ can insert into membranes and form pores (Dumenil *et al.*, 2004). Membrane insertion and the pore-forming activity of IcmQ is inhibited in the bacterial cytoplasm by its chaperone, IcmR (Dumenil and Isberg, 2001). In addition, IcmQ can be observed on the surface of *L. pneumophila* after exposure to macrophages (Dumenil *et al.*, 2004). Based on these findings, it has been proposed that IcmQ may have a role in forming the Dot/Icm secretion channel or in forming pores in the phagosomal membrane. However, several caveats to this model must be considered. Exposure of IcmQ during infection was observed in wild-type and *dot/icm* mutants, demonstrating that IcmQ exposure is not the result of secretion of IcmQ by the Dot/Icm complex. Like the surface exposure of DotO and DotH, surface exposure of IcmQ could be due to susceptibility of *L. pneumophila* cells to release of cellular contents under the fixation conditions used. Moreover, it is difficult to interpret the significance of characterization of IcmQ in the absence of IcmR. IcmR is not present in the closely related Dot/Icm T4SS found in *Coxiella burnetii*, although this system does encode an IcmQ homolog. However, the *C. burnetii* IcmQ protein has an amino-terminal extension that may eliminate the need for IcmR in this pathogen (Zamboni *et al.*, 2003). Thus, examination of the role of IcmQ in the absence of IcmR may not be physiologically relevant, as these proteins appear to be covalently linked in *C. burnetii* and may never exist separately in *L. pneumophila*. While it is clear that IcmQ is required for the function of the Dot/Icm complex, its specific role remains unclear.

IcmT

IcmT is a predicted inner membrane protein that is required for growth in macrophages and amoebae (Segal and Shuman, 1999). IcmT has a transmembrane domain at its amino terminus and is thought to be a component of the secretion apparatus. Mutations that result in carboxy-terminal truncations of IcmT have been identified that affect the ability of *L. pneumophila* to form pores and escape from host cells (Molmeret *et al.*, 2002a; Molmeret *et al.*, 2002b; Bitar *et al.*, 2005). These data demonstrate that IcmT is required for both secretion of substrates involved in formation of the *L. pneumophila* replicative phagosome and for lysis of host cells.

Remaining Dot/Icm proteins

Little is known about the remaining Dot/Icm proteins. DotK, like the core subcomplex proteins DotC and DotD, contains a lipobox motif that is essential for its function (Yerushalmi *et al.*, 2005). DotK is the only outer membrane component that is partially required for intracellular growth and its role in the T4SS is unclear (Segal and Shuman, 1999). Only one Dot/Icm protein localizes exclusively to the periplasm, IcmX (Matthews and Roy, 2000). Although this protein has been shown to be required for the function of the type IV secretion system, its specific role is also unknown.

Similarly, the roles of the remaining inner membrane proteins, DotE, DotI, DotJ, DotP, DotV and IcmV, remain uncharacterized. Three of these proteins, DotE, DotP, and DotV, exhibit a significant degree of homology, although they are not functionally redundant as each is required for intracellular growth (Segal *et al.*, 2005). Based on its size and predicted secondary structure, DotP is a candidate to form the Dot/Icm pilus, although this hypothesis has not been tested. According to this model, DotE and DotV may function akin to pseudopilins, but this also has not been tested. Interestingly, DotJ and DotV (like the cytoplasmic chaperone IcmR) do not have homologues in the *C. burnetii* Dot/Icm T4SS, indicating that their roles may be unique to *L. pneumophila* (Vogel, 2004).

The Dot/Icm complex consists of at least two critical subassemblies. The first is composed of the inner membrane proteins DotF and DotG and the outer membrane proteins DotC, DotD, and DotH. This subcomplex forms the 'core' of the T4SS, bridging the inner and outer membranes, and it was designated as the core-transmembrane complex (Vincent *et al.*, 2006b). The second subassembly includes the type IV coupling protein DotL, the inner membrane proteins DotM and DotN, and the cytoplasmic adaptor proteins IcmS and IcmW. This subcomplex is likely to be involved in the recognition of substrates by the secretion apparatus.

DotD/DotC/DotF/DotG/DotH core–transmembrane subcomplex

The core subcomplex of the Dot/Icm secretion apparatus is composed of the inner membrane proteins DotF and DotG and the outer membrane proteins DotC, DotD, and DotH (Fig. 9.1) (Vincent *et al.*, 2006b). One of the inner membrane components, DotF, was previously shown via a bacterial two-hybrid screen to interact with several Dot/Icm substrates in the cytoplasm (Luo and Isberg, 2004). Our findings show that DotF is required for the interactions between the inner membrane and outer membrane components of the core subcomplex. Together, these finding suggest that DotF plays an important role in the secretion apparatus. It is somewhat surprising, then, that a Δ*dotF* mutant is only partially attenuated for intracellular growth, demonstrating that DotF is not a strictly required component of the Dot/Icm T4SS (Purcell and Shuman, 1998). In light of this, it appears likely that DotF

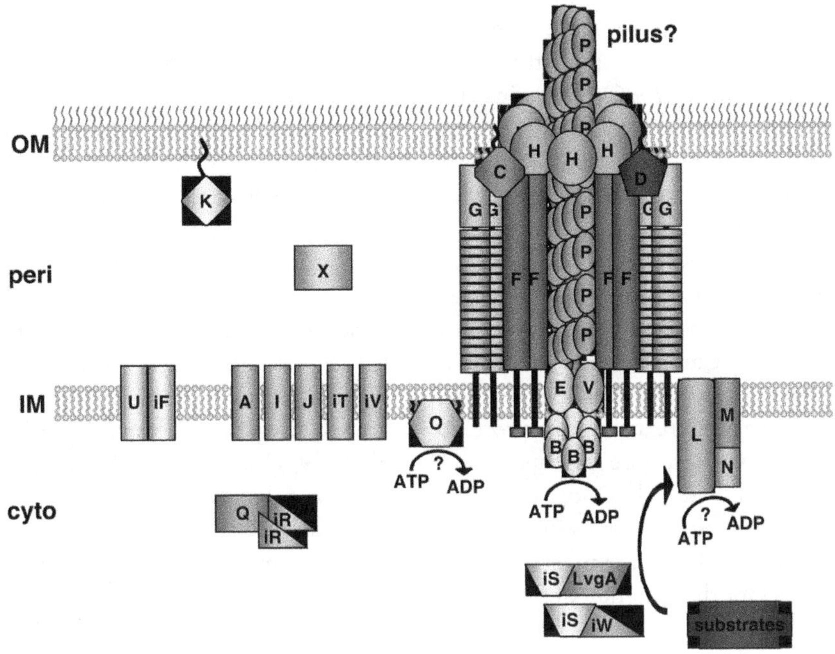

Figure 9.1 Model of the Dot/Icm type IVB secretion apparatus. Dot/Icm proteins are labelled using the last letter of their name. Dot proteins are indicated with a single letter whereas Icm proteins with the same last letter as a Dot protein are prefaced with the letter 'i'. The core subcomplex consists of the DotF, DotG, DotC, DotD, and DotH proteins. The coupling protein subcomplex contains DotL, DotM, DotN, and the secretion adaptor proteins IcmS and IcmW. The adaptor proteins are thought to mediate interactions between substrates and the type IV coupling protein DotL. Three proteins, DotJ, DotE, and IcmT, have not been localized experimentally but are predicted to be inner membrane proteins. DotP is shown as the Dot/Icm pilus, although this has not been experimentally confirmed.

functions with DotL (discussed below) either as a co-receptor for secreted substrates or as the protein to which DotL 'hands off' substrates in the process of translocation.

The other inner membrane component of the core subcomplex, DotG, exhibits limited homology over its carboxy-terminus to the *A. tumefaciens* protein VirB10. VirB10, a component of the *A. tumefaciens* core subcomplex, has been shown to function as an energy transducer that adopts different conformations in response to ATP hydrolysis by the VirB11 and VirD4 proteins (Cascales and Christie, 2004). In the presence of ATP, VirB10 interacts with the outer membrane component VirB9, but this interaction is not detected when cellular ATP levels are depleted. According to the current model, VirB10 responds to the energetic status of the complex by forming or disassembling a bridge between the inner and outer membrane components of the secretion system. It is likely that DotG performs the same role in the Dot/Icm T4SS. Supporting this notion, we have shown that DotG is an inner membrane protein with an identical topology to VirB10. In addition, we have shown that DotG interacts with the outer membrane components DotC, DotD, and DotH, similar to the interactions between VirB10 and the outer membrane proteins VirB7 and VirB9. Whether DotG adopts different conformations depending on ATP utilization by DotB and DotL remains to be experimentally determined.

The outer membrane components of the core subcomplex are DotC, DotD, and DotH. Two of these proteins, DotC and DotD, contain lipobox motifs and thus are likely to be covalently attached to a lipid that anchors them to the outer membrane (Yerushalmi *et al.*, 2005). DotC and DotD are required for insertion of DotH in the outer membrane, and in the absence of either of these lipoproteins, DotH accumulates in the periplasm. The dependence on a lipoprotein for the outer membrane insertion of DotH is reminiscent of secretins, which also require outer membrane lipoproteins called pilot proteins for insertion and oligomerization in the outer membrane (Bayan *et al.*, 2006). Secretins are found in a variety of secretion systems, including type II and type III secretion systems and type IV pilus biogenesis systems, although they have not been described in type IV secretion systems. In each of these diverse systems, secretins are though to perform a similar function as the outer membrane channel through which substrates are secreted. Although most secretins require a single pilot protein, the *Shigella* T3SS secretin MxiD is known to require two pilot proteins for outer membrane insertion (Schuch and Maurelli, 2001). Even though DotH does not exhibit detectable homology to the secretin family, we predict that is performs a similar role as an oligomeric channel for the secretion of substrates by the Dot/Icm T4SS.

DotL/DotM/DotN/IcmS/IcmW T4CP subcomplex

A second putative Dot/Icm subcomplex contains the inner membrane proteins DotL, DotM, and DotN and the cytoplasmic proteins IcmS and IcmW (C.D. Vincent and J.P. Vogel, unpublished). This subcomplex is of particular interest because of the key role proposed for DotL a member of the type IV coupling protein family (Buscher *et al.*, 2005). T4CPs function as inner membrane receptors for secreted substrates, linking the cytoplasmic substrates to the secretion machinery, and provide energy to the complex through ATP hydrolysis.

In support of DotL's predicted role in providing energy to the complex through ATP hydrolysis, we have observed that the DotL Walker box motif is essential for its function. In addition, expression of a DotL Walker box mutant inhibits the intracellular growth of *L. pneumophila* (C.D. Vincent and J.P. Vogel, unpublished) consistent with previous observations that other T4CPs form hexameric rings and display a dominant negative phenotype when their Walker box motif is mutated. In support of DotL's role as the inner membrane receptor for secreted substrates, DotL has been shown to interact with several components of the Dot/Icm secretion apparatus (C.D. Vincent and J.P. Vogel, unpublished). DotL was previously proposed to interact with DotM and DotN based on a shared phenotype of the *dotL*, *dotM*, and *dotN* mutants; all are essential genes but the lethality caused by their deletion can be suppressed by mutations in other *dot/icm* genes (Buscher *et al.*, 2005).

Although direct interactions between DotL and secreted substrates have not been reported, DotL has been found to interact with the secretion adaptor proteins IcmS and IcmW (C.D. Vincent and J.P. Vogel, unpublished). IcmS and IcmW are small, acidic, cytoplasmic proteins that are only partially required for intracellular growth (Segal and Shuman, 1997; Zuckman *et al.*, 1999; Coers *et al.*, 2000). Early evidence that IcmS and IcmW represented a unique class of Dot/Icm proteins included the observation that, unlike the majority of *dot/icm* mutants, *icmS* and *icmW* mutants still exhibit wild-type levels of contact-dependent cytotoxicity (Zuckman *et al.*, 1999; Coers *et al.*, 2000). In addition, *icmS* and *icmW* mutations do not suppress Δ*dotL* lethality (Buscher *et al.*, 2005). Because contact-dependent cytotoxicity and Δ*dotL* lethality require the presence of a functional secretion apparatus, these findings indicate that the Dot/Icm T4SS is assembled and functional in the *icmS* and *icmW* mutants. In spite of this, the mutants are severely attenuated in intracellular growth, suggesting that IcmS and IcmW may play a critical role in substrate secretion.

It was proposed that the role of IcmS and IcmW might be similar to type III secretion chaperones, proteins that bind to secreted substrates and facilitate their secretion (Page and Parsot, 2002; Parsot *et al.*, 2003). Consistent with this proposal, several reports have demonstrated interactions of IcmS and IcmW with substrates and a requirement for substrate export (Coers *et al.*, 2000; Bardill *et al.*, 2005; Ninio *et al.*, 2005). IcmS was shown to interact with the SidE family of secreted substrates, IcmW was found to interact with the substrates WipB and SidH, and both IcmS and IcmW were shown to bind to the substrates WipA and SidG (Bardill *et al.*, 2005; Ninio *et al.*, 2005). Notably, although IcmS and IcmW are required for the stable expression of each other, neither IcmS nor IcmW is required for the stable expression of their cognate secreted substrates. Therefore it is thought that the primary role of these proteins is to mediate interactions between substrates and the secretion apparatus. For these reasons, they have been termed 'adaptor proteins.'

Although several substrates require the adaptor proteins for secretion, at least one, RalF, does not require either IcmS or IcmW (Bardill *et al.*, 2005; Ninio *et al.*, 2005). This indicates that either some substrates are secreted by an adaptor-independent mechanism or that additional Dot/Icm adaptor proteins remain undiscovered. In fact, the LvgA protein appears to function as a third adaptor protein for the Dot/Icm T4SS (Vincent and Vogel, 2006). LvgA was originally identified as a virulence factor in an animal model of *L. pneumophila* infection, but its specific function was not determined (Edelstein *et al.*, 1999; Edelstein *et al.*, 2003). Subsequently, LvgA was shown to interact directly with IcmS and an Δ*lvgA* mutant was shown to phenotypically resemble a Δ*icmS* and Δ*icmW* mutant. In addition, LvgA was demonstrated to not be a component of the secretion apparatus nor be secreted by the Dot/Icm T4SS. As LvgA is more strictly required for growth in mouse bone marrow-derived macrophages than in human macrophages or in amoebae, LvgA appears to confer host specificity to *L. pneumophila*. Although a specific substrate(s) that requires LvgA for export has not yet been revealed, lvgA was identified in a screen for *L. pneumophila* mutants that were defective in the recruitment of Rab1 to the LCV (Murata *et al.*, 2006). The same report identified two secreted substrates, SidM and LidA, which are also required for recruitment of Rab1 to the LCV. It is tempting to speculate that the Rab1 recruitment defect of the *lvgA* mutant is due to a failure to secrete SidM and LidA, and that LvgA functions as an adaptor protein for one or both of these substrates.

Assembly factors for the Dot/Icm secretion apparatus

Considering that the Dot/Icm secretion apparatus is an extremely complicated multi-protein structure, it is not surprising that its assembly is not well understood. For example, the Dot/Icm complex likely requires factors to assist in the proper assembly of the secretion apparatus. In fact, two *L. pneumophila* proteins that are not components of the secretion apparatus, DjlA and LdsA, were recently shown to be required for the assembly of the Dot/Icm complex (Vincent *et al.*, 2006a). Although the *L. pneumophila djlA* and *ldsA* mutants still synthesize the Dot/Icm proteins, the mutants are unable to grow in macrophages and are defective for the phenotypes associated with an assembled, functional Dot/Icm secretion system (Vincent *et al.*, 2006a).

In *E. coli*, DjlA is an inner membrane protein that has homology to the heat shock protein DnaJ (Clarke *et al.*, 1996), is involved in the sensing of envelope stress and has been proposed to function as a chaperone for membrane proteins (Kelley and Georgopoulos, 1997). Like DnaJ, DjlA interacts with the chaperone DnaK (HSP70) and stimulates its ATPase activity (Wall *et al.*, 1994). DjlA was shown to be a virulence factor in the strain *L. dumoffii*, although its role in virulence is not understood (Ohnishi *et al.*, 2004). In *L. pneumophila*, DjlA is required for the assembly or activation of the Dot/Icm complex (Vincent *et al.*, 2006a). Based on the involvement of DjlA in assembly of the Dot/Icm complex, it is likely that the chaperone DnaK is also involved in folding and assembly of the Dot/Icm proteins.

In contrast to DjlA, the LdsA protein does not contain any clear functional motifs, thus obfuscating predictions of its function. LdsA appears to be a polytopic inner membrane protein, and although it is possible that this protein is part of the Dot/Icm secretion apparatus, it is unlikely because (1) it is not homologous to any known type IV secretion system proteins and (2) the *ldsA* gene is encoded distantly from the *dot/icm* pathogenicity islands. Instead LdsA may function in an accessory role perhaps as a specific chaperone for the assembly and/or activation of the Dot/Icm apparatus. Further characterization of the roles of DjlA and LdsA will likely provide insight into how the Dot/Icm system is assembled and activated.

Conclusions

A significant amount of progress has been made in understanding the *Legionella* Dot/Icm type IVB secretion system since the initial identification of DotA and IcmV/IcmW/IcmX in 1993 (Berger and Isberg, 1993; Sadosky *et al.*, 1993). The recent identification of two critical subcomplexes of the Dot/Icm complex exemplifies the progress in understanding how this secretion apparatus assembles and functions (Vincent *et al.*, 2006b). The first subcomplex is composed of five proteins (DotC, DotD, DotF, DotG, and DotH), functions as a molecular bridge between the inner and outer membranes, and has been designated as the 'core-transmembrane' subcomplex. In this subcomplex, DotH localization to the outer membrane requires DotC and DotD and DotH likely functions similarly to a secretin, forming a channel in the outer membrane. The inner membrane proteins DotF and DotG interact with each other and with the DotC/DotD/DotH proteins in the outer membrane, likely transducing energy from ATP hydrolysis to the outer membrane. The existence of a *Legionella* core-transmembrane complex demonstrates a functionally conserved core assembly in the evolutionarily diverse type IVA and type IVB secretion systems. The second Dot/Icm subcomplex contains DotL, DotM, DotN, IcmS, and IcmW. This subcomplex likely functions as the membrane receptor for the Dot/Icm substrates as DotL is related to type IV coupling proteins and IcmS and IcmW have been shown to interact with a number of the Dot/Icm substrates. In addition to these two subcomplexes, other subcomplexes likely consist of the remaining sixteen Dot/Icm proteins.

Finally, analysis of the *Legionella* Dot/Icm apparatus revealed a surprising difference between type IVA and type IVB secretion systems. Of the 26 *Legionella* Dot/Icm proteins, five are in the cytoplasm, one in the periplasm, and four in the outer membrane but sixteen are found associated with the inner membrane (Vincent *et al.*, 2006b). In contrast, type IVA systems consist of one cytoplasmic protein, one periplasmic protein, three outer membrane proteins, two pilus proteins and only five inner membrane proteins (Schroder and Lanka, 2005). Thus, it appears that type IVB systems have undergone a proliferation of non-conserved inner membrane proteins. Why the type IVB systems would require so many inner membrane components remains unclear. However, these 'extra' proteins do not appear to be accessory factors, because with the exception of DotU and IcmF, all are strictly required for the function of the Dot/Icm secretion apparatus. Moreover, the majority of these proteins are conserved in other type IVB systems, including the *C. burnetii* Dot/Icm secretion system and the plasmid transfer apparatus of the self-transmissible plasmids R64 and ColIb-P9, further supporting the idea that these proteins are essential for the function of the type IVB systems. Elucidation of the functions of these components will likely reveal important differences between the type IVA and type IVB secretion systems.

References

Andrews, H.L., Vogel, J.P., and Isberg, R.R. (1998). Identification of linked *Legionella pneumophila* genes essential for intracellular growth and evasion of the endocytic pathway. Infect. Immun. *66*, 950–958.

Atmakuri, K., Cascales, E., and Christie, P.J. (2004). Energetic components VirD4, VirB11 and VirB4 mediate early DNA transfer reactions required for bacterial type IV secretion. Mol. Microbiol. *54*, 1199–1211.

Atmakuri, K., Ding, Z., and Christie, P.J. (2003). VirE2, a type IV secretion substrate, interacts with the VirD4 transfer protein at cell poles of *Agrobacterium tumefaciens*. Mol. Microbiol. *49*, 1699–1713.

Balzer, D., Pansegrau, W., and Lanka, E. (1994). Essential motifs of relaxase (TraI) and TraG proteins involved in conjugative transfer of plasmid RP4. J. Bacteriol. *176*, 4285–4295.

Bardill, J.P., Miller, J.L., and Vogel, J.P. (2005). IcmS-dependent translocation of SdeA into macrophages by the *Legionella pneumophila* type IV secretion system. Mol. Microbiol. *56*, 90–103.

Bayan, N., Guilvout, I., and Pugsley, A.P. (2006). Secretins take shape. Mol. Microbiol. *60*, 1–4.

Berger, K. H., and Isberg, R. R. (1993). Two distinct defects in intracellular growth complemented by a single genetic locus in *Legionella pneumophila*. Mol. Microbiol. *7*, 7–19.

Bitar, D.M., Molmeret, M., and Kwaik, Y.A. (2005). Structure-function analysis of the C-terminus of IcmT of *Legionella pneumophila* in pore formation-mediated egress from macrophages. FEMS Microbiol. Lett. *242*, 177–184.

Burns, D.L. (2003). Type IV transporters of pathogenic bacteria. Curr. Opin. Microbiol. *6*, 29–34.

Buscher, B.A., Conover, G.M., Miller, J.L., Vogel, S.A., Meyers, S.N., Isberg, R.R., and Vogel, J.P. (2005). The DotL protein, a member of the TraG-coupling protein family, is essential for viability of *Legionella pneumophila* strain Lp02. J. Bacteriol. *187*, 2927–2938.

Cabezon, E., Lanka, E., and de la Cruz, F. (1994). Requirements for mobilization of plasmids RSF1010 and ColE1 by the IncW plasmid R388: *trwB* and RP4 *traG* are interchangeable. J. Bacteriol. *176*, 4455–4458.

Cabezon, E., Sastre, J.I., and de la Cruz, F. (1997). Genetic evidence of a coupling role for the TraG protein family in bacterial conjugation. Mol. Gen. Genet. *254*, 400–406.

Cascales, E., and Christie, P.J. (2003). The versatile bacterial type IV secretion systems. Nat. Rev. Microbiol. *1*, 137–149.

Cascales, E., and Christie, P.J. (2004). *Agrobacterium* VirB10, an ATP energy sensor required for type IV secretion. Proc. Natl. Acad. Sci. USA. *101*, 17228–17233.

Christie, P.J., Atmakuri, K., Krishnamoorthy, V., Jakubowski, S., and Cascales, E. (2005). Biogenesis, architecture, and function of bacterial type IV secretion systems. Annu. Rev. Microbiol. *59*, 451–485.

Christie, P.J., and Vogel, J.P. (2000). Bacterial type IV secretion: conjugation systems adapted to deliver effector molecules to host cells. Trends Microbiol. *8*, 354–360.

Clarke, D.J., Jacq, A., and Holland, I.B. (1996). A novel DnaJ-like protein in *Escherichia coli* inserts into the cytoplasmic membrane with a type III topology. Mol. Microbiol. *20*, 1273–1286.

Coers, J., Kagan, J.C., Matthews, M., Nagai, H., Zuckman, D.M., and Roy, C.R. (2000). Identification of Icm protein complexes that play distinct roles in the biogenesis of an organelle permissive for *Legionella pneumophila* intracellular growth. Mol. Microbiol. *38*, 719–736.

de Paz, H.D., Sangari, F.J., Bolland, S., Garcia-Lobo, J.M., Dehio, C., de la Cruz, F., and Llosa, M. (2005). Functional interactions between type IV secretion systems involved in DNA transfer and virulence. Microbiology *151*, 3505–3516.

Derre, I., and Isberg, R.R. (2004). *Legionella pneumophila* replication vacuole formation involves rapid recruitment of proteins of the early secretory system. Infect. Immun. *72*, 3048–3053.

Dumenil, G., and Isberg, R.R. (2001). The *Legionella pneumophila* IcmR protein exhibits chaperone activity for IcmQ by preventing its participation in high-molecular-weight complexes. Mol. Microbiol. *40*, 1113–1127.

Dumenil, G., Montminy, T.P., Tang, M., and Isberg, R.R. (2004). IcmR-regulated membrane insertion and efflux by the *Legionella pneumophila* IcmQ protein. J. Biol. Chem. *279*, 4686–4695.

Edelstein, P.H., Edelstein, M.A., Higa, F., and Falkow, S. (1999). Discovery of virulence genes of *Legionella pneumophila* by using signature tagged mutagenesis in a guinea pig pneumonia model. Proc. Natl. Acad. Sci. USA *96*, 8190–8195.

Edelstein, P.H., Hu, B., Higa, F., and Edelstein, M.A. (2003). *lvgA*, a novel *Legionella pneumophila* virulence factor. Infect. Immun. *71*, 2394–2403.

Fields, B.S., Benson, R.F., and Besser, R.E. (2002). *Legionella* and Legionnaires' disease: 25 years of investigation. Clin. Microbiol. Rev. *15*, 506–526.

Gilmour, M.W., Gunton, J.E., Lawley, T.D., and Taylor, D.E. (2003). Interaction between the IncHI1 plasmid R27 coupling protein and type IV secretion system: TraG associates with the coiled-coil mating pair formation protein TrhB. Mol. Microbiol. *49*, 105–116.

Gomis-Ruth, F.X., Moncalian, G., Perez-Luque, R., Gonzalez, A., Cabezon, E., de la Cruz, F., and Coll, M. (2001). The bacterial conjugation protein TrwB resembles ring helicases and F1-ATPase. Nature *409*, 637–641.

Hamilton, C.M., Lee, H., Li, P.L., Cook, D.M., Piper, K.R., von Bodman, S.B., Lanka, E., Ream, W., and Farrand, S.K. (2000). TraG from RP4 and TraG and VirD4 from Ti plasmids confer relaxosome specificity to the conjugal transfer system of pTiC58. J. Bacteriol. *182*, 1541–1548.

Horwitz, M.A. (1983a). Formation of a novel phagosome by the Legionnaires' disease bacterium (*Legionella pneumophila*) in human monocytes. J. Exp. Med. *158*, 1319–1331.

Horwitz, M.A. (1983b). The Legionnaires' disease bacterium (*Legionella pneumophila*) inhibits phagosome-lysosome fusion in human monocytes. J. Exp. Med. *158*, 2108–2126.

Horwitz, M.A., and Silverstein, S.C. (1980). Legionnaires' disease bacterium (*Legionella pneumophila*) multiples intracellularly in human monocytes. J. Clin. Invest. *66*, 441–450.

Kagan, J.C., and Roy, C.R. (2002). *Legionella* phagosomes intercept vesicular traffic from endoplasmic reticulum exit sites. Nat. Cell. Biol. *4*, 945–954.

Kagan, J.C., Stein, M.P., Pypaert, M., and Roy, C.R. (2004). *Legionella* subvert the functions of Rab1 and Sec22b to create a replicative organelle. J. Exp. Med. *199*, 1201–1211.

Kelley, W.L., and Georgopoulos, C. (1997). Positive control of the two-component RcsC/B signal transduction network by DjlA: a member of the DnaJ family of molecular chaperones in *Escherichia coli*. Mol. Microbiol. *25*, 913–931.

Kumar, R. B., and Das, A. (2002). Polar location and functional domains of the *Agrobacterium tumefaciens* DNA transfer protein VirD4. Mol. Microbiol. *43*, 1523–1532.

Llosa, M., Gomis-Ruth, F.X., Coll, M., and de la Cruz Fd, F. (2002). Bacterial conjugation: a two-step mechanism for DNA transport. Mol. Microbiol. *45*, 1–8.

Llosa, M., Zunzunegui, S., and de la Cruz, F. (2003). Conjugative coupling proteins interact with cognate and heterologous VirB10-like proteins while exhibiting specificity for cognate relaxosomes. Proc. Natl. Acad. Sci. USA *100*, 10465–10470.

Luo, Z.Q., and Isberg, R.R. (2004). Multiple substrates of the *Legionella pneumophila* Dot/Icm system identified by interbacterial protein transfer. Proc. Natl. Acad. Sci. USA. *101*, 841–846.

Matthews, M., and Roy, C.R. (2000). Identification and subcellular localization of the *Legionella pneumophila* IcmX protein: a factor essential for establishment of a replicative organelle in eukaryotic host cells. Infect. Immun. *68*, 3971–3982.

Molmeret, M., Alli, O.A., Radulic, M., Susa, M., Doric, M., and Kwaik, Y.A. (2002a). The C-terminus of IcmT is essential for pore formation and for intracellular trafficking of *Legionella pneumophila* within *Acanthamoeba polyphaga*. Mol. Microbiol. *43*, 1139–1150.

Molmeret, M., Alli, O.A., Zink, S., Flieger, A., Cianciotto, N.P., and Kwaik, Y.A. (2002b). icmT is essential for pore formation-mediated egress of *Legionella pneumophila* from mammalian and protozoan cells. Infect. Immun. *70*, 69–78.

Moncalian, G., Cabezon, E., Alkorta, I., Valle, M., Moro, F., Valpuesta, J.M., Goni, F.M., and de La Cruz, F. (1999). Characterization of ATP and DNA binding activities of TrwB, the coupling protein essential in plasmid R388 conjugation. J. Biol. Chem. *274*, 36117–36124.

Mougous, J.D., Cuff, M.E., Raunser, S., Shen, A., Zhou, M., Gifford, C.A., Goodman, A.L., Joachimiak, G., Ordonez, C. L., Lory, S., Walz, T., Joachimiak, A., and Mekalanos, J.J. (2006). A virulence locus of *Pseudomonas aeruginosa* encodes a protein secretion apparatus. Science *312*, 1526–1530.

Murata, T., Delprato, A., Ingmundson, A., Toomre, D.K., Lambright, D.G., and Roy, C.R. (2006). The *Legionella pneumophila* effector protein DrrA is a Rab1 guanine nucleotide-exchange factor. Nat. Cell. Biol. *8*, 971–977.

Nagai, H., and Roy, C.R. (2001). The DotA protein from *Legionella pneumophila* is secreted by a novel process that requires the Dot/Icm transporter. Embo J. *20*, 5962–5970.

Ninio, S., Zuckman-Cholon, D.M., Cambronne, E.D., and Roy, C.R. (2005). The *Legionella* IcmS-IcmW protein complex is important for Dot/Icm-mediated protein translocation. Mol. Microbiol. *55*, 912–926.

Ohnishi, H., Mizunoe, Y., Takade, A., Tanaka, Y., Miyamoto, H., Harada, M., and Yoshida, S. (2004). *Legionella dumoffii* DjlA, a member of the DnaJ family, is required for intracellular growth. Infect. Immun. *72*, 3592–3603.

Page, A.L., and Parsot, C. (2002). Chaperones of the type III secretion pathway: jacks of all trades. Mol. Microbiol. *46*, 1–11.

Parsot, C., Hamiaux, C., and Page, A.L. (2003). The various and varying roles of specific chaperones in type III secretion systems. Curr. Opin. Microbiol. *6*, 7–14.

Purcell, M., and Shuman, H.A. (1998). The *Legionella pneumophila icmGCDJBF* genes are required for killing of human macrophages. Infect. Immun. *66*, 2245–2255.

Roy, C.R., and Isberg, R.R. (1997). Topology of *Legionella pneumophila* DotA: an inner membrane protein required for replication in macrophages. Infect. Immun. *65*, 571–578.

Sadosky, A.B., Wiater, L.A., and Shuman, H.A. (1993). Identification of *Legionella pneumophila* genes required for growth within and killing of human macrophages. Infect. Immun. *61*, 5361–5373.

Sagulenko, E., Sagulenko, V., Chen, J., and Christie, P.J. (2001). Role of *Agrobacterium* VirB11 ATPase in T-pilus assembly and substrate selection. J. Bacteriol. *183*, 5813–5825.

Schroder, G., Krause, S., Zechner, E.L., Traxler, B., Yeo, H.J., Lurz, R., Waksman, G., and Lanka, E. (2002). TraG-like proteins of DNA transfer systems and of the *Helicobacter pylori* type IV secretion system: inner membrane gate for exported substrates? J. Bacteriol. *184*, 2767–2779.

Schroder, G., and Lanka, E. (2005). The mating pair formation system of conjugative plasmids-A versatile secretion machinery for transfer of proteins and DNA. Plasmid *54*, 1–25.

Schuch, R., and Maurelli, A.T. (2001). MxiM and MxiJ, base elements of the Mxi-Spa type III secretion system of *Shigella*, interact with and stabilize the MxiD secretin in the cell envelope. J. Bacteriol. *183*, 6991–6998.

Segal, G., Feldman, M., and Zusman, T. (2005). The Icm/Dot type-IV secretion systems of *Legionella pneumophila* and *Coxiella burnetii*. FEMS Microbiol. Rev. *29*, 65–81.

Segal, G., Purcell, M., and Shuman, H.A. (1998). Host cell killing and bacterial conjugation require overlapping sets of genes within a 22-kb region of the *Legionella pneumophila* genome. Proc. Natl. Acad. Sci. USA. *95*, 1669–1674.

Segal, G., and Shuman, H.A. (1997). Characterization of a new region required for macrophage killing by *Legionella pneumophila*. Infect. Immun. *65*, 5057–5066.

Segal, G., and Shuman, H.A. (1999). *Legionella pneumophila* utilizes the same genes to multiply within *Acanthamoeba castellanii* and human macrophages. Infect. Immun. *67*, 2117–2124.

Sexton, J.A., Miller, J.L., Yoneda, A., Kehl-Fie, T.E., and Vogel, J.P. (2004a). *Legionella pneumophila* DotU and IcmF are required for stability of the Dot/Icm complex. Infect. Immun. *72*, 5983–5992.

Sexton, J.A., Pinkner, J.S., Roth, R., Heuser, J.E., Hultgren, S.J., and Vogel, J.P. (2004b). The *Legionella pneumophila* PilT homologue DotB exhibits ATPase activity that is critical for intracellular growth. J. Bacteriol. *186*, 1658–1666.

Sexton, J.A., and Vogel, J.P. (2002). Type IVB secretion by intracellular pathogens. Traffic *3*, 178–185.

Sexton, J.A., Yeo, H.J., and Vogel, J.P. (2005). Genetic analysis of the *Legionella pneumophila* DotB ATPase reveals a role in type IV secretion system protein export. Mol. Microbiol. *57*, 70–84.

Sturgill-Koszycki, S., and Swanson, M.S. (2000). *Legionella pneumophila* replication vacuoles mature into acidic, endocytic organelles. J. Exp. Med. *192*, 1261–1272.

Swanson, M.S., and Isberg, R.R. (1995). Association of *Legionella pneumophila* with the macrophage endoplasmic reticulum. Infect. Immun. *63*, 3609–3620.

Szpirer, C.Y., Faelen, M., and Couturier, M. (2000). Interaction between the RP4 coupling protein TraG and the pBHR1 mobilization protein Mob. Mol. Microbiol. *37*, 1283–1292.

Tato, I., Zunzunegui, S., de la Cruz, F., and Cabezon, E. (2005). TrwB, the coupling protein involved in DNA transport during bacterial conjugation, is a DNA-dependent ATPase. Proc Natl. Acad. Sci. USA. *102*, 8156–8161.

VanRheenen, S.M., Dumenil, G., and Isberg, R.R. (2004). IcmF and DotU are required for optimal effector translocation and trafficking of the *Legionella pneumophila* vacuole. Infect. Immun. *72*, 5972–5982.

Vincent, C.D., Buscher, B.A., Friedman, J.R., Williams, L.A., Bardill, P., and Vogel, J.P. (2006a). Identification of non-dot/icm suppressors of the *Legionella pneumophila* D*dotL* lethality phenotype. J. Bacteriol. *188*, 8231–8243.

Vincent, C.D., Friedman, J.R., Jeong, K.C., Buford, E.C., Miller, J.L., and Vogel, J.P. (2006b). Identification of the core transmembrane complex of the *Legionella* Dot/Icm type IV secretion system. Mol. Microbiol. *62*, 1278–1291.

Vincent, C.D., and Vogel, J.P. (2006). The *Legionella pneumophila* IcmS-LvgA protein complex is important for Dot/Icm-dependent intracellular growth. Mol. Microbiol. *61*, 596–613.

Vogel, J.P. (2004). Turning a tiger into a house cat: using *Legionella pneumophila* to study *Coxiella burnetii*. Trends Microbiol *12*, 103–105.

Vogel, J.P., Andrews, H.L., Wong, S.K., and Isberg, R.R. (1998). Conjugative transfer by the virulence system of *Legionella pneumophila*. Science *279*, 873–876.

Vogel, J.P., Roy, C., and Isberg, R.R. (1996). Use of salt to isolate *Legionella pneumophila* mutants unable to replicate in macrophages. Ann. NY Acad. Sci. *797*, 271–272.

Walker, J.E., Saraste, M., Runswick, M.J., and Gay, N.J. (1982). Distantly related sequences in the alpha- and beta-subunits of ATP synthase, myosin, kinases and other ATP-requiring enzymes and a common nucleotide binding fold. Embo J. *1*, 945–951.

Wall, D., Zylicz, M., and Georgopoulos, C. (1994). The NH2-terminal 108 amino acids of the *Escherichia coli* DnaJ protein stimulate the ATPase activity of DnaK and are sufficient for lambda replication. J. Biol. Chem. *269*, 5446–5451.

Watarai, M., Andrews, H.L., and Isberg, R.R. (2001). Formation of a fibrous structure on the surface of *Legionella pneumophila* associated with exposure of DotH and DotO proteins after intracellular growth. Mol. Microbiol. *39*, 313–329.

Winans, S.C., Burns, D.L., and Christie, P.J. (1996). Adaptation of a conjugal transfer system for the export of pathogenic macromolecules. Trends Microbiol. *4*, 64–68.

Yerushalmi, G., Zusman, T., and Segal, G. (2005). Additive effect on intracellular growth by *Legionella pneumophila* Icm/Dot proteins containing a lipobox motif. Infect. Immun. *73*, 7578–7587.

Zamboni, D.S., McGrath, S., Rabinovitch, M., and Roy, C.R. (2003). *Coxiella burnetii* express type IV secretion system proteins that function similarly to components of the *Legionella pneumophila* Dot/Icm system. Mol. Microbiol. *49*, 965–976.

Zuckman, D.M., Hung, J.B., and Roy, C.R. (1999). Pore-forming activity is not sufficient for *Legionella pneumophila* phagosome trafficking and intracellular growth. Mol. Microbiol. *32*, 990–1001.

Mechanisms of Intracellular Survival and Replication of *Legionella pneumophila*

10

Tamara O'Connor*, Matthew Heidtman and Ralph R. Isberg*

Abstract

Upon uptake by phagocytic cells the *Legionella*-containing vacuole (LCV) evades targeting to and degradation by the endocytic pathway. Instead, mitochondria and early secretory vesicles are rapidly recruited to the LCV. Over a period of several hours, the LCV becomes studded with ribosomes and resembles an ER-like compartment. Within this organelle *L. pneumophila* is able to replicate to high numbers. Intracellular growth of *L. pneumophila* is dependent on the *dot/icm* genes. Many of the Dot/Icm proteins assemble into a large membrane-spanning complex that enables the delivery of bacterial proteins into the host cell cytosol. These translocated substrates are thought to modulate a variety of host cell processes to create an intracellular environment permissive for bacterial replication. The functions of most of these substrates are unknown, although at least some of them are known to modulate host cell vesicle trafficking and anti-apoptotic signalling pathways. Recent studies have sought to identify host cell processes important for *L. pneumophila* intracellular replication. The input of multiple host cell vesicle trafficking pathways, the ERAD system, anti-apoptotic signalling, autophagy, and host cell lipid metabolism all appear to be important to varying extents for efficient intracellular growth of *L. pneumophila*.

Modulation of host cell membrane trafficking pathways by *L. pneumophila*

Intracellular replication of *L. pneumophila* involves growth of the bacterium within a membrane-bound compartment that evades targeting into the endocytic pathway (Horwitz 1983a). Instead, nascent phagosomes harbouring virulent *L. pneumophila* associate with endoplasmic reticulum-derived membrane vesicles and mitochondria within minutes of phagocytosis (Horwitz, 1983a; Swanson and Isberg, 1995). Approximately 4–6 h after uptake, the *L. pneumophila* containing vacuole (LCV) is studded with ribosomes and closely resembles the rough ER in morphology. Within this ER-like replication compartment, the bacterium is able to grow to high numbers. Between 16–24 hours after the initiation of replication within mammalian cells the host cell lyses and large numbers of bacteria are released to the extracellular environment for further rounds of infection. Remodelling of the LCV into an ER-like organelle appears to be required for the bacterium to grow intracellularly, as *L. pneumophila* mutants defective for intracellular growth are targeted to endocytic compartments (Berger and Isberg, 1993; Sadosky *et al.*, 1993).

Central to the ability of *L. pneumophila* to modulate host cell membrane trafficking pathways is a set of 26 products of the dot (de-

*Corresponding authors.

fective in organelle trafficking)/icm (intracellular multiplication) genes (Marra *et al.*, 1992; Berger and Isberg, 1993). The vast majority of Dot/Icm proteins are thought to form a large integral membrane complex that allows the bacterium to inject proteins directly into the host cell cytoplasm (Vogel *et al.*, 1998; Vincent *et al.*, 2006a; Chapter 9). At least some of these translocated substrates are known to modulate the activities of host cell proteins involved in vesicular transport (Nagai *et al.*, 2002; Derre and Isberg, 2005; Machner and Isberg, 2006; Murata *et al.*, 2006). Through the action of these effectors it is believed that the bacterium is able to direct ER derived material to the LCV, ultimately remodelling this compartment into an organelle permissive for growth. Recent work has sought to understand the molecular mechanisms underlying LCV biogenesis and has focused on two overlapping areas of research: (1) identifying host cell proteins that are important for remodelling the LCV into a replicative organelle and (2) uncovering the molecular interactions between host cell factors and *L. pneumophila* effector proteins.

Subversion of early secretory pathway dynamics by *L. pneumophila*

Initial morphological and cytological studies suggested that the LCV acquires host cell membrane vesicles derived from the endoplasmic reticulum or other compartments of the early secretory pathway (Horwitz, 1983a; Swanson and Isberg, 1995). Therefore, it is important to understand some of the basic mechanisms underlying vesicular trafficking through the early limb of the eukaryotic secretory system as a prelude to a discussion regarding modulation of these pathways by *L. pneumophila* virulence factors. Specifically, vesicular traffic in the eukaryotic early secretory pathway relies upon the action of two heteromeric protein coat complexes, termed COPI (coat protein complex I) and COPII (coat protein complex II) (Bonifacino and Glick, 2004). COPII proteins bind to ER membranes and drive the formation of transport vesicles and membrane tubules containing secretory cargo (Gurkan *et al.*, 2006). The COPI complex associates with membranes of the ERGIC (ER-Golgi intermediate compartment) and Golgi complex. COPI function is required for ongoing anterograde transport between the ER and the Golgi complex (Rabouille and Klumperman, 2005), as well as retrograde transport within the Golgi and from the Golgi to the ER (Orci *et al.*, 1997; Letourneur *et al.*, 1994). Assembly of both coat complexes is regulated by small GTPases. COPI action is controlled by the GTPase ARF1 while COPII activity is directed by the GTPase Sar1 (Gurkan *et al.*, 2006). In the GTP bound state, Sar1 and ARF1 become activated and recruit their cognate coat proteins to membranes, initiating the assembly of coat complexes and ultimately stimulating the production of transport vesicles and/or tubules.

Recent studies have focused on defining the molecular pathway used by *L. pneumophila* to subvert host cell vesicular trafficking processes to create the LCV. Kagan and Roy, investigating the maturation of the LCV by fluorescence microscopy (Kagan and Roy, 2002), observed that p58, a protein that cycles between the ER and the Golgi within the lumen of COPI and COPII transport vesicles, was recruited to the majority of LCVs within 30 min of bacterial uptake into host cells. The early secretory vesicle marker YFP-KDEL, which has an ER-retrieval signal, displayed similar kinetics of association with the LCV. In contrast, ER resident proteins such as calnexin were detected at the LCV only at later time points in the infection process (~4 h). These observations suggest a biphasic model of LCV biogenesis in which early secretory vesicles are recruited to the LCV membrane first, followed by a second wave of membrane material containing ER resident proteins. This model is consistent with electron microscopy studies showing that the recruitment of small vesicles to the LCV surface occurs within minutes of uptake, while the appearance of ribosomes on the phagosome membrane occurs several hours later (Tilney *et al.*, 2001).

Functional studies gave support for the model that recruitment of early secretory vesicles controls formation of the LCV. Treatment of host cells with the fungal metabolite Brefeldin A (BFA) prior to *L. pneumophila* infection blocks the association of ER-to-Golgi vesicle proteins with the LCV. As BFA specifically inhibits the action of the COPI system, this indicates that vesicular traffic controlled by COPI system is important for LCV biogenesis. Similarly, production of the dominant interfering variants Sar1H79G and ARF1T31N in host cells during *L. pneumophila* infection significantly inhibits LCV formation (Kagan and Roy, 2002). As both Sar1 and ARF1 are key regulators of vesicle formation in the early secretory pathway, this observation again supports the hypothesis that ER-derived transport carriers are important for efficient LCV maturation. It is important to note that LCV formation was not completely blocked by the use of these inhibitors. This may indicate that *L. pneumophila* is able to acquire ER membranes from a route independent of the ARF1/COP1 and Sar1/COPII systems.

How might *L. pneumophila* hijack the activity of the ARF1/COPI and Sar1/COPII systems to create the LCV? An important clue to this question came with the identification of a translocated substrate of the Dot/Icm secretion system (Nagai *et al.*, 2002). Inspection of the *L. pneumophila* genome revealed the presence of an open reading frame predicted to encode a protein containing a region similar to Sec7 domains, which promote the exchange of GDP for GTP on ARF proteins. Biochemical experiments confirmed this prediction and the protein, termed RalF, was shown to be translocated into host cells dependent upon the Dot/Icm secretion system. Presumably the ARF exchange activity supports formation of the LCV, and the fact that ARF1 is recruited to a significant fraction of *L. pneumophila* phagosomes during the first two hours of infection is consistent with this model. Strikingly, the authors of this study demonstrated that *L. pneumophila* strains deleted for RalF were unable to recruit ARF1 to the LCV (Nagai *et al.*, 2002), although these strains did not display a growth defect in a variety of eukaryotic host cell types. One explanation for this result may be that *L. pneumophila* injects additional proteins into cells to exploit the ARF1/COPI system thereby enabling *L. pneumophila* to overcome the loss of RalF function in LCV biogenesis.

Vesicular transport between the ER and the Golgi complex defines the earliest limb of the secretory pathway in eukaryotic cells. In this process, transport vesicles carrying secretory cargo are generated from ER membranes by the action of the COPII protein coat (Gurkan *et al.*, 2006). These transport carriers are then efficiently targeted to the ERGIC and *cis*-Golgi membranes. The initial physical association of ER-derived transport carriers with target membranes is termed tethering, and is regulated by a family of small GTPases termed Rabs (Segev, 2001; Jahn *et al.*, 2003) and a diverse group of molecules referred to as tethering factors (Grosshans *et al.*, 2006; Sztul and Lupashin, 2006). Rab proteins are thought to act as molecular 'switches' to regulate the tethering of vesicles with the proper acceptor compartment. The exchange of GDP for GTP on Rabs allows for their activation, enabling Rabs to recruit various proteins necessary for vesicle tethering and fusion in a regulated temporal and spatial manner. Once tethering is complete, the vesicles fuse with the target membrane and deliver their contents.

Vesicle fusion depends upon a group of integral membrane proteins termed soluble *N*-ethylmalemide-sensitive factor attachment protein receptors (SNAREs) (Bonifacino and Glick, 2004). This process requires the interaction of SNAREs present in the vesicle membrane (termed a v-SNARE) with SNAREs in the acceptor compartment membrane (referred to as t-SNAREs) (Xu *et al.*, 2000). The tight association of v-SNAREs and t-SNAREs on apposing membranes is thought to supply the free energy needed to bring the membranes close enough to undergo fusion (Hanson *et al.*, 1997; Weber *et al.*, 1998; Chen and Scheller, 2001). In anterograde transport from the ER to the Golgi complex, the SNARE Sec22b is thought to function as a v-SNARE on ER-derived vesicles (Zhang *et al.*, 1999), while membrin, syntaxin 5, and Bet1 act as cognate t-SNAREs on pre-Golgi intermediate compartments (Xu *et al.*, 2000). In addition, recent evidence indicates ER-derived COPII vesicles can fuse with each other prior to their interaction with Golgi membranes, a process termed homotypic fusion (Xu and Hay, 2004; Bentley *et al.*, 2006). Homotypic fusion of ER-derived transport vesicles may allow *de novo* formation of ERGIC compartments, which then move through the Golgi stack by a process termed cisternal maturation (Losev *et al.*, 2006; Matsuura-Tokiata *et al.*, 2006; Pelham, 2006).

Recent studies have sought to define the manner in which early secretory vesicles are recruited, tethered and retained at the surface of the LCV. Two different groups showed that the host cell GTPase Rab1b is recruited to the LCV in a manner dependent upon a functional Dot/Icm translocation system (Derre and Isberg, 2004; Kagan *et al.*, 2004). Rab1b is a key regulator of the tethering and fusion of ER-derived vesicles with Golgi elements (Seabra *et al.*, 2002). Rab2 and Rab6, two Rab GTPases also known to function in ER-to-Golgi transport, were not observed to localize to the LCV (Kagan *et al.*, 2004). These findings suggest that *L. pneumophila* cells actively recruit Rab1b to the phagosome during infection, perhaps as a means to redirect secretory vesicles to the LCV surface. Consistent with this model, each group found that expression of a dominant interfering version of Rab1 (Rab1aS25N) interfered with efficient replication of *L. pneumophila* within host cells, although the observed effects were modest.

The observation that Rab1b is directed to the LCV suggested that SNARE proteins involved in ER-to-Golgi transport might also be involved in LCV biogenesis. Immunofluorescence microscopy revealed that the SNARE Sec22b is present at the LCV (Derre and Isberg, 2004; Kagan *et al.*, 2004). Interestingly, antibodies directed against the ER-to-Golgi SNARE proteins syntaxin 5, membrin, and Bet1 were also used as probes, but none of these proteins were found at the LCV. Immunogold labelling studies further revealed Sec22b staining of vesicles attached to the LCV and on the vacuole membrane surrounding *L. pneumophila* bacteria (Kagan *et al.*, 2004), suggesting that membrane fusion may occur between early secretory vesicles and the LCV. This hypothesis is further supported by recent electron microscopy studies utilizing immunohistochemistry and cytochemical techniques (Robinson and Roy, 2006), which indicate that vesicles attached to the LCV are derived from the ER based on the presence of the resident ER proteins glucose-6-phosphatase, protein disulphide isomerase (PDI) and proteins possessing the C-terminal KDEL ER-retention motif. At later time points of infection, electron microscopy analysis indicated that soluble ER markers were delivered to the lumen of the LCV, an observation that is consistent with the hypothesis that ER vesicles undergo fusion with the LCV (Robinson and Roy, 2006). Fusion of ER with the LCV may be important for the delivery of peptides, lipids and other nutrients into the lumen of the LCV that are utilized for bacterial replication.

Translocated substrates of the Dot/Icm system directly modulate Rab1, consistent with the model that most of the details of LCV formation are due to interaction of Dot/Icm substrates with early secretory system proteins. In one study, a visual screen was conducted to identify *L. pneumophila* mutants with defects in Rab1 recruitment to the LCV (Murata *et al.*, 2006). In a separate study, a biochemical approach was employed to identify potential Rab1 binding proteins from *L. pneumophila* lysates (Machner and Isberg, 2006). Both studies identified a protein termed SidM (substrate of the Icm/Dot system M)/DrrA (defective in Rab recruitment A). SidM/DrrA is capable of physically interacting with Rab1, and *L. pneumophila* strains deleted for SidM/DrrA fail to recruit Rab1 to the LCV at all time points of infection examined (Fig. 10.1; Machner and Isberg, 2006; Murata *et al.*, 2006). Furthermore, both groups showed that SidM/DrrA acts as a potent and specific Rab1 guanine nucleotide-exchange factor (GEF) *in vitro*. Rab1 activated in this fashion may recruit secretory vesicles to the LCV and facilitate their tethering to the vacuolar membrane.

LidA is a second substrate of the Dot/Icm system that appears to modulate Rab1 function. Mutations in *lidA* were originally isolated in a genetic screen designed to identify factors important for regulating the assembly and/or function of the Dot/Icm system (Conover *et al.*, 2003). Ectopic expression of LidA in eukaryotic cells results in fragmentation of the Golgi apparatus, a phenotype consistent with disruption of Rab1 function (Derre and Isberg, 2005). Furthermore, Machner and Isberg demonstrated that LidA physically interacts with Rab1 (Machner and Isberg, 2006). *L. pneumophila* strains deleted for LidA exhibited a defect in Rab1 recruitment to the LCV at early time points of infection (30 min) but displayed normal levels of Rab1 recruitment at later time points (2 h) (Machner and Isberg, 2006). Together, these data sug-

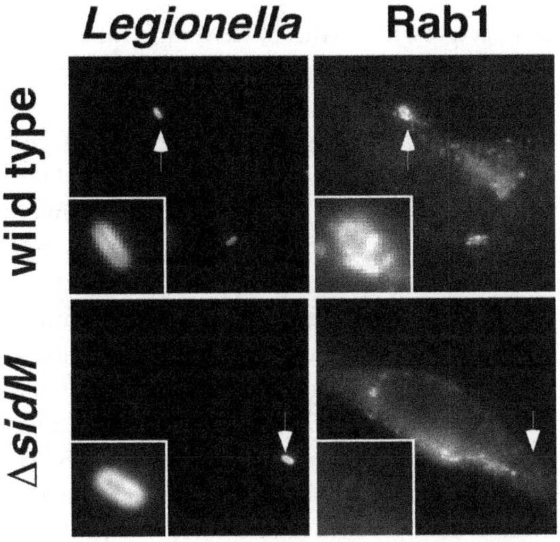

Figure 10.1 Recruitment of Rab1 to the LCV is mediated by the Dot/Icm substrate SidM/DrrA. In this experiment, mouse bone marrow-derived macrophages were infected with wild-type *L. pneumophila* or a *ΔsidM/drrA* mutant strain for 1 hour. Macrophages were then fixed and stained with antibodies directed against *L. pneumophila* or Rab1 and analysed by fluorescence microscopy. In the case of wild-type bacteria, note the robust Rab1 staining surrounding the *Legionella* containing vacuole (LCV). In contrast, *ΔsidM/drrA* mutants fail to recruit Rab1 to the LCV. (Images provided courtesy of M. Machner.)

gest a model where SidM/DrrA and LidA are translocated into host cells via the Dot/Icm translocon early in the infection process. SidM/DrrA then acts as a guanine nucleotide exchange factor to activate host cell Rab1, stimulating the recruitment of this protein to the LCV. LidA may collaborate with SidM/DrrA to support this process. Consistent with this hypothesis, beads coated with both purified SidM/DrrA and LidA bound significantly more ER-derived vesicles from host cell lysates than beads coated with each individual protein alone in an electron microscopy based assay (Machner and Isberg, 2006).

While the activities of SidM/DrrA and LidA are clearly important for integration of the LCV into the host secretory pathway, these proteins are apparently not absolutely required for this process. Similar to a *ΔralF* mutant, a *ΔsidM/drrA* mutant shows no intracellular growth defect, while *ΔlidA* mutants display only a mild defect in intracellular growth (Conover *et al.*, 2003; Machner and Isberg, 2006; Murata *et al.*, 2006). Additional translocated substrates other than RalF, SidM/DrrA, and LidA might be injected into host cells to direct ER-derived membranes to the LCV. Alternatively, there may be more than one source of vesicles trafficked to the LCV surface. This latter possibility is discussed below.

Multiple host membrane trafficking pathways are important for *L. pneumophila* intracellular replication

The absence of strong defects in *L. pneumophila* strains lacking individual translocated substrates suggest that there may be multiple strategies for generating the LCV. A recent study investigated this possibility by utilizing *Drosophila* tissue culture cells to identify host cell proteins that may facilitate *L. pneumophila* intracellular growth. This cell type is amenable to RNA interference, a technique used for inhibiting the expression of a specific target gene through the addition of exogenously added double-stranded RNA (dsRNA) (Hammond *et al.*, 2000). Furthermore, *Drosophila* cells offer the additional advantage of allowing 'knockdown' of multiple genes at once by the simultaneous addition of multiple dsRNAs (Boutros *et al.*, 2002). In this study, a bank of dsRNAs directed against various components of the secretory pathway, the endocytic pathway, and ER-related processes pathway was created, and knock-downs were screened visually for increased or decreased intracellular replication of *L. pneumophila*.

Treatment of cells with dsRNAs directed against Sar1 and ARF1 caused inhibition of *L. pneumophila* intracellular replication (Dorer *et al.*, 2006). This result is consistent with previous observations that expression of dominant interfering variants of these proteins also depresses *L. pneumophila* intracellular growth (Kagan and Roy 2002). Interestingly, treatment of host cells with most of the other dsRNAs directed against genes encoding early and late secretory factors did not depress *L. pneumophila* intracellular replication. For example, knockdown of the ER-to-Golgi SNARE protein Sec22 did not depress bacterial growth. This result is surprising as there is evidence that Sec22 is important for

L. pneumophila intracellular replication in CHO cells, where Sec22 function was disrupted by overexpression of the cognate SNARE membrin (Kagan *et al.*, 2004).

One explanation for the lack of effect by Sec22 knockdown is that multiple vesicle trafficking pathways flow into the LCV to facilitate vacuolar remodelling. In this model, severe defects in intracellular growth of the bacterium might only be observed when multiple trafficking pathways are simultaneously perturbed. Consistent with this idea, simultaneous knockdown of Sec22 and ARF1 severely inhibited *L. pneumophila* intracellular growth (Dorer *et al.*, 2006). Similarly, knockdown of Sec22 in combination with the early secretory factors Bet3 and Bet5, none of which show strong defects when performed individually, severely depressed intracellular growth of the bacterium. Bet3 and Bet5 are subunits of the multifunctional TRAPP (transport protein particle), a heteromeric multisubunit complex localized to the ERGIC and Golgi complex. Current data suggest that TRAPP plays an important role in the tethering of ER-derived vesicles to early Golgi elements as well as in the homotypic tethering of early secretory carriers (Loh *et al.*, 2005; Yu *et al.*, 2006; Kim *et al.*, 2006; Cai *et al.*, 2007).

These data are consistent with a model in which multiple transport pathways deliver membrane to the LCV. In this model, the disruption of a single input pathway fails to depress intracellular replication because of compensation by the action of additional input pathways providing the LCV with membrane material. An alternative explanation for these observations is that the double RNAi knockdowns cause additive, rather than synergistic effects.

A role for endoplasmic reticulum-associated degradation (ERAD) in *L. pneumophila* intracellular replication

While RNAi mediated knockdown of most single gene targets in *Drosophila* cells failed to cause significant defects in *L. pneumophila* intracellular growth, single knockdown of the ER associated ATPase Cdc48/p97 strongly depressed intracellular growth of the bacterium (Dorer *et al.*, 2006). Cdc48/p97 is a multifunctional protein thought to play roles in distinct biological pathways. Specifically, Cdc48/p97 function has been implicated in the fusion of mitotic Golgi fragments and the fusion of ER subdomains (Rabouille *et al.*, 1998; Roy *et al.*, 2000). Cdc48/p97 is also thought to play a key role in a process termed ERAD (ER associated degradation), which promotes the removal and degradation of misfolded proteins from the ER. Consistent with a role for Cdc48/p97 in ERAD, RNAi knockdown of factors that associate with Cdc48/p97 during ERAD caused defects in *L. pneumophila* intracellular growth. These factors include the proteins Npl4, Ufd1, and Ufd3. In contrast, knockdown of factors that bind to Cdc48/p97 in the context of membrane fusion pathways (such as p47) did not depress bacterial growth. Consistent with this observation, RNAi-mediated knockdown of Cdc48/p97 and cytosolic ERAD components in human tissue culture cells also strongly depressed *L. pneumophila* intracellular growth, indicating that knockdown in *Drosophila* cells is a valid predictive tool for *L. pneumophila* growth defects in mammalian cells (Dorer *et al.*, 2006). Furthermore, Cdc48/p97 function was shown to be specifically required for *L. pneumophila* growth, as knockdown of Cdc48/p97 did not affect the intracellular growth of two other pathogens, *Yersinia pseudotuberculosis* and *Salmonella typhimurium*.

Why is ERAD important for *L. pneumophila* intracellular growth? In host cells, the ERAD machinery functions to dislocate misfolded or otherwise damaged proteins from the ER and deliver them to the proteasome for ubiquitin-mediated degradation. Within this process, Cdc48/p97 is thought to act as a cytosolic chaperone and may provide force to remove misfolded proteins from ER membranes (Ye *et al.*, 2001; Jarosch *et al.*, 2002). By extension of this model, the ERAD machinery may therefore be exploited by *L. pneumophila* to assist in the removal of effector proteins from the Dot/Icm secretion channel. Specifically, the bacterium may recruit Cdc48/p97 and associated proteins to the nascent phagosome to assist in delivery of effector proteins into the host cell cytoplasm. Consistent with this idea, the authors show that Cdc48/p97 is present at the majority of LCVs early in the infection process (Dorer *et al.*, 2006). Cdc48/p97 appears to be important for the removal of the effector proteins LidA and SidC from the surface of the LCV (Dorer *et al.*, 2006). Alternatively, *L. pneumophila* may exploit the ERAD machinery to degrade effector proteins when their role in LCV maturation is fulfilled. Little is known about the manner in which the LCV is remodelled over the course of an infection, and efficient intracellular growth may require that certain effectors be eliminated once they have performed their function. Failure to regulate the activities of distinct groups of effectors at appropriate stages of the infection cycle may interrupt proper LCV maturation. Consistent with this idea, the majority of LCVs in mouse bone marrow derived macrophages positively stain for polyubiquitin via immunofluorescence microscopy throughout the infection cycle, although the identity of these polyubiquinated substrates is currently unknown.

Future studies should address the precise mechanisms by which the ERAD machinery functions to facilitate *L. pneumophila* intracellular growth. A key question concerns the identity of the polyubiquinated substrates observed at the surface of the LCV during the infection cycle. Furthermore, additional host cell factors important for *L. pneumophila* replication are likely to exist. Additional directed screens using RNAi in combination with *Drosophila* tissue culture cells should permit the identification of such factors.

Modulation of endosomal trafficking pathways by *L. pneumophila*

Recent studies support the hypothesis that *L. pneumophila* converts the nascent phagosome into an ER-derived organelle via the activities of Dot/Icm translocated substrates. It has also been hypothesized that some translocated effectors may inhibit host cell endocytic trafficking pathways. In this way, the bacterium may actively prevent fusion of the LCV with endocytic compartments and consequently block remodelling of the LCV into a lysosomal organelle restrictive for bacterial growth. Shuman and coworkers carried out a genetic screen employing the budding yeast *Saccharomyces cerevisiae* in an effort to identify translocated effectors that interfere with host cell endocytic function (Shohdy *et al.*, 2005). Yeast offers a well-characterized system in which to study organelle trafficking pathways (Lesser and Miller, 2001; Valdivia, 2004) by transforming cells with an *L. pneumophila* genomic library and screening for strains in which a secreted protein is missorted. In this case, the authors screened for missorting of a protein that normally targets to the vacuole (yeast lysosome), mimicking yeast mutants known as *vps* (vacuolar protein sorting) (Bankaitis *et al.*, 1986; Rothman and Stevens, 1986). *L. pneumophila* proteins that cause this phenotype were considered to be candidates that modulate host cell endocytic processes during infection.

The three different *L. pneumophila* proteins that caused a *vps* phenotype (VipA, VipD and VipF) were shown to be translocated into host cells during infection using an enzymatic reporter assay (Shohdy *et al.*, 2005). Closer examination of the consequences of the expression of these proteins in yeast revealed mechanistic differences in the manner in which the *vps* phenotype appeared to be generated. For example, expression of VipD caused a defect in transport of proteins between the ER and the Golgi complex as well as perturbing proper trafficking to the vacuole. Interestingly, *L. pneumophila* possesses three other genes with significant homology to *vipD*. Individual deletions of the Vip proteins failed to produce a significant growth defect during infection (Shohdy *et al.*, 2005; VanRheenen *et al.*, 2006), although a bacterial strain lacking VipD and its three paralogues displayed a growth advantage in the amoebal host *Dictyostelium discoideum*, suggesting this protein family may effect intracellular growth in a complex manner. It is likely that additional translocated effectors which modulate endosomal trafficking exist which have yet to be identified and characterized. Fig. 10.2 depicts a working model for the interaction of the LCV with host membrane trafficking pathways and the ERAD system, as well as the role particular Dot/Icm translocated substrates may play in this process.

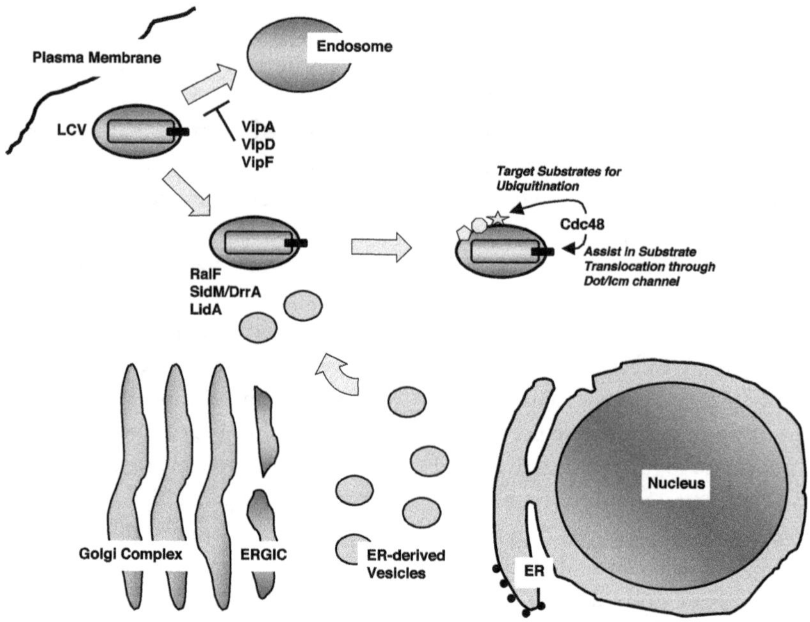

Figure 10.2 Working model for modulation of host cell processes by *L. pneumophila* to establish and maintain infection. During phagocytosis by host cells, *L. pneumophila* deposits effector proteins into the host cell via the Dot/Icm secretion system. Some of these proteins, such as VipA, VipD and VipF may function to block entry of the LCV (*Legionella* containing vacuole) into the endocytic pathway by targeting as yet unidentified host cell factors. Other effectors, such as RalF, SidM/DrrA and LidA modulate the action of host cell ER-to-Golgi vesicular trafficking factors ARF1 and Rab1. Modulation of ARF1 and Rab1 function by these translocated substrates results in the recruitment of ER-derived vesicles to the LCV, facilitating the remodelling of this compartment into a replicative organelle. Cdc48/p97 is recruited to the LCV by an unknown mechanism. Cdc48/p97 may act to assist the removal of additional translocated substrates from the Dot/Icm channel. Alternatively, Cdc/p97 may work in conjunction with the ubiquitin/proteasome system to target effectors for degradation once their functions in LCV biogenesis have been fulfilled.

Interaction of *L. pneumophila* with host cell phosphoinositide signalling and autophagy pathways

An additional critical aspect of host cell membrane traffic pathways is the modulation of lipid dynamics and composition. In particular, the metabolism of phosphoinositides is known to be a key event in many membrane trafficking events including phagoctyosis, endocytosis and exocytosis (Gillooly *et al.*, 2001; De Matteis and Godi, 2004). The local levels of specific phosphoinositides can be rapidly altered by a wide array of kinases and phosphatases, allowing temporal and spatial control of signalling events to promote the fidelity of membrane traffic (Behnia and Munro, 2005). Phosphoinositide metabolism appears to affect intracellular growth of *L. pneumophila*. *L. pneumophila* replicates more robustly within a *D. discoideum* strain lacking the class I phosphoinositol (3) kinases PI3K-1 and –2, which phosphorylate the D-*myo*-inositol ring on phosphoinositide (PI) at position 3 to generate phosphoinositide 3-phosphate (PI(3)P) (Weber *et al.*, 2006). This observation suggests that the activity of PI3K1/2 in some way restricts intracellular growth of the bacterium. Consistent with this idea, pharmacological inhibition of PI3K1/2 activity also results in enhanced bacterial growth in *D. discoideum*. However, treatment of a mammalian macrophage-like cell line with the same inhibitor did not result in increased *L. pneumophila* replication, suggesting these findings may be specific to *D. discoideum*.

The finding that alterations in host cell phosphoinositide metabolism can cause changes in intracellular replication suggests *L. pneumophila* may interact with components of these pathways. In fact, the translocated substrate SidC and its paralogue SdcA specifically bound phosphoinositol 4-phosphate (PI(4)P) *in vitro*. Interestingly, both SidC and SdcA lack classical PH or other phosphoinositol-binding domains suggesting their interaction with PI(4)P is mediated by a novel protein motif yet to be characterized (Weber *et al.*, 2006). Further, PI(4)P is present on LCVs isolated from infected *D. discoideum* and mammalian cells (Weber *et al.*, 2006). These observations suggest that *L. pneumophila* may exploit host cell PI(4)P as a means of anchoring translocated substrates to the surface of the replicative vacuole. Further, the PI3K1/2 deletion strain displays increased SidC staining about the LCV as compared to a wild-type host, consistent with the PI3K1/2 deletion strain containing more PI(4)P compared to wild-type amoebae at steady state (Zhou *et al.*, 1998). Future questions that extend from this work include asking how PI(4)P is generated and maintained at the LCV membrane. Further, the identification of additional translocated substrates that interact with phosphoinositide derivatives should help elucidate the molecular mechanisms by which these molecules impact intracellular replication.

Autophagy is another host cell protein trafficking pathway that has been suggested to play a role in the remodelling of the LCV into a replicative organelle (Swanson and Fernandez-Moreira, 2002). This hypothesis extends in part from the observation that the LCV bears some morphological resemblance to an organelle termed an autophagosome (Swanson and Isberg, 1995; Amer and Swanson, 2005). Furthermore, in permissive A/J mouse macrophages, some studies indicate that the ER-derived LCV slowly merges with lysosomes, a maturation process that apparently favours bacterial replication (Sturgill-Koszycki and Swanson, 2000). However, other work has shown that phagosomal acidification is not a prerequisite for intracellular multiplication of *L. pneumophila* in human monocytes (Horwitz and Maxfield, 1984; Wieland *et al.*, 2004). Autophagy is a conserved membrane trafficking pathway by which eukaryotic cells encapsulate cytoplasmic molecules for delivery to lysosomes and degradation as a response to nutrient starvation (Shintani and Klionsky, 2004; Eskelinen, 2005). During maturation of the double membrane-bound autophagosome it acquires lysosomal makers and eventually become a degradative organelle (Dunn, 1990a). The origin of the membrane material comprising the nascent phagosome remains unclear, although different studies have proposed input from the ER, a post-Golgi compartment, or a novel organelle called a phagophore (Dunn, 1990b; Ueno *et al.*, 1991; Stromhaug *et al.*, 1998; Rich *et al.*, 2003).

To investigate the hypothesis that the host cell autophagy machinery may account for the engulfment of the *L. pneumophila* phagosome by secretory membranes, different experimental approaches have been undertaken. In one study, *D. discoideum* strains deficient for particular autophagy genes were found to support efficient *L. pneumophila* replication (Otto *et al.*, 2004). However, in mouse macrophages the autophagy proteins Atg7 and Atg8 are recruited to the LCV during infection (Amer and Swanson, 2005). In addition, this study showed that LCVs in infected macrophages derived from a strain of mice resistant to *L. pneumophila* infection appeared to acquire autophagy markers more quickly as compared to macrophages from permissive mice. Based on these and other experiments, the authors propose that macrophages elevate autophagy as a barrier to infection (Swanson, 2006). According to this model, host factors initiate recruitment of secretory membranes to the bacterial vacuole, whereas certain effectors of Type IV secretion stabilize these interactions to retard delivery of *L. pneumophila* to autophagolysosomes. Further experiments are necessary to elucidate the precise mechanistic contributions of autophagy to the ability of *L. pneumophila* to replicate within eukaryotic hosts.

Modulation of host cell death signalling pathways by *L. pneumophila* to promote intracellular survival

In addition to sequestering and killing invading bacteria through the endocytic pathway, the presence of bacteria in the host cell elicits a number of alternative defences including the activation of programmed cell death. Thus, in addition to preventing fusion of the LCV with lysosomes, the intracellular survival of *L. pneumophila* also depends on its ability to modulate host cell death signalling pathways that are initiated in response to infection (Muller *et al.*, 1996; Gao and Abu Kwaik, 1999a; Gao and Abu Kwaik, 1999b). Apoptotic cell death involves regulated catabolism of the cell and is mediated by a series of cysteine proteases called caspases. This form of programmed cell death is characterized by nuclear condensation and DNA fragmentation in the presence of an intact membrane (Kerr *et al.*, 1995; McGahon *et al.*, 1995). Gao and Abu Kwaik (1999a, 1999b) demonstrated that caspase-dependent apoptotic signalling pathways were activated in response to *L. pneumophila* within hours of infection however, the majority of infected cells did not exhibit the classic signs of apoptosis at this early time point. Instead, nuclear condensation and DNA fragmentation occurred much later in the infection cycle suggesting that *L. pneumophila* was able to inhibit or at least delay this process until replication was complete (Muller *et al.*, 1996; Laguna *et al.*, 2006).

Activation of host apoptotic cell death pathways by *L. pneumophila*

To better appreciate how *L. pneumophila* is able to modulate host cell death pathways is it important to briefly review the pathways that mediate the apoptotic death response (for review see Allen *et al.*, 1998; Chen and Wang, 2002) and the mechanisms by which the bacterium activates these processes. The extrinsic death receptor pathway involves signalling through tumour necrosis factor receptor 1 (TNFR-1) and CD95/Fas/Apo-1 receptor in response to the extracellular ligands TNFα and Fas, respectively (Fig. 10.3). The activated receptors recruit their corresponding adaptor proteins TNF-associated death domain (TNFADD) and Fas-associated death domain (FADD) which in turn activate the initiator caspase-8. The intrinsic mitochondrial death pathway is initiated by damage from UV and γ-irradiation, staurosporin, cycloheximide and anticancer drugs which cause release of cytochrome *c* from mitochondria. This is facilitated by the pro-apoptotic proteins Bak/Bax and inhibited by the anti-apoptotic protein Bcl-2. Cytochrome *c* in the cytoplasm interacts with Apaf-1 which together activate the initiator

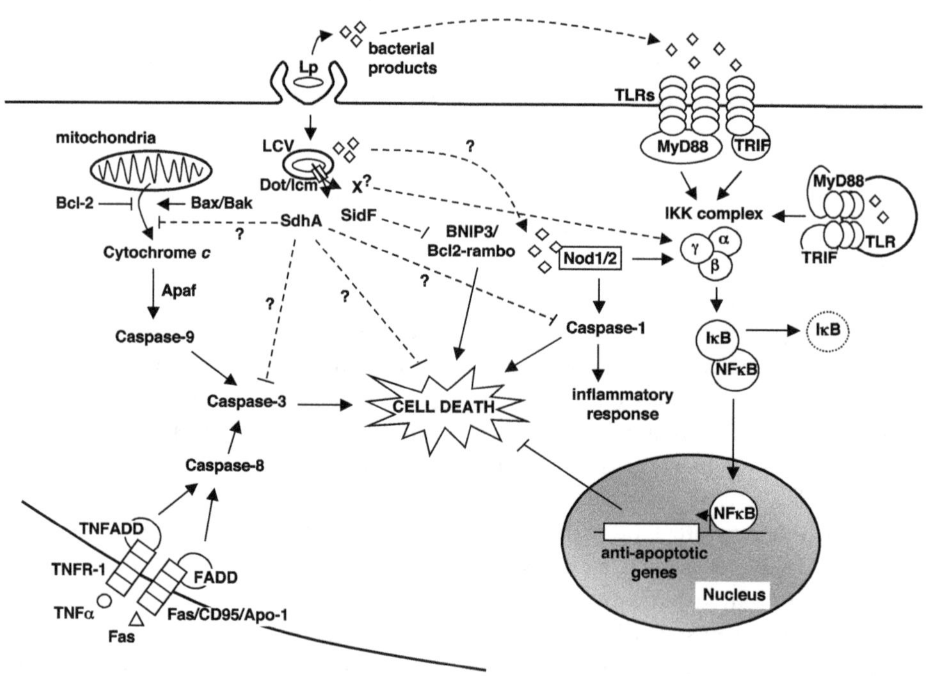

Figure 10.3 Modulation of host cell death signalling pathways by *L. pneumophila*. *L. pneumophila* employs at least two mechanisms to prevent host apoptotic cell death during infection: the first involves the function of the Dot/Icm secreted effector SdhA which is important for preventing cytochrome *c* release from mitochondria and inhibits caspase activity however, the precise targets of SdhA and the molecular mechanism by which it accomplishes this are unknown; the second involves increased expression of anti-apoptotic genes through the activation of NFκB. This could be mediated by another, yet unknown translocated effector (X), by the host cell in response to various bacterial macromolecules or by other mechanisms as yet undetermined.

caspase-9. Activated caspase-8 or caspase-9 in turn activates the executioner caspases-3, 6 and 7 which then initiate the process of programmed cell death.

L. pneumophila infection of both mammalian macrophages and alveolar epithelial cells has been shown by Gao and Abu Kwaik (1999b) to cause cleavage of PARP, a endogenous substrate of caspase-3 suggesting a role for caspase-3 in mediating the apoptotic response of the cell. Indeed, treatment of cells prior to infection with a caspase-3 specific inhibitor, z-DEVD-fmk prevented both production of the PARP cleavage product and the death of infected cells associated with apoptosis (Gao and Abu Kwaik, 1999b). To better define the pathway involved in caspase-3-mediated cell death induced by *L. pneumophila*, three independent groups tested whether caspase-3 activation was dependent on various components of both the intrinsic and extrinsic death signalling pathways (Neumeister *et al.*, 2002; Molmeret *et al.*, 2004; Fischer *et al.*, 2006).

In Jurkat cells infected with *L. pneumophila* at a high multiplicity of infection, cytoplasmic cytochrome *c* as well as the proteolytic processing of caspase-8, caspase-9 and caspase-3 to their active forms has been observed (Neumeister *et al.*, 2002; Fischer, 2006). Neumeister *et al.*, (2002) demonstrated that *L. pneumophila*-induced activation of caspase-8, caspase-9, caspase-3 and cleavage of PARP could be blocked either by the overexpression of Bcl-2 or by treating cells with the broad spectrum caspase inhibitor z-VAD-fmk. Furthermore, *L. pneumophila* induced similar levels of DNA fragmentation and PARP cleavage in Jurkat cells lacking either the death receptor CD95 or its adapter protein FADD compared to wild-type cells (Neumeister *et al.*, 2002). In addition, treatment of HL-60 cells with a TNFR-1 inhibitor pentoxifylline or anti-TNFα had no effect on the ability of *L. pneumophila* to induce apoptosis (Hagele *et al.*, 1998; Arakaki *et al.*, 2002). Collectively these results suggested that *L. pneumophila* activate apoptotic cell death through the intrinsic mitochondrial death pathway.

Although the activation of caspase-8 supports a role for death receptor mediating apoptosis, caspase-8$^{-/-}$ Jurkat cells infected with *L. pneumophila* exhibited both caspase-3 processing and cleavage of PARP, although to a lesser extent than in the parental strain (Neumeister *et al.*, 2002). These results suggested that although caspase-8 is not required for *L. pneumophila*-induced apoptotic cell death, it may contribute via a non-classical death receptor-independent pathway. Indeed, it has been demonstrated that caspase-8 can be induced by anticancer drugs independent of death receptor function (Ferrari *et al.*, 1998; Wesselborg *et al.*, 1999) and can also act downstream of caspase-9 during activation of apoptosis through the intrinsic death pathway (Chinnaiyan *et al.*, 1997; Slee *et al.*, 1999; Engels *et al.*, 2000).

Molmeret *et al.*, (2004) have proposed that at very early stages of an infection, *L. pneumophila* may activate caspase-3 directly as a critical step in supporting replication. In this study, the authors reported that at early time points, *L. pneumophila*-infected U937 cells exhibited proteolytic processing of caspase-3 but not caspase-8 or caspase-9, nor did the authors observe cytochrome *c* release from mitochondria (Molmeret *et al.*, 2004). Moreover, although caspase-3 activation was detected in cells expressing a dominant negative variant of FADD consistent with the results obtained by Neumeister *et al.*, (2002), overexpression of Bcl-2 in human prostate carcinoma cells did not block *L. pneumophila*-induced activation of caspase-3. It is possible that the discrepancies in these results may involve differences in the strains, cell lines and experimental procedures used in the two studies. The authors proposed a role for caspase-3 during the initial stages of infection in preventing fusion of the LCV with endocytic vesicles based on two observations: 1) *L. pneumophila* infection resulted in cleavage of Rabaptin-5, an event known to result from caspase-3 cleavage, which may interfere with endosomal fusion events (Stenmark *et al.*, 1995; Cosulich *et al.*, 1997; Swanton *et al.*, 1999) and 2) treating cells with a caspase-3 specific inhibitor restricted *L. pneumophila* growth and caused wild-type bacteria to co-localize with the late lysosomal marker LAMP-2, although the presence of the early endosomal marker LAMP-1 about the LCV was not observed. Thus, direct activation of caspase-3 at early time points may initially benefit the bacterium by enabling it to escape lysosomal degradation (Molmeret *et al.*, 2004).

Interestingly, Gao and Abu Kwaik (1999b) reported that *L. pneumophila* infected cells treated with caspase-3 inhibitor began to show signs of apoptosis 12 hours post infection despite the presence of the inhibitor throughout the infection. Thus, although inhibition of caspase activity prevented cell death at early stages of the infection, it was not sufficient for sustained protection against *L. pneumophila*-induced cytopathogenicity (Gao and Abu Kwaik, 1999b). This suggested the existence of an alternative, caspase-3-independent pathway for activating apoptosis in response to *L. pneumophila* infection. Indeed, *L. pneumophila* also activate a caspase-1-dependent signalling pathway. This involves recognition of the bacterial flagellar protein flagellin in the host cytoplasm by the Nod-like protein Naip5 and the subsequent induction of caspase-1 (Ren *et al.*, 2006; Molofsky *et al.*, 2006; see Chapter 5). Although the activity of casapase-1 is typically associated with the activation of a proinflammatory response to bacterial invasion, recent work has suggested an additional role for capase-1 in inducing apoptosis in response to intracellular pathogens. Indeed, Derre and Isberg (2005) showed that in a subpopulation of restrictive C57B/6 macrophages in which *L. pneumophila* had successfully initiated replication, there were signs of advanced apoptosis. That is, if the bacterium is able to establish a replicative vacuole in spite of the restrictive nature of the host macrophage, the cell may induce apoptosis as a second line of defence.

It has been shown that at high multiplicities of infection, *L. pneumophila* cause rapid cytotoxic killing of host cells (Keen *et al.*, 1989; Szeto *et al.*, 1990; Wintermeyer *et al.*, 1991; Husmann *et al.*, 1994). This necrotic form of cell death is attributed to the Dot/Icm-dependent pore-forming activity of *L. pneumophila* that punctures holes in the host plasma membrane causing the cell to rupture (Kirby *et al.*, 1998). The apoptotic response described above was similarly shown to be dependent on various components of the Dot/Icm machinery (Gao and Abu Kwaik, 1999b; Molmeret *et al.*, 2004). This presents a serious problem for the bacterium because although a functional Dot/Icm secretion system is essential for the intracellular survival of *L. pneumophila*, this complex is also responsible for inducing host cell death (Watarai *et al.*, 2001; Diez *et al.*, 2003; Wright *et al.*, 2003; Derre and Isberg, 2004; Zamboni *et al.*, 2006) which is antagonistic to its survival. Therefore, the ability to interfere with apoptotic signalling pathways is necessary for permitting intracellular growth of the bacterium. In support of this, Abu-Zant *et al.*, (2007) demonstrated that infection of human monocyte-derived macrophages (hMDM) with *L. pneumophila* could protect the cells from death induced by the exogenous addition of the apoptosis-stimulating agent staurosporin, indicating that a bacterial product may interfere with cell death pathways activated in response to infection (Abu-Zant *et al.*, 2007). The bacterial products required for this interference appear to be Dot/Icm substrates (Abu-Zant *et al.*, 2007).

SidF-mediated protection against host cell death

SidF, originally identified amongst several Dot/Icm substrates using an inter-bacterial transfer assay (Luo and Isberg, 2004), has been shown to play a role in protecting the host cells against *L. pneumophila*-induced cell death. (Banga *et al.*, 2007). Indeed, Banga *et al.*, (2007) reported that a Δ*sidF* mutant exhibits a slight defect in intracellular growth in murine derived bone marrow macrophages but grows as proficiently as the wild-type strain in both human-derived macrophage-like U937 cells and the amoebae *D. discoideum*. Furthermore, this mutant is able to avoid delivery to lysosomes and is able to generate a vacuole that supports bacterial replication. However, the authors also observed a significant increase in the number of infected host cells undergoing apoptosis when infected with the Δ*sidF* mutant compared to wild type or a Δ*dotA* mutant that is defective for Dot/Icm-dependent translocation. Moreover, the absence of SidF rendered *L. pneumophila* less capable of protecting the host against staurosporine-induced apoptosis compared to wild-type cells whereas the overexpression of *sidF* in either HeLa or U937 cells resulted in an increased resistance to staurosporine-induced cell death (Banga *et al.*, 2007).

Immunofluorescence analysis demonstrated that SidF was translocated to the host cell cytoplasm in a Dot/Icm-dependent manner (Banga *et al.*, 2007). Subsequent yeast two-hybrid experiments identified two host cell targets of SidF from the Bcl2 family of pro-apoptotic proteins, Bcl-rambo and BNIP3 (Banga *et al.*, 2007). MCF-7 cells transfected to express both Bcl-rambo and *sidF* showed significantly less host cell apoptosis than in cells expressing Bcl-rambo alone demonstrating that SidF is capable of protecting host cells against Bcl-rambo mediated death (Banga *et al.*, 2007). Future characterization of the mechanism by which SidF manipulates these Bcl2 family proteins will provide insight as to how *L. pneumophila* is able to modulate host cell death pathways to promote its intracellular survival.

SdhA-mediated protection against host cell death

To identify genes encoding Dot/Icm effectors that are required for intracellular growth, Laguna *et al.*, (2006) performed a screen to isolate *L. pneumophila* transposon mutants that were defective in intracellular growth in U937 macrophages-like cells but which had an intact Dot/Icm apparatus (Conover *et al.*, 2003; Laguna *et al.*, 2006). This work identified a novel gene designated *sdhA* whose encoded protein could be detected in the cytoplasm of U937 cells infected with wild-type *L. pneumophila* but not the Δ*dotA* mutant. These results demonstrated that SdhA was important for growth in macrophages and deposited into the host cell by the Dot/Icm secretion system during an infection.

An *sdhA* null mutant exhibits over a 100-fold defect in intracellular growth in bone-marrow derived mouse macrophages compared to wild-type bacteria (Laguna *et al.*, 2006). Despite this severe growth defect, the absence of SdhA caused no apparent defect in intracellular trafficking of the LCV (Laguna *et al.*, 2006). Instead, cells showed increased amounts of DNA fragmentation, aberrant cytochrome *c* staining, condensed cell morphology with extensive membrane ruffling and significant increases in caspase activity, classic indicators of apoptosis (Laguna *et al.*, 2006). Interestingly, treatment of host cells with a broad specificity caspase inhibitor rescued the growth defect of the Δ*sdhA* mutant (Laguna *et al.*, 2006). These data indicated that SdhA functions to regulate the progression of the host cell death response of infected cells by interfering with caspase-dependent apoptotic signalling events (Laguna *et al.*, 2006). The mechanism by which SdhA achieves this is unknown. There are no significant SdhA homologues present in the GenBank NR database however, this protein is predicted to have several coiled-coil protein binding domains and a putative GRIP domain that is typically responsible for targeting proteins to the trans-Golgi or Golgi membrane and mediating their interaction with Arl1 (Arf1-like) small GTPases (Lu *et al.*, 2006). Future characterization of the molecular mechanism by which SdhA exerts its effect on the host will be important for understanding how *L. pneumophila* is able to modulate cell death signalling pathways.

Based on this work, there was significant evidence to suggest that *L. pneumophila*-induced host cell death occurs via multiple pathways. First, caspase inhibition in host cells was not sufficient to fully restore growth of the Δ*sdhA* mutant to wild-type levels (Laguna *et al.*, 2006). Second, although caspase inhibitor treated cells infected with the Δ*sdhA* mutant showed no DNA degradation, a subpopulation of infected cells still exhibited abnormal cytochrome *c* staining and aberrant cell morphology (Laguna *et al.*, 2006). These results suggested that SdhA inhibits host cell death by multiple mechanisms. Third, the growth of wild-type bacteria within the first 48 hours of infection was enhanced in cells treated with caspase inhibitor compared to that in untreated cells (Laguna *et al.*, 2006). This suggested that SdhA alone was not sufficient to completely inhibit caspase-dependent cell death pathways. Fourth, inhibition of caspase activity only protected against host cell death at early time points during infection with an Δ*sdhA* mutant as cells eventually succumbed to apoptotic death as the infection progressed (Laguna *et al.*, 2006). This observation was consistent with those reported by Gao and Abu Kwaik (1999b) in which treatment of infected cells with caspase inhibitor could not provide prolonged protection against apoptotic host cell death.

Although the alternative pathways involved in mediating *L. pneumophila*-induced cell death are not known, there are several possibilities. First, Dot/Icm effector proteins may be toxic to the host cell and inadvertently induce programmed cell death. Second, in addition to activating anti-apoptotic genes, *L. pneumophila* induced translocation of NFκB to the nucleus (see below) also results in the upregulation of a number of pro-apoptotic genes (Losick *et al.*, 2006; Abu-Zant *et al.*, 2007). The balance between these two opposing pathways appears to favour cell survival at early stages of infection although, activation of pro-apoptotic signalling pathways may contribute to the eventual death of the host cell at late stages of the infection cycle. Third, it has recently been shown that *L. pneumophila* encodes a glycosyltransferase protein, Lpg1369 designated Lgt1 that may function to glycosylate the eukaryotic protein EF1A (Belyi *et al.*, 2006). Modification of EF1A inhibits eukaryotic protein synthesis which eventually results in host cell death. Conceivably, Lgt1 activity may inadvertently promote host cell death as modification of EF1A may have some other, as yet undetermined role in the infectious process for which the induction of apoptosis is an indirect consequence. Indeed, EF1A also functions to regulate actin cytoskeletal rearrangements (Demma *et al.*, 1990; Shiina *et al.*, 1994; Edmonds *et al.*, 1995), lipid metabolism (Yang *et al.*, 1993) and has been suggested to play role in regulating apoptosis (Duttaroy *et al.*, 1998) and ubiquitin-dependent protein degradation (Gonen *et al.*, 1994; Gonen *et al.*, 1996). Finally, both necrotic and/or autophagic cell death pathways may be activated (Vandenabeele *et al.*, 2006). Indeed, it has been proposed that autophagy may be activated as a means to clear the bacteria from infected cells (Amer and Swanson, 2005). If so, *L. pneumophila* must be properly equipped to modulate these pathways in order to maintain host cell viability until the bacteria are ready to exit the cell.

Induction of host anti-apoptotic signalling by *L. pneumophila* through the activation of NFκB

In addition to bacterial proteins that interfere with cell death pathways, *L. pneumophila* also induce the expression of anti-apoptotic host proteins during intracellular growth (Losick *et al.*, 2006; Abu-Zant *et al.*, 2007). In the U937 cell line, there is upregulation of the dual specificity phosphatase genes *dusp1*, *dusp2* and *dusp6* (Losick *et al.*, 2006) whose encoded proteins function to limit host cell death by inhibiting mitogen-activated protein kinase and c-Jun N-terminal kinase activity (Kamata *et al.*, 2005). In addition, *L. pneumophila* infection also increases expression of anti-apoptotic genes regulated by the transcription factor NFκB including *ciap1*, *ciap2* (Losick *et al.*, 2006), which inhibit apoptosis by directly binding caspases (Deveraux *et al.*, 1999), *pai-2* and *bcl2* (Losick *et al.*, 2006), which blocks cytochrome *c* release from mitochondria (Werner *et al.*, 2002). At least one of the encoded proteins, PAI-2 plays a direct role in protecting cells from death as $pai2^{-/-}$ mouse macrophages showed increased cell death in response to *L. pneumophila* (Losick *et al.*, 2006). In an independent study, Abu-Zant *et al.*, (2007) also reported an increase in expression of several host anti-apoptotic genes in hMDM cells challenged with *L. pneumophila* including two NFκB-regulated genes, *bcl2* and *xiap*, a potent inhibitor of caspase-3, 7 and 9 (Chai *et al.*, 2001; Riedl *et al.*, 2001; Shiozaki *et al.*, 2003). These studies clearly indicated there is an induction of host anti-apoptotic genes in response to *L. pneumophila* and that NFκB plays a potential role in mediating this process.

Due to the enhanced expression of genes regulated by NFκB, both groups tested whether NFκB was translocated to the nucleus in response to *L. pneumophila* infection. At a low multiplicity of infection (one bacterium per cell), wild-type *L. pneumophila* induce rapid, sustained translocation of the p65 subunit of NFκB to the nucleus whereas strains lacking a functional Dot/Icm complex were unable to do so at this low bacterial dose (Losick *et al.*, 2006). At a higher multiplicity of infection (10–50 bacteria per cell), nuclear localization of NFκB was observed in both wild-type and Δ*dotA* mutant infected cells. Unlike wild-type bacteria, the Δ*dotA* mutant was unable to maintain this effect (Losick *et al.*, 2006; Abu-Zant *et al.*, 2007). *L. pneumophila*-induced translocation of NFκB to the nucleus was shown to be dependent on the IKK kinase complex that functions to activate NFκB in response to cellular signalling events

(Abu-Zant *et al.*, 2007). These results suggested that *L. pneumophila* possess at least two mechanisms, one Dot/Icm-dependent mechanism and one Dot/Icm-independent mechanism for modulating NFκB activity.

Consistent with a role for NFκB in preventing apoptotic cell death, blocking NFκB translocation by treating cells with caffeic acid phenethyl ester (CAPE) or by overexpressing a dominant negative variant of IκB resulted in increased numbers of infected cells exhibiting condensed nuclei characteristic of apoptosis (Losick *et al.*, 2006). Furthermore, CAPE treated cells exhibiting apoptotic morphology were restrictive for the growth of wild-type bacteria (Losick *et al.*, 2006), although this was not observed in human monocyte-derived macrophages (Abu-Zant *et al.*, 2007). In addition, *L. pneumophila* were unable to protect mouse embryonic fibroblasts deficient in p65 against death induced by staurosporin, a phenomenon that was dependent on the IKK kinase complex (Abu-Zant *et al.*, 2007). Thus, NFκB activation by *L. pneumophila* plays a role in promoting host cell survival during infection.

NFκB activation can occur via both toll-like receptor (TLR) signalling or the cytoplasmic surveillance proteins Nod1/2 (for review see Philpott and Girardin, 2004). To determine if either pathway was involved in mediating the NFκB-dependent response to *L. pneumophila*, Losick *et al.*, (2006) tested whether macrophages deficient in either the TLR broad specificity adapter protein MyD88 or Nod1 exhibited NFκB translocation to the nucleus upon infection. At low multiplicities of infection, NFκB activation by wild-type bacteria did not require MyD88-dependent TLR signalling or Nod1 (Losick *et al.*, 2006). In contrast, NFκB activation by the Δ*dotA* mutant at higher multiplicities of infection was abolished in the absence of MyD88 indicating that activation of NFκB via the Dot/Icm-independent mechanism requires signalling through a TLR pathway (Losick *et al.*, 2006). Since the absence of MyD88 does not abolish TLR4 signalling it was possible that activation of NFκB may be mediated by a response to lipopolysaccharide (LPS). However, Abu-Zant *et al.*, (2007) demonstrated that formalin-killed bacteria were unable to induce translocation of NFκB to the nucleus suggesting that LPS was not likely responsible for this phenomenon.

Bacterial egress

Survival of *L. pneumophila* during infection not only requires the formation of a replicative niche that supports growth but also the ability to escape the host cell once replication is complete. Gao and Abu Kwaik (1999a) hypothesized that the necrotic killing of host cells at late stages of infection is important for bacterial egress and that mutants lacking the pore formation-associated cytotoxic activity (Husmann *et al.*, 1994; Kirby *et al.*, 1998; Alli *et al.*, 2000) would remain trapped in host cells at the end of the replication cycle. To test this theory, the authors screened a library of *L. pneumophila* transposon insertion mutants for those that exhibited defects in cytotoxic killing of amoebae and U937 macrophages-like cells but appeared to grow normally within these cells. From this work they identified six mutants designated release of intracellular bacteria (*rib*) mutants. Subsequent studies revealed that all six *rib* mutants have a frame shift mutation in *icmT* which results in the expression of a C-terminally truncated version of the IcmT protein (Molmeret *et al.*, 2002) and thus may be partial loss of function mutants.

More recently, it has been demonstrated that disruption and eventual degradation of the phagosomal membrane in both mammalian and protozoan cells may begin as soon as 12 hr post infection and that *L. pneumophila* are released into the host cell cytoplasm where they reside, and potentially replicate further, prior to the disruption of the host cell membrane and their escape into the extracellular environment (Molmeret *et al.*, 2004). It is possible that bacterial egress from the replicative vacuole and release from the host cell cytosol into the surrounding environment may be two distinct processes. Regardless, coordinating the end of replication with host cell death would allow egress of the bacteria at a point when they are highly virulent. This would promote the survival of the bacteria by exposing them to the extracellular milieu only when they are most infectious and thus most capable of initiating a second round of infection.

There is some evidence to suggest that *L. pneumophila* is capable of exiting protozoan hosts by a route that exploits the host exocytic pathway (Chen *et al.*, 2004). Chen *et al.*, (2004) predicted a role for two *L. pneumophila* genes *lepA* and *lepB* (*Legionella* effector protein) in modulating vesicle trafficking events during an infection based on their similarity to mammalian EEA1 and yeast USO1 vesicle tethering proteins. Both proteins are translocated into the host cell cytoplasm by the Dot/Icm secretion system during an infection and in the absence of both LepA and LepB, the number of bacteria released from amoebae into the supernatant during a 72 hour infection was significantly less compared to wild-type bacteria (Chen *et al.*, 2004). Transmission electron microscopy demonstrated that at 14 hours post infection, *Acanthamoebae castellanii* cells infected with the Δ*lepA*, Δ*lepB* double mutant showed numerous bacteria-containing vesicles associated with the remains of lysed cells (Chen *et al.*, 2004). Since the double mutant is able to induce high multiplicity killing of amoebae similar to the wild-type strain, the authors concluded that LepA and LepB mediate bacterial release from the host cell. It has previously been demonstrated that *L. pneumophila* are released from amoebae enclosed in intact vesicles (Berk *et al.*, 1998). The homology LepA and LepB share with SNARE proteins and their translocation to the host cytoplasm suggest they may function in trafficking the LCV via the exocytic pathway into the extracellular environment. Interestingly, none of the three Δ*lep* strains were defective in their ability to escape from mammalian cells, suggesting that neither gene was important for egress from mammalian cells (Chen *et al.*, 2004).

Identification and characterization of additional Dot/Icm translocated effectors

To better understand how *L. pneumophila* is able to manipulate the host cell, it is crucial to identify translocated Dot/Icm substrates and elucidate the mechanisms by which they function during infection. Unfortunately, this has been difficult using classical mutant screens that are based on defects in intracellular growth or strong phenotypes associated with null mutations as chromosomal deletion mutations in the majority of effector-encoding genes cause only slight, if any, measurable phenotype. As a result, a series of more direct approaches has been devised to identify substrates of the Dot/Icm secretion system.

Sid family proteins

A large family of putative effectors was identified from a screen in which a Cre/*loxP* reporter system was used to directly measure Dot/Icm-dependent protein translocation from one bacterium to another (Luo *et al.*, 2004). This strategy was based on the finding that the Dot/Icm complex was able to promote conjugative transfer of DNA between bacterial strains, a process that involves translocation of protein:DNA substrates (Vergunst *et al.*, 2000). Candidates for this screen were pre-selected based on their ability to interact with DotF, a component of the Dot/Icm machinery using a bacterial two hybrid assay (Luo *et al.*, 2004). From this work, Luo *et al.*, (2004) identified eight Dot/Icm translocated proteins which were designated substrates of Icm/Dot or Sid proteins (Table 10.1). One of these proteins, SidC, was shown to be translocated by the Dot/Icm system into the cytoplasm of mouse bone-marrow derived macrophages during an infection with wild-type *L. pneumophila*. Subsequent work has demonstrated that several other proteins identified by this screen, including the SidE paralogue SdeA (Bardill *et al.*, 2005), SidF (Banga *et al.*, 2007), SidG (Ninio *et al.*, 2005), SidH (Ninio *et al.*, 2005) and SidJ (Liu and Luo, 2007), are also translocated by the Dot/Icm system into the host cell cytoplasm during an infection.

Several of the *sid* genes including *sidB*, *sidC*, *sidE* and *sidH* have one or more paralogues within the genome (Table 10.1) (Luo *et al.*, 2004). In some cases, sets of paralogues are highly homologous throughout the entire protein while others share similarities in only specific regions or domains and the degree of similarity in these regions can range from almost identical to weakly homologous. Unlike the *dot* and *icm* genes that are organized into two main loci (Marra *et al.*, 1992; Berger and Isberg, 1993; Brand *et al.*, 1994), the majority of genes encoding known Dot/Icm substrates are individually scattered throughout the genome or in small clusters of genes (Fig. 10.4). As an exception, one cluster contains three *sidE* paralogues *sdeA*, *sdeB*

Table 10.1 Substrates of the Dot/Icm type IV secretion system

Protein	Phila-1	Paris	Lens	Paralogues	Domain/function	Evidence for Translocation*	Reference
Effectors							
LidA	lpg0940	lpp1002	lpl0971	–	Coiled-coil/ Rab1 recruitment	IF, PNS	Conover *et al*., 2003
RalF	lpg1950	lpp1932	lpl1919	–	sec7 homology domain/ Arf1 GEF	CA, IF	Nagai *et al*., 2002; Nagai *et al*., 2005
SdhA	lpg0376	lpp443	lpl419	SdhB, SidH	GRIP domain, coiled-coil/ anti-apoptosis	SE	Laguna *et al*., 2006
SidC	lpg2511	lpp2579	lpl2433	SdcA	PI(4)P binding domain	IF, PNS, ST	Luo and Isberg, 2004
SidF	lpg2584	lpp2637	lpl2507	–	Bcl2-rambo/BNIP3 binding domain	IF	Banga *et al*., 2007
SidJ	lpg2155	NH	lpl2053	SdjA	ER recruitment	SE, ST	Liu and Luo, 2007
SidM/DrrA	lpg2464	NH	lpl2384	–	Rab1 GEF	CA, IF, PNS	Machner and Isberg, 2006; Murata *et al*., 2006
Predicted effectors: contain signal for Dot/Icm-dependent translocation							
Ceg10	lpg0284	lpp360	lpl336	–	PmrA regulon	CA	Zusman *et al*., 2007
Ceg23	lpg1621	lpp1591	lpl1402	–	PmrA regulon	CA	Zusman *et al*., 2007
Ceg29	lpg2409	lpp2476	lpl2332	–	PmrA regulon	CA	Zusman *et al*., 2007
LegL3	lpg1660	lpp1631	lpl1625	–	Leucine-rich repeat	CA	deFelipe *et al*., 2005
LegLC4	lpg1948	NH	NH	LegL2, LegL5, LegL6, LegL7	Leucine-rich repeat, coiled-coil	CA	deFelipe *et al*., 2005
LegL5	lpg1958	pp1940	NH	LegC4, LegL2, LegL6, LegL7	Leucine-rich repeat	CA	deFelipe *et al*., 2005
LegC5	lpg1488	lpp1488	lpl1540	LegC8, Lpg1368	Coiled-coil	CA	deFelipe *et al*., 2005
LegC8	lpg2862	NH	NH	LegC5, Lpg1368	Coiled-coil	CA	deFelipe *et al*., 2005
LegG2	lpg0276	lpp0350	lpl0328	–	Ras GEF	CA	deFelipe *et al*., 2005

Table 10.1 *continued*

Protein	Phila-1	Paris	Lens	Paralogues	Domain/function	Evidence for Translocation*	Reference
LepA	lpg2793	lpp2839	lpl2708	LepB, Lpg2620	homology to EEA1, USO1 SNAREs/ bacterial egress	CA	Chen *et al.*, 2003
LepB	lpg1871	lpp1836	lpl1833	LepA, Lpg2620	homology to EEA1, USO1 SNAREs/ bacterial egress	CA	Chen *et al.*, 2003
SidA	lpg0621	lpp675	lpl658	–	–	IT	Luo and Isberg, 2004
SidB	lpg1642	NH	lpl1384	SdbA, SdbB, SdbC	Rtx toxin/lipase	IT	Luo and Isberg, 2004
SidD	lpg2465	NH	lpl2385	–	–	IT	Luo and Isberg, 2004
SidE	lpg0234	lpp0304	lpl0288	SdeA, SdeB, SdeC, SdeD	(vesicle trafficking)	IT	Luo and Isberg, 2004
SdeA	lpg257	lpp2096	lpl2085	SidE, SdeB, SdeC, SdeD	–	IT	Luo and Isberg, 2004
SdeC	lpg2157	lpp2096	lpl2085	SidE, SdeA, SdeB, SdeD	–	IT	Luo and Isberg, 2004
SidF	lpg2584	lpp2637	lpl2507	–	–	IT	Luo and Isberg, 2004
SidG	lpg1355	lpp1309	NH	–	–	CA, IT	Luo and Isberg, 2004; Ninio *et al.*, 2005
SidH	lpg2829	NH	NH	SdhA, SdhB	–	CA, IT	Luo and Isberg, 2004
SdjA	lpg2508	lpp2576	lpl2430		–	ST	Liu and Luo, 2007
VipA	lpg0390	lpp0457	lpl0433	–	(vesicle trafficking)	CA	Shohdy *et al.*, 2005
VipD	lpg2831	lpp2888	NH	VpdA, VpdB, VpdC	patatin-like phospholipase/(vesicle trafficking)	CA, ST	Shohdy *et al.*, 2005; VanRheenen *et al.*, 2006
VpdA	lpg2410	lpp2479	lpl2334	VipD, VpdB, VpdC	patatin-like phospholipase	ST	VanRheenen *et al.*, 2006
VpdB	lpg1227	lpp1235	lpl1235	VipD, VpdA, VpdC	patatin-like phospholipase	ST	VanRheenen *et al.*, 2006

VipF	lpg0103	lpp0117	lpl0103	–	Acetyltransferase (vesicle trafficking)	CA	Shohdy *et al*., 2005
WipA	lpg2718	lpg2775	lpl2646	WipB, WipC	–	CA	Ninio *et al*., 2005
WipB	lpg0642	lpg0696	lpl0679	WipA, WipC	–	CA	Ninio *et al*., 2005
YlfA/LegC7	lpg2298	lpp2246	lpl2217	YlfB/LegC2	Coiled-coil/(vesicle trafficking)	CA	Campondonico *et al*., 2005; deFelipe *et al*., 2005
YlfB/LegC2	lpg1884	lpp1848	lpl1845	YlfA/LegC7	Coiled-coil	CA	Campondonico *et al*., 2005; deFelipe, 2005
Putative effectors: paralogues of known or putative effectors							
LegL2	lpg1602	lpp1567	lpl1423	LegC4, LegL5, LegL6, LegL7	Leucine-rich repeat	ND	deFelipe *et al*., 2005
LegL6	lpg2392	lpp2459	lpl2316	LegC4, LegL2, LegL5, LegL7	Leucine-rich repeat	ND	deFelipe *et al*., 2005
SdbA	lpg0275	lpp0349	lpl0327	SidB, SdbB, SdbC	Rtx toxin/lipase	ND	Luo and Isberg, 2004
SdbB	lpg2482	lpp2546	lpl2402	SidB, SdbA, SdbC	Rtx toxin/lipase	ND	Luo and Isberg, 2004
SdbC	lpg2391	lpp2458	lpl2315	SidB, SdbA, SdbB	Rtx toxin/lipase	ND	Luo and Isberg, 2004
SdcA	lpg2510	lpp2578	lpl2432	SidC	PI(4)P binding domain/(vesicle trafficking)	ND	Luo and Isberg, 2004; Weber *et al*., 2005
SdeB	lpg2156	lpp2095	lpl2084	SidE, SdeA, SdeC, SdeD	–	ND	Luo and Isberg, 2004
SdeD	lpg2509	lpp2577	lpl2431	SidE, SdeA, SdeB, SdeC	–	ND	Luo and Isberg, 2004
SdhB	lpg0135	lpp0150	lpl0135	SidH, SdhA	GRIP domain, coiled-coil	ND	Laguna *et al*., 2006
VipE	lpg2813	lpp2865	lpl2728	–	–	ND	Shohdy *et al*., 2005

*Assays to test for Dot/Icm-dependent translocation: IT, inter-bacterial transfer; IF: immunofluorescence microscopy; ST, SidC-based translocation assay; CA,: cya-fusion assay; SE, saponin extraction; PNS, presence of protein on phagosomes isolated from postnuclear supernatant of infected cells; ND, not determined. Functions listed in parentheses indicate putative roles for proteins based on the disruption of vesicle trafficking when overexpressed in yeast.

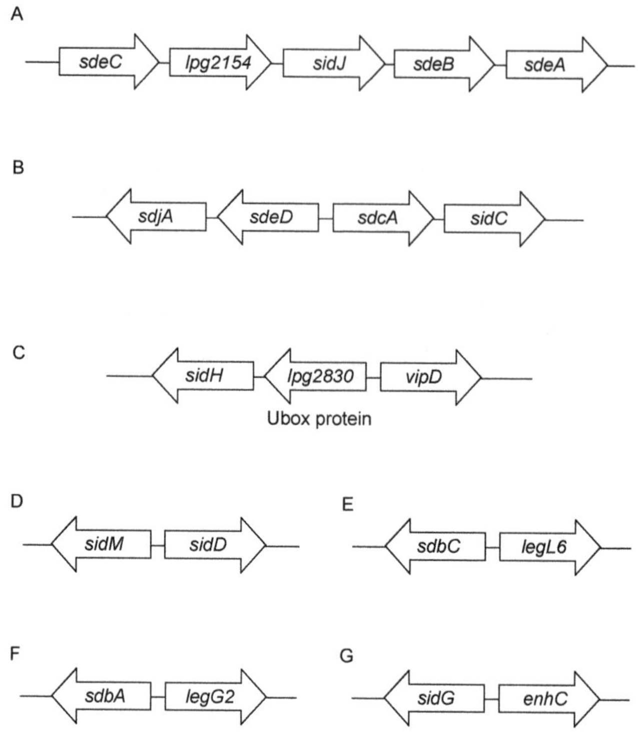

Figure 10.4 Genomic clustering of *L. pneumophila* virulence genes.

and *sdeC* that flank two additional open reading frames, *lpg2154* and *sidJ*. A mutant strain of *L. pneumophila* deleted for this region showed a partial defect in intracellular growth in mouse macrophages that could be rescued by *sidJ*, demonstrating a role for its encoded protein in mediating infection (Liu and Luo, 2007). Similar to other Dot/Icm substrate-encoding genes such as *ralF* and *lidA*, *sidC* and *sidE* have been shown to be upregulated at post-exponential phase of growth ((Byrne and Swanson, 1998; Hammer *et al.*, 1999; Nagai *et al.*, 2002; Conover *et al.*, 2003; Luo *et al.*, 2004). Whether these and other *sid* genes are coordinately regulated in an RpoS-dependent manner (Hales *et al.*, 1999; Bachman *et al.*, 2001; Zusman *et al.*, 2002) similar to other virulence determinants (Byrne and Swanson, 1998; Hammer *et al.*, 1999) remains to be determined.

The lack of strong defects in intracellular growth resulting from mutations in the *sid* genes has made it difficult to determine the role of these genes in *L. pneumophila* pathogenicity, a problem that is exacerbated by the lack of Sid protein homologues in any other organism. One exception to this is SdbA, a paralogue of SidB, which contains a putative lipase domain with homology in some regions to the Rtx toxin of *Vibrio cholera* (Lin *et al.*, 1999). A deletion mutation in this gene partially limits replication in mouse bone-marrow derived macrophages (Luo *et al.*, 2004). Deletion of all four *sidB* paralogues in a single strain does not further attenuate intracellular growth, suggesting that either the SidB family of proteins plays only a minor role in mediating *L. pneumophila* pathogenicity in this cell type or that the activities of these proteins are not redundant.

A mutation in *sdeC*, one of five paralogues of *sidE*, causes a partial defect in growth in mouse bone-marrow derived macrophages and the amoebae *Dictyostelium discoideum*, and this defect can be complemented by *sdeC* (Luo *et al.*, 2004). In subsequent work Bardill *et al.*, (2005), showed that the partial growth defect observed for a Δ*sidE*, Δ*sdeA*, Δ*sdeB*, Δ*sdeC* quadruple mutant in the protozoan host *A. castellanii* could be fully complemented by another *sidE* family paralogue, *sdeA*. Thus, at least in the case of SdeA and SdeC, one paralogue is sufficient for promoting growth in the absence of the other. Alternatively, it is possible that different paralogues are important in different host species or strains, SdeC in mammalian cells and *D. discoideum*, SdeA in *A. castellanii*.

Of the SidE family of proteins, SidE, SdeA, SdeB and SdeC have all been demonstrated to interact with the Dot/Icm chaperone protein IcmS (Bardill *et al.*, 2005) suggesting they may all be substrates of the type IV secretion system. Three of these proteins, SdeA, SdeB and SdeC, are expressed in *L. pneumophila* and demonstrate an increase in production during the transition from exponential to stationary phase growth (Bardill *et al.*, 2005), consistent with other known substrates of the Dot/Icm complex. Although the function of proteins in the SidE family is unknown, SidE has been shown to inhibit growth when overproduced in yeast (Campodonico *et al.*, 2005), suggesting that this protein is also capable of manipulating an essential host cellular process.

In contrast to other Dot/Icm substrates whose corresponding genes are upregulated during post-exponential phase (Nagai *et al.*, 2002; Conover *et al.*, 2003; Luo *et al.*, 2004), members of the SidJ family of proteins, SidJ and SdjA are produced at similar levels throughout the life cycle of the bacterium (Liu and Luo, 2007). Both proteins have been shown to be translocated into the host cell cytoplasm in a Dot/Icm dependent manner and are detectable on the surface of the LCV in the first hour of infection of macrophages (Liu and Luo, 2007). However, unlike SidE, whose presence in the host does not persist to later stage of the infection, SidJ is still detectable after 14 hours in the cytoplasm of differentiated U937 cells (Liu and Luo, 2007). These results suggest that SidJ levels in the host are maintained throughout the infection process which could be achieved by either its constant secretion into the host or its stability within the host cytoplasm (Liu and Luo, 2007).

Interestingly, the Δ*sidJ* and Δ*sdjA* mutant strains show differential growth defects in both mouse bone marrow-derived macrophages and the amoebae *D. discoideum* (Liu and Luo, 2007). The Δ*sidJ* mutant exhibits a moderate defect during growth in macrophages relative to the wild-type strain at all time points tested in a 72 hour period. In amoebae, this mutant shows no replication within the first 24 hours of infection and, although growth was detectable at 48 and 72 hours, the rate of growth was much less than that of the parental strain (Liu and Luo, 2007). In marked contrast, the Δ*sdjA* mutant grew as efficiently as the wild-type strain in mouse bone marrow-derived macrophages and in amoebae during the first 48 hours of growth. Only at the later time point of 72 hours did the Δ*sdjA* mutant begin to show a defect in replication in amoebae (Liu and Luo, 2007). The Δ*sidJ* mutant efficiently escapes trafficking to the lysosomal compartment and initiates replication at a similar rate as the wild-type strains in macrophages (Liu and Luo, 2007). However, by 14 hours post infection, macrophages harbouring the Δ*sidJ* mutant contained only a small fraction of vacuoles with large numbers of bacteria relative to the parental strain (Liu and Luo, 2007). Furthermore, in both macrophages and amoebae, recruitment of ER proteins to the LCV by the Δ*sidJ* mutant was not as efficient as observed for wild-type bacteria (Liu and Luo, 2007). This suggests that the lack of growth exhibited by the Δ*sidJ* mutant results from an inability to properly remodel the phagosome into an environment that supports growth. The exact role SidJ plays in the acquisition of ER vesicles to the phagosome remains to be determined.

Leg family proteins

de Felipe *et al.*, (2005) used a bioinformatics-based screen to identify *L. pneumophila* substrates based on the prediction that they would have protein domains that are widespread throughout eukaryotic species but not typically found in bacteria. These included proteins with coiled-coil domains, ankyrin-rich repeats and leucine-rich repeats as well as proteins predicted to have enzymatic activities more typically found in eukaryotes that in prokaryotes. This approach identified several known Dot/Icm substrate-encoding genes including *ralF*, *lepA*, *lepB* and several *sid* gene family members as well as 48 previously uncharacterized genes that were designated *Legionella* effector genes (*leg*) (de Felipe *et al.*, 2005). Interestingly, many of these domains are conserved across different isolates of *L. pneumophila* (de Felipe *et al.*, 2005).

Although the host cell targets of many of these proteins have yet to be determined, the diversity of the putative functions of their corresponding domains demonstrates the versatility of effector proteins at the disposal of *L. pneumophila* and the multiple mechanisms *L. pneumophila* may employ to modulate host cellular processes.

To date, eight Leg proteins have been shown by an enzymatic reporter assay to be translocated by the Dot/Icm apparatus during infection of *D. discoideum* (Table 10.1) (de Felipe *et al.*, 2005). These represent three classes of Leg proteins: those that contain coiled-coil domains (LegC family proteins), leucine-rich repeats (LegL family proteins) or guanosine nucleotide exchange factor domains (LegG family proteins). Two of the Leg proteins, LegC2 and LegC7, were also isolated in an independent study for proteins that inhibit growth when overexpressed in yeast: these were designated YlfB and YlfA, respectively (see below, Campodonico *et al.*, 2005).

The presence of proteins that contain functional domains typically found exclusively in eukaryotic proteins suggests that many of the proteins may have been acquired by horizontal gene transfer (de Felipe *et al.*, 2005). This has previously been predicted for RalF (Nagai *et al.*, 2002) and is thought to be the source of many of the Leg proteins as they all differ in both overall G+C content and positional G+C content relative to the rest of the genes encoded in the genome (de Felipe *et al.*, 2005). Thus, it is possible that over time *L. pneumophila* has randomly acquired an assortment of protein motifs and functional domains from its protozoan and eukaryotic hosts that allow it to manipulate the host cell environment to promote its survival or replication. A more detailed analysis of the *L. pneumophila* genome, its features and trends are covered in detail in Chapter 7.

Wip family proteins

Based on the observations that the Dot/Icm proteins IcmS and IcmW form a complex that resides in the bacterial cytosol (Ninio *et al.*, 2005) and that it is predicted to facilitate the translocation of effectors through the Dot/Icm complex (Coers *et al.*, 2000), Ninio *et al.*, (2005) used a yeast two-hybrid genetic screen to identify proteins that interact with one of these proteins, IcmW. This work resulted in the identification of the previously isolated *sidG* and *sidH* genes and an uncharacterized gene that was designated IcmW interacting protein A or *wipA*. The WipA protein has two paralogues encoded in the genome, designated WipB and WipC which share approximately 30% identity and 40% similarity with WipA (Ninio *et al.*, 2005). Consistent with other known effectors, these proteins do not share homology with proteins of any other organism nor do they contain any recognizable domains of predicted function. Neither WipA nor WipB are required for intracellular growth of *L. pneumophila* in *A. castellanii*, as replication of a double mutant lacking both *wipA* and *wipB* is indistinguishable from that of the wild-type strain (Ninio *et al.*, 2005).

Ylf family proteins

In an effort to identify substrates of the Dot/Icm secretion system, Campodonico *et al.*, (2005) tested for genes that inhibit growth when overexpressed in yeast. Based on this screen, two previously identified Dot/Icm substrates were isolated, SidE and SdcA as well as a protein of unknown function whose corresponding gene was named yeast lethal factor A (*ylfA*) (Campodonico *et al.*, 2005). YlfA is predicted to have a single N-terminal transmembrane domain and two C-terminal coiled-coil domains (Campodonico *et al.*, 2005). Of these, the transmembrane domain in combination with the N-terminal coiled-coil domain was shown to be necessary and sufficient to inhibit growth when overexpressed in yeast (Campodonico *et al.*, 2005). Translocation of the YlfA protein has been shown by an enzymatic reporter assay to begin within the first hour of infection but its presence on the LCV was undetectable by immunofluorescence microscopy at this early time point (Campodonico *et al.*, 2005). This was due to the trafficking of YlfA to other subcellular locations in the host as YlfA staining at later time points was visibly associated with both the LCV and structures that did not contain *L. pneumophila* (Campodonico *et al.*, 2005). Furthermore, YlfA staining showed extensive overlap with the sec61-associated TRAP complex and the KDEL protein motif, two markers of the ER (Campodonico *et al.*, 2005). Collectively these results suggested that, upon translocation into the host cell, YlfA not only decorates the surface of the LCV but also infiltrates host vesicle trafficking pathways. In spite of this, a Δ*ylfA* mutant was able to recruit ER-derived vacuoles to the LCV as well as the parental strain (Campodonico *et al.*, 2005) and neither YlfA nor its homologue YlfB were required for growth in *A. castellanii* or primary bone-marrow macrophages (Campodonico *et al.*, 2005).

Ceg family proteins

Bioinformatic analysis of the regulatory regions upstream of Dot/Icm substrate-encoding genes has revealed a conserved regulatory element (cTTAATatT) in the promoter region of several of these genes (Zusman *et al.*, 2007). The consensus sequence is recognized by the *L. pneumophila* protein, PmrA, a two-component response regulator, based on the observations that a Δ*pmrA* mutant exhibits decreases levels of expression of the predicted target genes *in vivo* and that PmrA directly binds DNA containing the consensus sequence *in vitro* (Zusman *et al.*, 2007). The importance of this regulator for virulence was demonstrated by the defect in intracellular growth exhibited by a Δ*pmrA* mutant in both HL-60-derived human macrophages and the amoebae *A. castellanii* (Zusman *et al.*, 2007). This work suggests that several effector-encoding genes are coregulated and members of a PmrA regulon that is activated during the transition from exponential to stationary phase growth and the onset of virulence trait expression.

Based on the conservation of the PmrA recognition sequence, 35 additional putative PmrA target genes were identified in the *L. pneumophila* genome which have been designated coregulated with the effector encoding genes (*ceg*) (Zusman *et al.*, 2007). Several of these genes are clustered with other known Dot/Icm substrate-encoding genes, and many have one or more paralogues encoded within the *L. pneumophila* genome. The majority of the *ceg* genes encode proteins of no known function and have no homologues in other organisms, although a small number of the Ceg proteins are predicted to contain domains or motifs typically found in eukaryotic proteins such as ankyrin rich repeats and coiled-coil domains (Zusman *et al.*, 2007). One of these genes, *ceg12*, was previously identified by de Felipe *et al.*, (2005) as *legA9*. Of the five *ceg* genes tested, the expression of all five was shown to be dependent on PmrA (Zusman *et al.*, 2007). Furthermore, three of these, *ceg10*, *ceg23* and *ceg29*, were shown to be translocated into the host during infection of HL-60-derived macrophages in a Dot/Icm-dependent manner (Zusman *et al.*, 2007). Future characterization of the roles of these genes in *L. pneumophila* pathogenicity and identification of the stimuli to which PmrA responds and activates these genes will prove valuable in elucidating how Dot/Icm substrate-encoding genes are regulated.

Functional redundancy

Individual mutations in any of the *dot* or *icm* genes completely abolish intracellular growth of *L. pneumophila* in virtually all host cell types (Marra *et al.*, 1992; Brand *et al.*, 1994; Roy *et al.*, 1998; Matthews *et al.*, 2000; VanRheenen *et al.*, 2004). In contrast, when genes encoding substrates of the Dot/Icm secretion system are deleted, the resulting strains typically have only partial, if any, defect in intracellular replication (Nagai *et al.*, 2002; Luo *et al.*, 2004; VanRheenen *et al.*, 2006; Machner and Isberg, 2006; Murata *et al.*, 2006). Although this may allude to a less important role for these proteins in mediating the infection process, it seems more likely that the lack of any one effector can be compensated for by the activity of another, thereby defining a series of functionally redundant sets of proteins.

One potential source of functional redundancy is the presence of one or more paralogues of any given effector encoded within the genome (Table 10.1). Consistent with this idea, it has been shown that the partial defect exhibited by a Δ*sidE*, Δ*sdeA*, Δ*sdeB*, Δ*sdeC* quadruple mutant can be fully rescued by the overexpression of either *sdeA* or *sdeC* (Luo *et al.*, 2004; Bardill *et al.*, 2005). In addition, the effects of the Δ*lepA* and Δ*lepB* single mutations on the ability of *L. pneumophila* to escape from the host appear to have additive effects when combined in a Δ*lepA*, Δ*lepB* double mutant (Chen *et al.*, 2004). Thus, although one paralogue may play a more predominant role than other members of the same family, the absence of any one paralogue can be sufficiently compensated for by one of its siblings. Consistent with this idea, a number of translocated substrates are not conserved in other strains of virulent *L. pneumophila* (Table 10.1). For example, *sidB* is present in the Philadelphia-1 and Paris strains but not Lens whereas, *vipD* is present in Philadelphia-1 and Lens but not Paris and *sidH* is absent from both Paris and Lens.

In many cases however, when combinations of paralogues from a single family are deleted simultaneously, the corresponding mutants still show no significant defects in intracellular growth: this has been demonstrated for the *vipD, sidB, sidC, sidE* and *sidJ* families of genes. These results imply that seemingly unrelated proteins may perform similar functions. Indeed, this may explain the lack of phenotype observed for strains with mutations in *ralF* and *legG2*, genes that have no paralogues within the genome. As well, there are 305 and 293 genes in the Paris and Lens genomes respectively that are not present in the Philadelphia-1 strain which may encode alternative virulence factors. Collectively, these observations suggest the existence of a more complex layer of functional redundancy in which the bacterium employs multiple overlapping mechanisms to ensure its survival within the host.

L. pneumophila may deploy multiple proteins that target a single cellular pathway. For example, *L. pneumophila* employs RalF, SidM/DrrA, LidA and possibly others to target host proteins involved in early secretory vesicle trafficking to remodel the LCV to an ER-studded vacuole. Although deletions in *ralF* and *sidM/drrA* abolish Arf1 and Rab1 recruitment to the LCV respectively, neither mutant is severely defective in intracellular growth. This suggests that the activity of one may be sufficient to generate a replicative vacuole in the absence of the other. Alternatively, there may be other proteins that also target this pathway. Indeed, YlfA and SidJ have been shown to interact with early secretory vesicles and to recruit ER proteins to the LCV respectively. In addition, SdcA and SidE have been shown to disrupt cellular process in yeast that are important for growth. It is possible that these two proteins may also play roles in remodelling the LCV.

Another strategy employed by *L. pneumophila* is to target multiple host proteins, some in seemingly unrelated pathways to accomplish the same feat. For example, in addition to recruiting vesicles from the early secretory pathway, remodelling of the LCV may also involve membrane assembly events common to the biogenesis of autophagic compartments. This is based on the observations that early stages in replicative vacuole biogenesis morphologically resemble early autophagosomal compartments (Swanson and Isberg 1995) and that the autophagy proteins Atg7 and Atg8 are recruited to the LCV at early stages during its formation (Amer and Swanson, 2005). Furthermore, both LidA and SidM/DrrA demonstrated some affinity for Rab6 and Rab8 (LidA) (Machner and Isberg, 2006) and Rab35 (SidM/DrrA) (Murata *et al.*, 2006) which may provide the bacterium with alternative mechanisms for membrane vesicle recruitment in the absence of Rab1. Consistent with the extent to which functional redundancy appears to play a role in pathogenesis, depletion of any single component involved in host vesicle transport has only modest or no effect on the intracellular growth of the bacterium (Dorer *et al.*, 2006). Only when individual components are depleted in combination are any effects on intracellular replication observed (Dorer *et al.*, 2006).

Generating and maintaining a replicative vacuole is central to the survival of the bacterium within the host. Thus, it is not surprising that during its co-evolution with phagocytic amoebae and protozoa, *L. pneumophila* has accumulated multiple overlapping mechanisms for ensuring the successful trafficking of the LCV during infection. By hijacking or perturbing numerous vesicle trafficking pathways that function between the ER and the Golgi, *L. pneumophila* employs a functionally redundant system that is advantageous in many ways. First, by exporting numerous substrates of similar function, *L. pneumophila* ensures its survival in the event that any one of these substrates is lost or ineffective. Second, by targeting multiple components of the host vesicle trafficking pathways, *L. pneumophila* ensures that upon perturbation of any one of these pathways or its constituents it is still able to form a replicative niche through alternative means. Finally, by hijacking multiple pathways, *L. pneumophila* ensures that no one pathway vital to the host cell is completely disrupted due to depletion of any one component. By sustaining host cell viability throughout the replicative phase of the infection cycle, the bacterium indirectly contributes to its own survival.

L. pneumophila infects and survives in a multitude of distinct host cell types ranging from amoebae and ciliated protozoa to eukaryotic macrophage and alveolar epithelial cells. Within each host, the pathogen manipulates similar cellular processes to establish a replicative niche able to support growth (Horwitz, 1983a; Horwitz, 1983b; Swanson and Isberg, 1995; Abu Kwaik, 1996; Bozue *et al.*, 1996). This has been demonstrated by the essentiality of the Dot/Icm secretion system for infecting both macrophages and amoebae as well as the similar morphology of the LCVs in both cell types (Segal and Shuman, 1997). Thus, it appears *L. pneumophila* employ a core set of virulence factors that are necessary for growth and survival in two evolutionarily distinct cells types (Gao *et al.*, 1997; Segal and Shuman, 1999).

In contrast, *L. pneumophila* also encode a subset of virulence determinants that are important only under specific circumstances or in specific species. For example, a number of genes have been isolated that are important for the infection of macrophages but not amoebae, including the *mil* (macrophage-specific infectivity loci) genes (Gao *et al.*, 1998), *htrA*, which codes for a homologueof the HtrA/DegP family of stress induced chaperone/proteases (Pedersen *et al.*, 2001), and *lvgA*, which encodes a putative chaperone protein that function in conjunction with IcmS (Vincent *et al.*, 2006b). Furthermore, work by Segal and Shuman (1999) has demonstrated that some of the *icm* genes are more important in one species than another. For instance, whereas *dotK/icmN* is not required for growth of *L. pneumophila* in HL-60 derived macrophages, a strain containing an insertion mutation in this gene shows a partial growth defect in amoebae. Similarly, insertion mutations in *icmS* or *dotF/icmG* completely abolish growth in amoebae but only cause partial phenotypes in macrophages. There also appears to be a species-specific requirement for some members of a paralogous families over others as described above for the Δ*sidJ* and Δ*sdjA* mutants (Liu and Luo, 2007), the Δ*sdhA* mutant (Laguna *et al.*, 2006) and the Δ*lepA* and Δ*lepB* mutants (Chen *et al.*, 2004). Thus, *L. pneumophila* has acquired subsets of species-specific genes that allow them to compensate for species-specific variation in host cellular processes.

In addition to a role for functional redundancy in facilitating the successful infection of a specific host, it also appears that functionally redundant mechanisms have evolved to ensure the survival of *L. pneumophila* in a broad range of cell types. In the environment, *L. pneumophila* randomly encounter any number of protozoan species within a given body of water which at any time can be inhaled by humans into the lung where the bacteria are exposed to a completely different set of evolutionarily distinct cells, in particular macrophages and epithelial cells. Thus, *L. pneumophila* must be adequately equipped with the proper tools to invade and survive within any one of them. Subtle but significant differences between strains and species-to-species variation may account for the multiple variants of any given substrate *L. pneumophila* encodes or the multiple mechanisms utilized to mediate the same event during an infection. As a result, although the different forms of any given effector or sets of functionally redundant proteins may function in the same capacity during infection, they may be more effective in one host cell type compared to another. Thus, as a consequence of its promiscuity, *L. pneumophila* has likely acquired, through random mutation or horizontal gene transfer, a multitude of virulence factors. The random yet equal probability of encountering evolutionarily distinct hosts provides a selective pressure to maintain within the genome all variations of effector proteins and functionally redundant genes sets which together, promote the survival of the bacterium.

Future directions

Several genetic, biochemical and bioinformatic-based screens have successfully identified approximately 40 known or potential Dot/Icm substrates. However, current efforts to define the complete set of *L. pneumophila* effectors predict that the total number may include as many as 150 proteins (Huang, O'Connor, Chen and Isberg, unpublished data). Significant efforts have been made to characterize the function of many of these proteins. For a few, their host targets and the mechanisms by which they exert their effects have been elucidated in considerable detail. Nevertheless, the roles of the majority of effectors during an infection remain elusive. Future biochemical and molecular biological studies to further define the targets of these proteins and their modes of action will be necessary to understand how *L. pneumophila* manipulates host cellular processes. Indeed, the diversity of the effectors identified to date demonstrate that *L. pneumophila* is capable of influencing an assortment of host cell functions including vesicle trafficking, lipid metabolism, protein synthesis, autophagy and apoptosis.

Classical screens to identify mutants based on strong phenotypic defects are not as effective for studying *L. pneumophila* type IV secreted substrates due to the presence of multiple variants of effectors encoded in the genome and the overlapping functions that seemingly unrelated proteins play during infection. Thus, alternative approaches that address the functional redundancy problem inherent to *L. pneumophila* will be necessary to identify the full repertoire of effectors and how they are utilized by the bacterium to generate an environment that supports bacterial growth. Identifying sources of functional redundancy will also aid in understanding how *L. pneumophila* infects and survives within a diverse array of evolutionarily distinct cells types.

References

Abu Kwaik, Y. (1996). The phagosome containing *Legionella pneumophila* within the protozoan *Hartmannella vermiformis* is surrounded by the rough endoplasmic reticulum. Appl. Environ. Microbiol. *62*, 2022–2028.

Abu-Zant, A., Jones, S., Asare, R., Suttles, J., Price, C., Graham, J., and Kwaik, Y.A. (2007). Anti-apoptotic signalling by the Dot/Icm secretion system of *L. pneumophila*. Cell. Microbiol. *9*, 246–264.

Allen, R.T., Cluck, M.W., and Agrawal, D.K. (1998). Mechanisms controlling cellular suicide: role of Bcl-2 and caspases. CMLS Cell. Mol. Life Sci. *54*, 427–445.

Alli, O.A., Gao, L.Y., Pedersen, L.L., Zink, S., Radulic, M., Doric, M., and Abu Kwaik, Y. (2000). Temporal pore formation-mediated egress from macrophages and alveolar epithelial cells by *Legionella pneumophila*. Infect. Immun. *68*, 6431–6440.

Amer, A.O. and Swanson, M.S. (2005). Autophagy is an immediate macrophage response to *Legionella pneumophila*. Cell. Microbiol. *7*, 765–778.

Arakaki, N., Higa, F., Koide, M., Tateyama, M., and Saito, A. (2002). Induction of apoptosis of human macrophages *in vitro* by *Legionella longbeachae* through activation of the caspase pathway. J. Med. Microbiol. *51*, 159–168.

Bachman, M.A. and Swanson, M.S. (2001). RpoS co-operates with other factors to induce *Legionella pneumophila* virulence in the stationary phase. Mol. Microbiol. *40*, 1201–1214.

Banga, S., Gao, P., Shen, X., Fiscus, V., Zong, W., Chen, L., and Luo, Z. (2007). *Legionella pneumophila* inhibits macrophage apoptosis by targeting pro-death members of the Bcl2 protein family. Proc. Natl. Acad. Sci. USA. *104*, 5121–5126.

Bankaitis, V.A., Johnson, L.M., and Emr, S.D. (1986). Isolation of yeast mutants defective in protein targeting to the vacuole. Proc. Natl. Acad. Sci. USA *83*, 9075–9079.

Bardill, J.P., Miller, J.L., and Vogel, J.P. (2005). IcmS-dependent translocation of SdeA into macrophages by the *Legionella pneumophila* type IV secretion system. Mol. Microbiol. *56*, 90–103.

Behnia, R. and Munro, S. (2005). Organelle identity and the signposts for membrane traffic. Nature *438*, 597–604.

Belyi, Y., Niggeweg, R., Opitz, B., Vogelsgesang, M., Hippenstiel, S., Wilm, M., and Aktories, K. (2006). *Legionella pneumophila* glucosyltransferase inhibits host elongation factor 1A. Proc. Natl. Acad. Sci. *103*, 16953–16958.

Bentley, M., Liang, Y., Mullen, K., Xu, D., Sztul, E., and Hay, J.C. (2006). SNARE Status Regulates Tether Recruitment and Function in Homotypic COPII Vesicle Fusion. J. Biol. Chem. *281*, 38825–38833.

Berger, K.H. and Isberg, R.R. (1993). Two distinct defects in intracellular growth complemented by a single genetic locus in *Legionella pneumophila*. Mol. Microbiol. *7*, 7–19.

Berk, S.G., Ting, R.S., Turner, G.W., and Ashburn, R.J. (1998). Production of respirable vesicles containing live *Legionella pneumophila* cells by two *Acanthamoeba* spp. Appli. Environ. Microbiol. *64*, 279–286.

Bonifacino, J.S. and Glick, B.S. (2004). The mechanisms of vesicle budding and fusion. Cell. *116*, 153–166.

Boutros, M., Agaisse, H., and Perrimon, N. (2002). Sequential activation of signaling pathways during innate immune responses in *Drosophila*. Dev. Cell. *3*, 711–722.

Bozue, J.A. and Johnson, W. (1996). Interaction of *Legionella pneumophila* with *Acanthamoeba castellanii*: uptake by coiling phagocytosis and inhibition of phagosome-lysosome fusion. Infect. Immun. *64*, 668–673.

Brand, B.C., Sadosky, A.B., and Shuman, H.A. (1994). The *Legionella pneumophila icm* locus: a set of genes required for intracellular multiplication in human macrophages. Mol. Microbiol. *14*, 797–808.

Byrne, B. and Swanson, M.S. (1998). Expression of *Legionella pneumophila* virulence traits in response to growth conditions. Infect. Immun. *66*, 3029–3034.

Campodonico, E.M., Chesnel, L., and Roy, C.R. (2005). A yeast genetic system for the identification and characterization of substrate proteins transferred into host cells by the *Legionella pneumophila* Dot/Icm system. Mol. Microbiol. *56*, 918–933.

Cai, H., Yu, S., Menon, S., Cai, Y., Lazarova, D., Fu, C., Reinisch, K., Hay, J.C., and Ferro-Novick, S. (2007). TRAPPI tethers COPII vesicles by binding the coat subunit Sec23. Nature *445*, 941–944.

Chai, J., Shiozaki, E., Srinivasula, S.M., Wu, Q., Datta, P., Alnemri, E.S., and Shi, Y. (2001). Structural basis of caspase-7 inhibition by XIAP. Cell. *104*, 769–780.

Chen, J., de Felipe, K.S., Clarke, M., Lu, H., Anderson, O.R., Segal, G., and Shuman, H.A. (2004). *Legionella* effectors that promote nonlytic release from protozoa. Science. *303*, 1358–1361.

Chen, Y.A. and Scheller, R.H. (2001). SNARE-mediated membrane fusion. Nat. Rev. Mol. Cell. Biol. *2*, 98–106.

Chen, M. and Wang, J. (2002). Initiator caspases in apoptosis singaling pathways. Apoptosis *7*, 313–319.

Chinnaiyan, A.M., O'Rourke, K., Lane, B.R., and Dixit, V.M. (1997). Interaction of CED-4 with CED-3 and CED-9: a molecular framework for cell death. Science. *275*, 1122–1126.

Coers, J., Kagan, J.C., Matthews, M., Nagai, H., Zuckman, D.M., and Roy, C.R. (2000). Identification of Icm protein complexes that play distinct roles in the biogenesis of an organelle permissive for *Legionella pneumophila* intracellular growth. Mol. Microbiol. *38*, 719–736.

Conover, G.M., Derre, I., Vogel, J.P., and Isberg, R.R. (2003). The *Legionella pneumophila* LidA protein: a translocated substrate of the Dot/Icm system associated with maintenance of bacterial integrity. Mol. Microbiol. *48*, 305–321.

Cosulich, S.C., Horiuchi, H., Zerial, M., Clarke, P.R., and Woodman, P.G. (1997). Cleavage of rabaptin-5 blocks endosome fusion during apoptosis. EMBO J. *16*, 6182–6191.

de Felipe, K.S., Pampou, S., Jovanovic, O.S., Pericone, C.D., Ye, S.F., Kalachikov, S., and Shuman, H.A. (2005). Evidence for acquisition of *Legionella* type IV secretion substrates via interdomain horizontal gene transfer. J. Bacteriol. *187*, 7716–7726.

De Matteis, M.A., and Godi, A. (2004). PI-loting membrane traffic. Nat. Cell. Biol. *6*, 487–492.

Demma, M., Warren, V., Hick, R., Dharmawardhane, S., and Condeelis, J. (1990). Isolation of an abundant 50,000-dalton actin filament bundling protein from *Dictyostelium amoebae*. J. Biol. Chem. *265*, 2286–2291.

Derre, I. and Isberg, R.R. (2004). *Legionella pneumophila* replication vacuole formation involves rapid recruitment of proteins of the early secretory system. Infect. Immun. *72*, 3048–3053.

Derre, I. and Isberg, R.R. (2005). LidA, a translocated substrate of the *Legionella pneumophila* type IV secretion system, interferes with the early secretory pathway. Infect. Immun. *73*, 4370–4380.

Deveraux, Q.L. and Reed, J.C. (1999). IAP family proteins–suppressors of apoptosis. Genes Dev. *13*, 239–252.

Diez, E., Lee, S.H., Gauthier, S., Yaraghi, Z., Tremblay, M., Vidal, S., and Gros, P. (2003). Birc1e is the gene within the Lgn1 locus associated with resistance to *Legionella pneumophila*. Nat. Genet. *33*, 55–60.

Dorer, M.S., Kirton, D., Bader, J.S., and Isberg, R.R. (2006). RNA interference analysis of *Legionella* in *Drosophila* cells: exploitation of early secretory apparatus dynamics. PLoS Pathog. *2*, e34.

Dunn, W.A., Jr. (1990a). Studies on the mechanisms of autophagy: maturation of the autophagic vacuole. J. Cell. Biol. *110*, 1935–1945.

Dunn, W.A., Jr. (1990b). Studies on the mechanisms of autophagy: formation of the autophagic vacuole. J. Cell. Biol. *110*, 1923–1933.

Duttaroy, A., Bourbeau, D., Wang, X.L., and Wang, E. (1998). Apoptosis rate can be accelerated or decelerated by overexpression or reduction of the level of elongation factor-1 alpha. Exp. Cell. Res. *238*, 168–176.

Edmonds, B.T., Murray, J., and Condeelis, J. (1995). pH regulation of the F-actin binding properties of *Dictyostelium* elongation factor 1alpha. J. Biol. Chem. *270*, 15222–15230.

Engels, I.H., Stepczynska, A., Stroh, C., Lauber, K., Berg, C., Schwenzer, R., Wajant, H., Janicke, R.U., Porter, A.G., Belka, C., Gregor, M., Schulze-Osthoff, K., and Wesselborg, S. (2000). Caspase-8/FLICE functions as an executioner caspase in anticancer drug-induced apoptosis. Oncogene. *19*, 4563–4573.

Eskelinen, E.L. (2005). Maturation of autophagic vacuoles in Mammalian cells. Autophagy *1*, 1–10.

Ferrari, D., Stepczynska, A., Los, M., Wesselborg, S., and Schulze-Osthoff, K. (1998). Differential regulation and ATP requirement for caspase-8 and caspase-3 activation during CD95- and anticancer drug-induced apoptosis. J. Exp. Med. *188*, 979–984.

Fischer, S.F., Vier, J., Muller-Thomas, C., and Hacker, G. (2006). Induction of apoptosis by *Legionella pneumophila* in mammalian cells requires the mitochondrial pathway for caspase activation. Microbes Infect. *8*, 662–669.

Gao, L.Y. and Abu Kwaik, Y. (1999a). Apoptosis in macrophages and alveolar epithelial cells during early stages of infection by *Legionella pneumophila* and its role in cytopathogenicity. Infect. Immun. *67*, 862–870.

Gao, L.Y. and Abu Kwaik, Y. (1999b). Activation of caspase 3 during *Legionella pneumophila*-induced apoptosis. Infect. Immun. *67*, 4886–4894.

Gao, L.Y., Harb, O.S., and Abu Kwaik, Y. (1997). Utilization of similar mechanisms by *Legionella pneumophila* to parasitize two evolutionarily distant host cells, mammalian macrophages and protozoa. Infect. Immun. *65*, 4738–4746.

Gao, L.Y., Harb, O.S., and Abu Kwaik, Y.A. (1998). Identification of macrophage-specific infectivity loci (*mil*) of *Legionella pneumophila* that are not required for infectivity of protozoa. Infect. Immun. *66*, 883–892.

Gillooly, D.J., Simonsen, A., and Stenmark, H. (2001). Phosphoinositides and phagocytosis. J. Cell. Biol. *155*, 15–17.

Gonen, H., Dickman, D., Schwartz, A.L., and Ciechanover, A. (1996). Protein synthesis elongation factor EF-1 alpha is an isopeptidase essential for ubiquitin-dependent degradation of certain proteolytic substrates. Adv. Exp. Med. Biol. *389*, 209–219.

Gonen, H., Smith, C.E., Siegel, N.R., Kahana, C., Merrick, W.C., Chakraburtty, K., Schwartz, A.L., and Ciechanover, A. (1994). Protein synthesis elongation factor EF-1alpha is essential for ubiquitin-dependent degradation of certain N-acetylated proteins and may be substituted by the bacterial elongation factor EF-TU. Proc. Natl. Acad. Sci. USA *91*, 7648–7652.

Grosshans, B.L., Ortiz, D., and Novick, P. (2006). Rabs and their effectors: achieving specificity in membrane traffic. Proc Natl Acad Sci USA *103*, 11821–11827.

Gurkan, C., Stagg, S.M., Lapointe, P., and Balch, W.E. (2006). The COPII cage: unifying principles of vesicle coat assembly. Nat. Rev. Mol. Cell. Biol. *7*, 727–738.

Hagele, S., Hacker, J., and Brand, B.C. (1998). *Legionella pneumophila* kills human phagocytes but not protozoan host cells by inducing apoptotic cell death. FEMS Micrbiol. Lett. *169*, 51–58.

Hales, L.M. and Shuman, H.A. (1999). The *Legionella pneumophila rpoS* gene is required for growth within *Acanthamoeba castellanii*. J. Bacteriol. *181*, 4879–4889.

Hammer, B.K. and Swanson, M.S. (1999). Co-ordination of *Legionella pneumophila* virulence with entry into stationary phase by ppGpp. Mol. Microbiol. *33*, 721–731.

Hammond, S.M., Bernstein, E., Beach, D., and Hannon, G.J. (2000). An RNA-directed nuclease mediates post-transcriptional gene silencing in *Drosophila* cells. Nature. *404*, 293–296.

Hanson, P.I., Roth, R., Morisaki, H., Jahn, R., and Heuser, J.E. (1997). Structure and conformational changes in NSF and its membrane receptor complexes visualized by quick-freeze/deep-etch electron microscopy. Cell. *90*, 523–535.

Horwitz, M.A. (1983a). Formation of a novel phagosome by the Legionnaires' disease bacterium (*Legionella pneumophila*) in human monocytes. J. Exp. Med. *158*, 1319–1331.

Horwitz, M.A. (1983b). The Legionnaires' disease bacterium (*Legionella pneumophila*) inhibits phagosome-lysosome fusion in human monocytes. J. Exp. Med. *158*, 2108–2126.

Horwitz, M. A. and Maxfield, F. R. (1984). *Legionella pneumophila* inhibits acidification of its phagosome in human monocytes. J. Cell Biol. *99*, 1936–1943.

Husmann, L.K. and Johnson, W. (1994). Cytotoxicity of extracellular *Legionella pneumophila*. Infect. Immun. *62*, 2111–2114.

Jahn, R., Lang, T., and Sudhof, T.C. (2003). Membrane fusion. Cell. *112*, 519–533.

Jarosch, E., Taxis, C., Volkwein, C., Bordallo, J., Finley, D., Wolf, D.H., and Sommer, T. (2002). Protein dislocation from the ER requires polyubiquitination and the AAA-ATPase Cdc48. Nat. Cell. Biol. *4*, 134–139.

Kagan, J.C. and Roy, C.R. (2002). *Legionella* phagosomes intercept vesicular traffic from endoplasmic reticulum exit sites. Nat. Cell. Biol. *4*, 945–954.

Kagan, J.C., Stein, M.P., Pypaert, M., and Roy, C.R. (2004). *Legionella* subvert the functions of Rab1 and Sec22b to create a replicative organelle. J. Exp. Med. *199*, 1201–1211.

Kamata, H., Honda, S., Maeda, S., Chang, L., Hirata, H., and Karin, M. (2005). Reactive oxygen species promote TNFα-induced death and sustained JNK activation by inhibiting MAP kinase phosphatases. Cell. *120*, 649–661.

Keen, M.G. and Hoffman, P.S. (1989). Characterization of a *Legionella pneumophila* extracellular protease exhibiting hemolytic and cytotoxic activities. Infect. Immun. *57*, 732–738.

Kerr, J.F., Gobe, G.C., Winterford, C.M., and Harmon, B.V. (1995). Anatomical methods in cell death. Methods Cell. Biol. *46*, 1–27.

Kim, Y.G., Raunser, S., Munger, C., Wagner, J., Song, Y.L., Cygler, M., Walz, T., Oh, B.H., and Sacher, M. (2006). The architecture of the multisubunit TRAPP I complex suggests a model for vesicle tethering. Cell. *127*, 817–830.

Kirby, J.E., Vogel, J.P., Andrews, H.L., and Isberg, R.R. (1998). Evidence for pore-forming ability by *Legionella pneumophila*. Mol. Microbiol. *27*, 323–336.

Laguna, R.K., Creasey, E.A., Li, Z., Valtz, N., and Isberg, R.R. (2006). A *Legionella pneumophila*-translocated substrate that is required for growth within macrophages and protection from host cell death. Proc. Natl. Acad. Sci. USA *103*, 18745–18750.

Lesser, C.F. and Miller, S.I. (2001). Expression of microbial virulence proteins in *Saccharomyces cerevisiae* models mammalian infection. EMBO J. *20*, 1840–1849.

Letourneur, F., Gaynor, E.C., Hennecke, S., Demolliere, C., Duden, R., Emr, S.D., Riezman, H., and Cosson, P. (1994). Coatomer is essential for retrieval of dilysine-tagged proteins to the endoplasmic reticulum. Cell *79*, 1199–1207.

Lin, W., Fullner, K.J., Clayton, R., Sexton, J.A., Rogers, M.B., Calia, K.E., Calderwood, S.B., Fraser, C., and Mekalanos, J.J. (1999). Identification of a *Vibrio cholerae* RTX toxin gene cluster that is tightly linked to the cholera toxin prophage. Proc. Natl. Acad. Sci. USA *96*, 1071–1076.

Liu, Y. and Luo, Z.Q. (2007). The *Legionella pneumophila* Effector SidJ Is required for efficient recruitment of endoplasmic reticulum proteins to the bacterial phagosome. Infect. Immun. *75*, 592–603.

Loh, E., Peter, F., Subramaniam, V.N., and Hong, W. (2005). Mammalian Bet3 functions as a cytosolic factor participating in transport from the ER to the Golgi apparatus. J. Cell. Sci. *118*, 1209–1222.

Losev, E., Reinke, C.A., Jellen, J., Strongin, D.E., Bevis, B.J., and Glick, B.S. (2006). Golgi maturation visualized in living yeast. Nature. *441*, 1002–1006.

Losick, V.P. and Isberg, R.R. (2006). NF-kappaB translocation prevents host cell death after low-dose challenge by *Legionella pneumophila*. J. Exp. Med. *203*, 2177–2189.

Lu, L., Tai, G., Wu, M., Song, H., and Hong, W. (2006). Multilayer interactions determine the Golgi localization of GRIP golgins. Traffic. *7*, 1399–1407.

Luo, Z.Q. and Isberg, R.R. (2004). Multiple substrates of the *Legionella pneumophila* Dot/Icm system identified by interbacterial protein transfer. Proc. Natl. Acad. Sci USA *101*, 841–846.

Machner, M.P. and Isberg, R.R. (2006). Targeting of host Rab GTPase function by the intravacuolar pathogen *Legionella pneumophila*. Dev. Cell. *11*, 47–56.

Marra, A., Blander, S.J., Horwitz, M.A., and Shuman, H.A. (1992). Identification of a *Legionella pneumophila* locus required for intracellular multiplication in human macrophages. Proc. Natl. Acad. Sci. USA *89*, 9607–9611.

Matsuura-Tokita, K., Takeuchi, M., Ichihara, A., Mikuriya, K., and Nakano, A. (2006). Live imaging of yeast Golgi cisternal maturation. Nature. *441*, 1007–1010.

Matthews, M. and Roy, C.R. (2000). Identification and subcellular localization of the *Legionella pneumophila* IcmX protein: a factor essential for establishment of a replicative organelle in eukaryotic host cells. Infect. Immun. *68*, 3971–3982.

McGahon, A.J., Martin, S.J., Bissonnette, R.P., Mahboubi, A., Shi, Y., Mogil, R.J., Nishioka, W.K., and Green, D.R. (1995). The end of the (cell) line: methods for the study of apoptosis in vitro. Methods Cell. Biol. *46*, 153–185.

Molmeret, M., Alli, O.A., Zink, S., Flieger, A., Cianciotto, N.P., and Abu Kwaik, Y.A. (2002). *icmT* is essential for pore formation-mediated egress of *Legionella pneumophila* from mammalian and protozoan cells. Infect. Immun. *70*, 69–78.

Molmeret, M., Zink, S.D., Han, L., Abu-Zant, A., Asari, R., Bitar, D.M., and Abu Kwaik, Y. (2004). Activation of caspase-3 by the Dot/Icm virulence system is essential for arrested biogenesis of the *Legionella*-containing phagosome. Cell. Microbiol. *6*, 33–48.

Molofsky, A.B., Byrne, B.G., Whitfield, N.N., Madigan, C.A., Fuse, E.T., Tateda, K., and Swanson, M.S. (2006). Cytosolic recognition of flagellin by mouse macrophages restricts *Legionella pneumophila* infection. J. Exp. Med. *203*, 1093–1104.

Muller, A., Hacker, J., and Brand, B.C. (1996). Evidence for apoptosis of human macrophage-like HL-60 cells by *Legionella pneumophila* infection. Infect. Immun. *64*, 4900–4906.

Murata, T., Delprato, A., Ingmundson, A., Toomre, D.K., Lambright, D.G., and Roy, C.R. (2006). The *Legionella pneumophila* effector protein DrrA is a Rab1 guanine nucleotide-exchange factor. Nat. Cell. Biol. *8*, 971–977.

Nagai, H., Kagan, J.C., Zhu, X., Kahn, R.A., and Roy, C.R. (2002). A bacterial guanine nucleotide exchange factor activates ARF on *Legionella* phagosomes. Science. *295*, 679–682.

Neumeister, B., Faigle, M., Lauber, K., Northoff, H., and Wesselborg, S. (2002). *Legionella pneumophila* induces apoptosis via the mitochondrial death pathway. Microbiol. *148*, 3639–3650.

Ninio, S., Zuckman-Cholon, D.M., Cambronne, E.D., and Roy, C.R. (2005). The *Legionella* IcmS–IcmW protein complex is important for Dot/Icm-mediated protein translocation. Mol. Microbiol. *55*, 912–926.

Orci, L., Stamnes, M., Ravazzola, M., Amherdt, M., Perrelet, A., Sollner, T.H., and Rothman, J.E. (1997). Bidirectional transport by distinct populations of COPI-coated vesicles. Cell. *90*, 335–349.

Otto, G.P., Wu, M.Y., Clarke, M., Lu, H., Anderson, O.R., Hilbi, H., Shuman, H.A., and Kessin, R.H. (2004). Macroautophagy is dispensable for intracellular replication of *Legionella pneumophila* in *Dictyostelium discoideum*. Mol. Microbiol. *51*, 63–72.

Pedersen, L.L., Radulic, M., Doric, M., and Abu Kwaik, Y. (2001). HtrA homologue of *Legionella pneumophila*: an indispensable element for intracellular infection of mammalian but not protozoan cells. Infect. Immun. *69*, 2569–2579.

Pelham, H.R. (2006). Maturation of Golgi cisternae directly observed. Trends Biochem Sci. *31*, 601–604.

Philpott, D.J. and Girardin, S.E. (2004). The role of Toll-like receptors and Nod proteins in bacterial infection. Mol. Immunol. *41*, 1099–1108.

Rabouille, C. and Klumperman, J. (2005). Opinion: The maturing role of COPI vesicles in intra-Golgi transport. Nat. Rev. Mol. Cell. Biol. *6*, 812–817.

Rabouille, C., Kondo, H., Newman, R., Hui, N., Freemont, P., and Warren, G. (1998). Syntaxin 5 is a common component of the NSF- and p97-mediated reassembly pathways of Golgi cisternae from mitotic Golgi fragments *in vitro*. Cell. *92*, 603–610.

Ren, T., Zamboni, D.S., Roy, C.R., Dietrich, W.F., and Vance, R.E. (2006). Flagellin-deficient *Legionella* mutants evade caspase-1- and Naip5-mediated macrophage immunity. PLoS Pathog. *2*, e18.

Rich, K.A., Burkett, C., and Webster, P. (2003). Cytoplasmic bacteria can be targets for autophagy. Cell. Microbiol. *5*, 455–468.

Riedl, S.J., Renatus, M., Schwarzenbacher, R., Zhou, Q., Sun, C., Fesik, S.W., Liddington, R.C., and Salvesen, G.S. (2001). Structural basis for the inhibition of caspase-3 by XIAP. Cell. *104*, 791–800.

Robinson, C.G. and Roy, C.R. (2006). Attachment and fusion of endoplasmic reticulum with vacuoles containing *Legionella pneumophila*. Cell. Microbiol. *8*, 793–805.

Rothman, J.H. and Stevens, T.H. (1986). Protein sorting in yeast: mutants defective in vacuole biogenesis mislocalize vacuolar proteins into the late secretory pathway. Cell. *47*, 1041–1051.

Roy, C.R., Berger, K.H., and Isberg, R.R. (1998). *Legionella pneumophila* DotA protein is required for early phagosome trafficking decisions that occur within minutes of bacterial uptake. Mol. Microbiol. *28*, 663–674.

Roy, L., Bergeron, J.J., Lavoie, C., Hendriks, R., Gushue, J., Fazel, A., Pelletier, A., Morre, D.J., Subramaniam, V.N., Hong, W., and Paiement, J. (2000). Role of p97 and syntaxin 5 in the assembly of transitional endoplasmic reticulum. Mol. Biol. Cell. *11*, 2529–2542.

Sadosky, A.B., Wiater, L.A., and Shuman, H.A. (1993). Identification of *Legionella pneumophila* genes required for growth within and killing of human macrophages. Infect. Immun. *61*, 5361–5373.

Seabra, M.C., Mules, E.H., and Hume, A.N. (2002). Rab GTPases, intracellular traffic and disease. Trends Mol. Med. *8*, 23–30.

Segal, G. and Shuman, H.A. (1997). Characterization of a new region required for macrophage killing by *Legionella pneumophila*. Infect. Immun. *65*, 5057–5066.

Segal, G. and Shuman, H.A. (1999). *Legionella pneumophila* utilizes the same genes to multiply within *Acanthamoeba castellanii* and human macrophages. Infect. Immun. *67*, 2117–2124.

Segev, N. (2001). Ypt and Rab GTPases: insight into functions through novel interactions. Curr. Opin. Cell. Biol. *13*, 500–511.

Shiina, N., Gotoh, Y., Kubomura, N., Iwamatsu, A., and Nishida, E. (1994). Microtubule severing by elongation factor 1alpha. Science *266*, 282–285.

Shintani, T. and Klionsky, D.J. (2004). Autophagy in health and disease: a double-edged sword. Science. *306*, 990–995.

Shiozaki, E.N., Chai, J., Rigotti, D.J., Riedl, S.J., Li, P., Srinivasula, S.M., Alnemri, E.S., Fairman, R., and Shi, Y. (2003). Mechanism of XIAP-mediated inhibition of caspase-9. Mol. Cell. *11*, 519–527.

Shohdy, N., Efe, J.A., Emr, S.D., and Shuman, H.A. (2005). Pathogen effector protein screening in yeast identifies *Legionella* factors that interfere with membrane trafficking. Proc. Natl. Acad. Sci. *102*, 4866–4871.

Slee, E.A., Harte, M.T., Kluck, R.M., Wolf, B.B., Casiano, C.A., Newmeyer, D.D., Wang, H.G., Reed, J.C., Nicholson, D.W., Alnemri, E.S., Green, D.R., and Martin, S.J. (1999). Ordering the cytochrome c-initiated caspase cascade: hierarchical activation of caspases-2, -3, -6, -7, -8, and -10 in a caspase-9-dependent manner. J. Cell Biol. *144*, 281–292.

Stenmark, H., Vitale, G., Ullrich, O., and Zerial, M. (1995). Rabaptin-5 is a direct effector of the small GTPase Rab5 in endocytic membrane fusion. Cell. *83*, 423–432.

Stromhaug, P.E., Berg, T.O., Fengsrud, M., and Seglen, P.O. (1998). Purification and characterization of autophagosomes from rat hepatocytes. Biochem. J. *335*, 217–224.

Sturgill-Koszycki, S. and Swanson, M. S. (2000). *Legionella pnuemophila* replication vacuoles mature in acidic, endocytic organelles. J Exp. Med. *192*, 1261–1272

Swanson, M. S. (2006). Autophagy: eating for good health. J. Immunol. *177*, 4945–4951.

Swanson, M.S. and Fernandez-Moreira, E. (2002). A microbial strategy to multiply in macrophages: the pregnant pause. Traffic *3*, 170–177.

Swanson, M.S. and Isberg, R.R. (1995). Association of *Legionella pneumophila* with the macrophage endoplasmic reticulum. Infect. Immun. *63*, 3609–3620.

Swanton, E., Bishop, N., and Woodman, P. (1999). Human rabaptin-5 is selectively cleaved by caspase-3 during apoptosis. J. Biol. Chem. *274*, 37583–37590.

Szeto, L. and Shuman, H.A. (1990). The *Legionella pneumophila* major secretory protein, a protease, is not required for intracellular growth or cell killing. Infect. Immun. *58*, 2585–2592.

Sztul, E. and Lupashin, V. (2006). Role of tethering factors in secretory membrane traffic. Am. J. Physiol. Cell. Physiol. *290*, C11–26.

Tilney, L.G., Harb, O.S., Connelly, P.S., Robinson, C.G., and Roy, C.R. (2001). How the parasitic bacterium *Legionella pneumophila* modifies its phagosome and transforms it into rough ER: implications for conversion of plasma membrane to the ER membrane. J Cell. Sci. *114*, 4637–4650.

Ueno, T., Muno, D., and Kominami, E. (1991). Membrane markers of endoplasmic reticulum preserved in autophagic vacuolar membranes isolated from leupeptin-administered rat liver. J. Biol. Chem. *266*, 18995–18999.

Valdivia, R.H. (2004). Modeling the function of bacterial virulence factors in *Saccharomyces cerevisiae*. Eukaryot. Cell. *3*, 827–834.

Vandenabeele, P., Vanden Berghe, T., and Festjens, N. (2006). Caspase inhibitors promote alternative cell death pathways. Sci STKE. *2006*, pe44.

VanRheenen, S.M., Dumenil, G., and Isberg, R.R. (2004). IcmF and DotU are required for optimal effector translocation and trafficking of the *Legionella pneumophila* vacuole. Infect. Immun. *72*, 5972–5982.

VanRheenen, S.M., Luo, Z.Q., O'Connor, T.J., and Isberg, R.R. (2006). Members of a *Legionella pneumophila* family of proteins with ExoU (phospholipase A) active sites are translocated to target cells. Infect. Immun. *74*, 3597–3606.

Vergunst, A.C., Schrammeijer, B., den Dulk-Ras, A., de Vlaam, C.M., Regensburg-Tuink, T.J., and Hooykaas, P.J. (2000). VirB/D4-dependent protein translocation from *Agrobacterium* into plant cells. Science *290*, 979–982.

Vincent, C.D., Friedman, J.R., Jeong, K.C., Buford, E.C., Miller, J.L., and Vogel, J.P. (2006a). Identification of the core transmembrane complex of the *Legionella* Dot/Icm type IV secretion system. Mol. Microbiol. *62*, 1278–1291.

Vincent, C.D. and Vogel, J.P. (2006b). The *Legionella pneumophila* IcmS-LvgA protein complex is important for Dot/Icm-dependent intracellular growth. Mol. Microbiol. *61*, 596–613.

Vogel, J.P., Andrews, H.L., Wong, S.K., and Isberg, R.R. (1998). Conjugative transfer by the virulence system of *Legionella pneumophila*. Science. *279*, 873–876.

Watarai, M., Derre, I., Kirby, J., Growney, J.D., Dietrich, W.F., and Isberg, R.R. (2001). *Legionella pneumophila* is internalized by a macropinocytotic uptake pathway controlled by the Dot/Icm system and the mouse Lgn1 locus. J. Exp. Med. *194*, 1081–1096.

Weber, S.S., Ragaz, C., Reus, K., Nyfeler, Y., and Hilbi, H. (2006). *Legionella pneumophila* exploits PI(4)P to anchor secreted effector proteins to the replicative vacuole. PLoS Pathog. *2*, e46.

Weber, T., Zemelman, B.V., McNew, J.A., Westermann, B., Gmachl, M., Parlati, F., Sollner, T.H., and Rothman, J.E. (1998). SNAREpins: minimal machinery for membrane fusion. Cell *92*, 759–772.

Werner, A.B., de Vries, E., Tait, S.W., Bontjer, I., and Borst, J. (2002). Bcl-2 family member Bfl-1/A1 sequesters truncated bid to inhibit is collaboration with pro-apoptotic Bak or Bax. J. Biol. Chem. *277*, 22781–22788.

Wesselborg, S., Engels, I.H., Rossmann, E., Los, M., and Schulze-Osthoff, K. (1999). Anticancer drugs induce caspase-8/FLICE activation and apoptosis in the absence of CD95 receptor/ligand interaction. Blood *93*, 3053–3063.

Wieland, H., Goetz, F., and Neumeister, B. (2004). Phagosomal acidification is not a prerequisite for intracellular multiplication of *Legionella pneumophila* in human monocytes. J Infect. Dis. *189*, 1610–1614.

Wintermeyer, E., Rdest, U., Ludwig, B., Debes, A., and Hacker, J. (1991). Characterization of legiolysin (lly), responsible for haemolytic activity, colour production and fluorescence of *Legionella pneumophila*. Mol. Microbiol. 5, 1135–1143.

Wright, E.K., Goodart, S.A., Growney, J.D., Hadinoto, V., Endrizzi, M.G., Long, E.M., Sadigh, K., Abney, A.L., Bernstein-Hanley, I., and Dietrich, W.F. (2003). Naip5 affects host susceptibility to the intracellular pathogen *Legionella pneumophila*. Curr. Biol. *13*, 27–36.

Xu, D. and Hay, J.C. (2004). Reconstitution of COPII vesicle fusion to generate a pre-Golgi intermediate compartment. J. Cell. Biol. *167*, 997–1003.

Xu, D., Joglekar, A.P., Williams, A.L., and Hay, J.C. (2000). Subunit structure of a mammalian ER/Golgi SNARE complex. J. Biol. Chem. *275*, 39631–39639.

Yang, W., Burkhart, W., Cavallius, J., Merrick, W.C., and Boss, W.F. (1993). Purification and characterization of a phosphatidylinositol 4-kinase activator in carrot cells. J. Biol. Chem. *268*, 392–398.

Ye, Y., Meyer, H.H., and Rapoport, T.A. (2001). The AAA ATPase Cdc48/p97 and its partners transport proteins from the ER into the cytosol. Nature. *414*, 652–656.

Yu, S., Satoh, A., Pypaert, M., Mullen, K., Hay, J.C., and Ferro-Novick, S. (2006). mBet3p is required for homotypic COPII vesicle tethering in mammalian cells. J. Cell. Biol. *174*, 359–368.

Zamboni, D.S., Kobayashi, K.S., Kohlsdorf, T., Ogura, Y., Long, E.M., Vance, R.E., Kuida, K., Mariathasan, S., Dixit, V.M., and Flavell, R.A., Dietrich W.F., and Roy, C.R. (2006). The Birc1e cytosolic pattern-recognition receptor contributes to the detection and control of *Legionella pneumophila* infection. Nat. Immunol 7, 318–325.

Zhang, T., Wong, S.H., Tang, B.L., Xu, Y., and Hong, W. (1999). Morphological and functional association of Sec22b/ERS-24 with the pre-Golgi intermediate compartment. Mol. Biol. Cell. *10*, 435–453.

Zhou, K., Pandol, S., Bokoch, G., and Traynor-Kaplan, A.E. (1998). Disruption of *Dictyostelium* PI3K genes reduces [32P]phosphatidylinositol 3,4 bisphosphate and [32P]phosphatidylinositol trisphosphate levels, alters F-actin distribution and impairs pinocytosis. J. Cell. Sci. *111 (Pt 2)*, 283–294.

Zusman, T., Aloni, G., Halperin, E., Kotzer, H., Degtyar, E., Felman, M., and Segal, G. (2007). The response regulator PmrA is a major regulator of the *icm/dot* type IV secretion system in *Legionella pneumophila* and *Coxiella burnetii*. Mol. Microbiol. *63*, 1508–1523.

Zusman, T., Gal-Mor, O., and Segal, G. (2002). Characterization of a *Legionella pneumophila relA* insertion mutant and toles of RelA and RpoS in virulence gene expression. J. Bacteriol. *184*, 67–75.

Nutrient Acquisition and Assimilation Strategies of *Legionella pneumophila*

11

Maris V. Fonseca, John-Demian Sauer and Michele S. Swanson*

Abstract

The fitness of the opportunistic pathogen *L. pneumophila* is a result of selective pressures in its natural freshwater environments. *L. pneumophila* has thus developed a sophisticated morphogenetic code that equips it to dramatically alter its developmental state in response to changes in its microenvironment. It is now clear that the cellular and molecular transformations that occur during the *L. pneumophila* life cycle are integrated with its metabolic state. This is exemplified by its reciprocal expression of traits that promote replication within host cells and those that promote transmission to a new host or its differentiation into cyst-like forms of greatly reduced metabolic activity. Insight into some aspects of the metabolic machinery that equip *L. pneumophila* to satisfy its nutritional requirements have been gained through studies of the molecular mechanisms that promote its differentiation. Availability of the genome sequences of *L. pneumophila* strains Philadelphia-1, Paris and Lens has also increased our power to decipher the strategies *L. pneumophila* employs to acquire and assimilate nutrients from its environment. By integrating knowledge gained from genetic, biochemical, molecular and computational approaches, this chapter presents our current view of the metabolic capabilities of *L. pneumophila*, focusing on its key metabolic features, nutritional requirements and specialized nutrient acquisition systems.

Introduction

Following the initial outbreak of Legionnaires' disease and the identification of *L. pneumophila* as its aetiological agent, several research groups undertook the task of identifying the nutritional and physiological growth requirements of this bacterium in broth culture. From these studies, valuable information was obtained regarding optimal *in vitro* growth conditions, carbon and energy source utilization and absolute nutritional requirements, such as amino acids and trace metals. This knowledge has led to the development of defined and complex media that have been used extensively both for routine culturing of *L. pneumophila* in the laboratory and for analysing the circuitry that controls its developmental pathways (Pine *et al.*, 1979; Weiss *et al.*, 1980; Reeves *et al.*, 1981; Ristroph *et al.*, 1981; Tesh *et al.*, 1983; Hoffman, 1984; Molofsky and Swanson, 2004).

Analysis of the *L. pneumophila* life cycle has been greatly facilitated by the ability to mimic in synchronous broth cultures many aspects of its differentiation into replicative and transmissive forms (Molofsky and Swanson, 2004). Studies of broth-grown *L. pneumophila* led to the discovery that multiple traits associated with virulence are regulated by growth phase, including motility, cytotoxicity and evasion of lysosomes (Byrne and Swanson, 1998; Heuner *et al.*, 1999). Moreover, in macrophage cultures, a similar expression pattern has been observed for several of these traits (Molofsky and Swanson, 2004). The modelling of the *L. pneumophila* life cycle in broth cultures also enabled the use of genetic and phenotypic screens to identify key regulators of differentiation as well as factors important for nutrient acquisition (Molofsky and Swanson, 2004; Sauer *et al.*, 2005).

*Corresponding author.

Metabolic capabilities and nutritional requirements of *L. pneumophila*

Carbohydrates

Initial studies by Pine and coworkers concluded that *L. pneumophila* is a strict aerobe (Pine *et al.*, 1979; Hoffman, 1984), suggesting that catabolism of carbohydrates would proceed by their oxidation using oxygen as an electron acceptor rather than by their fermentation (Pine *et al.*, 1979; Hoffman, 1984). Consistent with these findings, broth cultures of *L. pneumophila* do not contain detectable levels of glucose fermentation products (Weaver and Feeley, 1979).

The use of glucose as a carbon and energy source by *L. pneumophila* was examined in more detail by several investigators (Weiss *et al.*, 1980; Tesh *et al.*, 1983). When cultured *in vitro*, *L. pneumophila* do not catabolize substantial amounts of glucose, as measured by radiorespirometric assays. The low amount of CO_2 generated from its catabolism strongly suggested that the glycolytic pathway is not a preferential carbon and energy generation route for *L. pneumophila* (Tesh *et al.*, 1983). Although the Entner–Doudoroff and pentose phosphate pathways were suggested as the routes for *L. pneumophila* to catabolize glucose (Tesh *et al.*, 1983), direct measurements of key enzyme activities of the catabolic and anabolic pathways of carbohydrate metabolism did not detect components of Entner-Doudoroff (EDD) pathway in strains Philadelphia-1 and Knoxville-1 (Hoffman, 1984; Keen and Hoffman, 1984). Nonetheless, genome annotation analyses of the *L. pneumophila* Philadelphia-1 sequence (Keseler *et al.*, 2005) indicate open-reading frames (ORFs) predicted to encode the phosphogluconate dehydratase (Edd) and 2-keto-3-deoxy-6-phosphogluconate aldolase (Eda) EDD pathway enzymes (Fig. 11.1). Furthermore, global gene expression analyses indicate that *L. pneumophila* strains Philadelphia-1, Paris, and Lens transcribe these ORFs during the replicative phase both *in vivo* and *in vitro* (Brüggemann *et al.*, 2006). Perhaps the ability to detect the activity of these enzymes in cell-free extracts of *L. pneumophila* requires particular culture conditions, or the detected transcripts are not translated into functional proteins.

A *L. pneumophila* pentose phosphate pathway is suggested by the glucose-6-phosphate dehydrogenase (Zwf) and 6-phosphogluconate dehydrogenase (Gnd) activities detected in cell extracts (Fig. 11.2). However, the genome sequences of strains Philadelphia-1, Paris or Lens do not contain an obvious homologue of the Gnd enzyme (Keseler *et al.*, 2005), suggesting that another enzyme may be performing this reaction in *L. pneumophila*. Nevertheless, the slow metabolism of glucose observed in radiotracer studies, as judged by the amount of CO_2 release from cells grown in its presence, suggests that the enzyme activities observed may be inefficient or nonspecific (Tesh *et al.*, 1983). The transaldolase and transketolase activities that are also key to this metabolic route were not assayed, but assumed to be present (Hoffman, 1984). Furthermore, genome sequence analysis of *L. pneumophila* strain Philadelphia-1 identifies a transketolase I (TktA) orthologue, but not transaldolase A or B orthologues (Chien *et al.*, 2004; Keseler *et al.*, 2005). Because the pentose phosphate pathway requires Gnd and transaldolase activities, at this time there is not sufficient evidence to support the canonical pentose phosphate pathway in *L. pneumophila*. Perhaps *L. pneumophila* is equipped with alternative metabolic pathways for generation of precursors required for the synthesis of coenzymes, nucleotides and amino acids such as histidine and tryptophan (Neidhardt *et al.*, 1990). Such mechanisms would not be unprecedented, as alternative metabolic routes to the pentose phosphate pathway have been documented in some methanogenic archaea (Tumbula *et al.*, 1997; Grochowski *et al.* 2005).

The glycolytic pathway appears to be complete in *L. pneumophila* strains, as judged by genome sequence analyses (Cazalet *et al.*, 2004; Chien *et al.*, 2004) (Fig. 11.1). Nevertheless, results from both enzymatic assays and radiorespirometric studies suggest very low, if any, metabolic flux through this route (Tesh *et al.*, 1983; Hoffman, 1984). Instead, *L. pneumophila* preferentially incorporates glucose into lipid components, as demonstrated by radiorespirometric

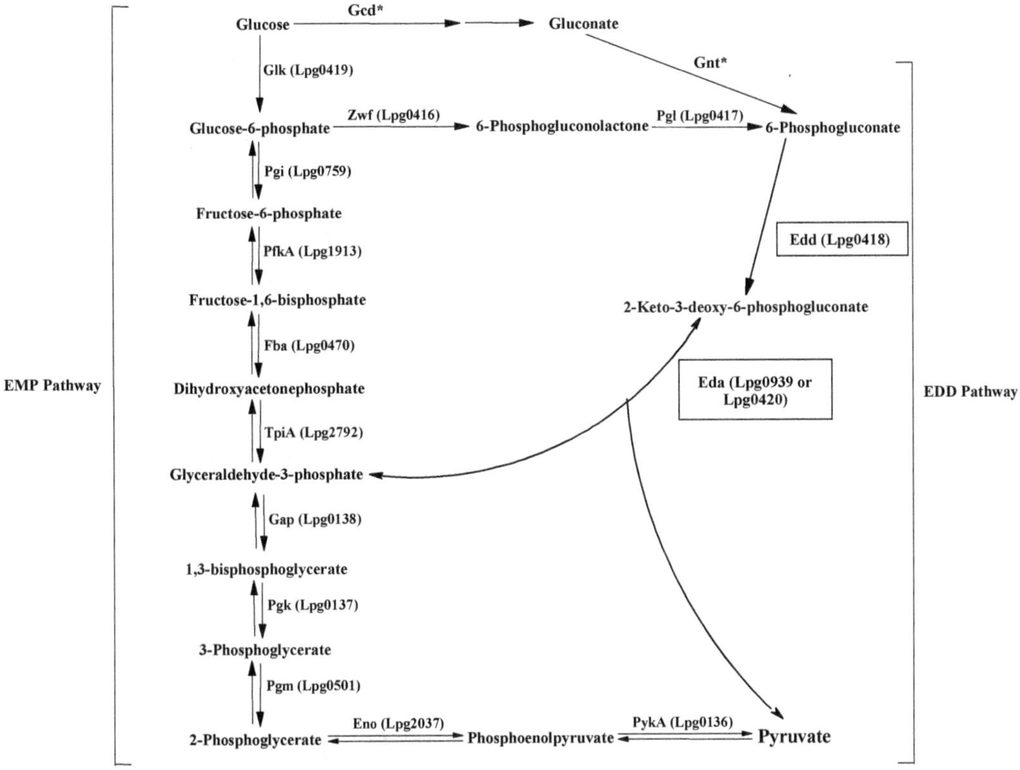

Figure 11.1 Proposed Embden–Meyerhof–Parnas (EMP) and Entner–Doudoroff pathways in *L. pneumophila*. Annotated functions in the *L. pneumophila* strain Philadelphia-1 genome are shown. Encoding open-reading frames (ORFs) are specified in parenthesis. Functions marked with an asterisk have no annotated homologues in the *L. pneumophila* genome. Abbreviations for putative EMP enzymes: Gcd, glucose dehydrogenase; Gnt, gluconate transporter; Glk, glucokinase; Zwf, glucose-6-phosphate-1-dehydrogenase; Pgl, 6-phosphogluconolactonase; Pgi, phosphoglucose isomerase; PfkA, pyrophosphate-dependent phosphofructokinase; Fba, fructose bisphosphate aldolase (Class I); TpiA, triosephosphate isomerase; Gap, glyceraldehyde-3-phosphate dehydrogenase; Pgk, phosphoglycerate kinase; Pgm, phosphoglycerate mutase; Eno, enolase; PykA, pyruvate kinase II. Abbreviations for EDD enzymes (in boxes): Edd, 6-phosphogluconate dehydratase; Eda, 2-keto-3-deoxy-phosphogluconate aldolase.

studies (Tesh *et al.*, 1983). Thus, *L. pneumophila* likely has an anabolic use for glucose rather than a catabolic one.

Further supporting that the bacterium may favour formation of glucose via anaplerotic reactions rather than its uptake and utilization are results from glucose uptake studies that strongly suggest that glucose is not transported by *L. pneumophila*, but rather may enter the cell by diffusion or via non-specific binding (Keen and Hoffman, 1984). Indeed, *L. pneumophila* extracts contain readily detectable gluconeogenic enzyme activities, namely phosphoenolpyruvate (PEP) carboxylase and PEP carboxykinase (Hoffman, 1984). Interestingly, while putative orthologues of the inorganic phosphate-dependent PEP carboxylase (EC 4.1.1.31) can be found within the genome sequences of *L. pneumophila* strains Philadelphia-1, Paris and Lens, orthologues of the ATP-dependent PEP carboxykinase (EC 4.1.1.49) cannot (Keseler *et al.*, 2005). Therefore, the source of the PEP carboxykinase activity detected in cell extracts may be the putative PEP carboxylase enzyme itself: ATP hydrolysed by other enzyme activities in the extracts could generate the inorganic phosphate this enzyme needs to produce PEP from oxaloacetate. Another plausible explanation is that the PEP carboxykinase activity observed is generated by an as yet unidentified function.

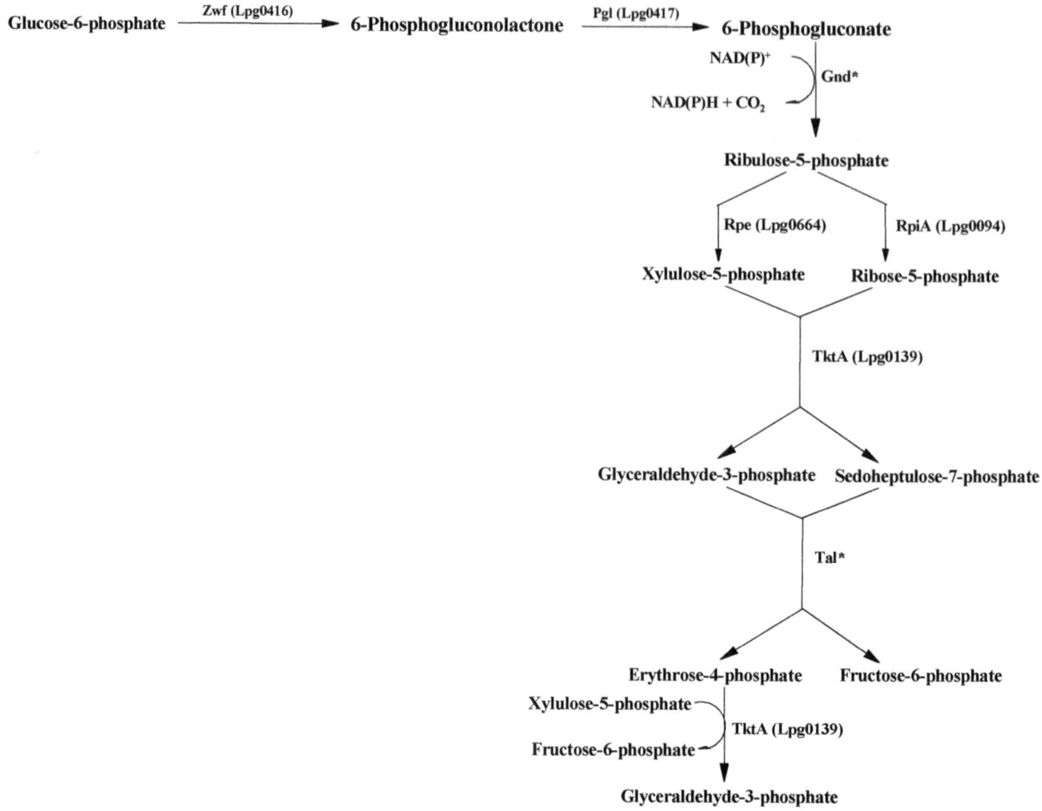

Figure 11.2 Canonical pentose phosphate pathway. Annotated ORFs in the *L. pneumophila* strain Philadelphia-1 genome are shown in parenthesis. Functions marked with an asterisk have no annotated homologues in the *L. pneumophila* genome. Abbreviations for proposed functions: Zwf, glucose-6-phosphate-1-dehydrogenase; Pgl, 6-phosphogluconolactonase; Gnd, 6-phosphogluconate dehydrogenase; Rpe, ribulose-5- phosphate-3-epimerase; RpiA, ribose-5-phosphate isomerase; TktA, transketolase; Tal, transaldolase.

Also detected in *L. pneumophila* cell-free extracts were activities for additional components of anabolic routes to glucose such as malate dehydrogenase (NAD^+ and $NADP^+$-linked), pyruvate carboxylase, PEP carboxylase and glyceraldehyde-3-phosphate dehydrogenase. Consistent with the readily detected activities is the presence of orthologues for these components in the genomes of *L. pneumophila* strains (Chien *et al.*, 2004; Cazalet *et al.*, 2004; Keseler *et al.*, 2005). The apparent high functional redundancy in routes leading to PEP are suggestive of the importance of the gluconeogenic pathway to *L. pneumophila* (Fig. 11.3).

Interestingly, based solely on computational analyses, the BioCyc database has assigned the fructose 1,6-bisphosphatase function required for generation of fructose-6-phosphate to ORF Lpg0469, encoding the putative phenol hydroxylase enzyme PoxF (Green and Karp, 2004). The putative function shares homology with ferredoxin-$NADP^+$ reductase enzymes which contain NAD^+- and FAD-binding domains (Altschul *et al.*, 1990). It is difficult to envision how a protein with oxidoreductase activity may function as a phosphatase. Perhaps this assignment was made based on the close proximity of the proposed fructose bisphosphate aldolase function in the *L. pneumophila* genome, a feature that the methodology employed takes into consideration (Green and Karp, 2004). An interesting possibility for the otherwise missing function is raised by the homology of the putative phosphofructokinase PfkA function in *L. pneumophila* to eukaryotic

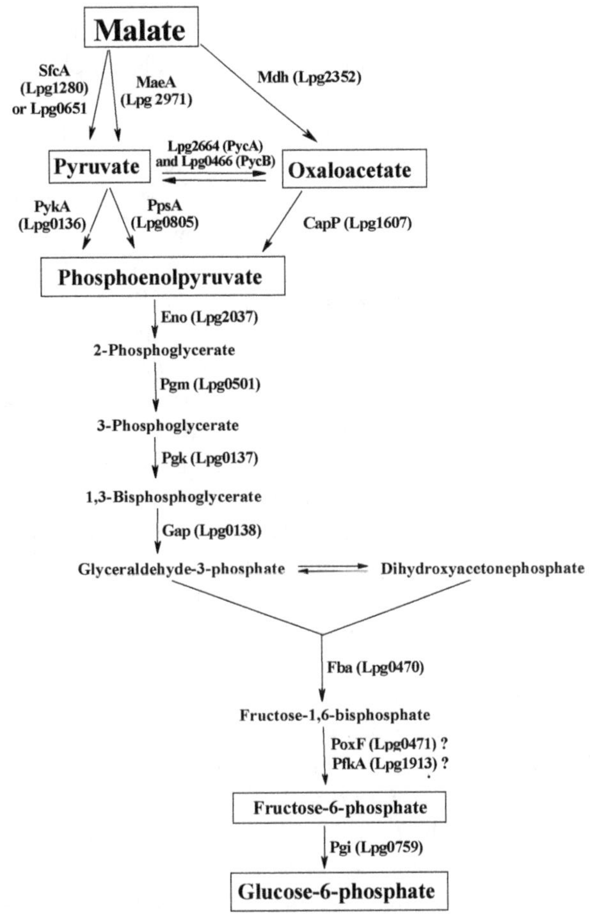

Figure 11.3 Proposed gluconeogenic pathway in *L. pneumophila*. Annotated functions in the *L. pneumophila* strain Philadelphia-1 genome are shown. Encoding open-reading frames (ORFs) are specified in parenthesis. Functions marked with a question mark have no clear annotation in the *L. pneumophila* genome. Proposed key intermediates are shown in boxes. Abbreviations for proposed functions: SfcA, malate oxidoreductase; Lpg0651, malate oxidoreductase; MaeA, malate dehydrogenase (NAD-linked; a.k.a. malic enzyme); Lpg2664 (PycA), pyruvate carboxylase, subunit A; Lpg0466 (PycB), pyruvate carboxylase subunit B; Mdh, malate dehydrogenase; PykA, pyruvate kinase II; PpsA, phosphoenolpyruvate synthase; CapP, phosphoenolpyruvate carboxylase; Eno, enolase; Pgm, phosphoglycerate mutase; Pgk, phosphoglycerate kinase; Gap, glyceraldehyde-3-phosphate dehydrogenase; Fba, fructose bisphosphate aldolase (Class I); PoxF, phenol hydroxylase; PfkA, pyrophosphate-dependent phosphofructokinase; Pgi, phosphoglucose isomerase.

pyrophosphate-dependent phosphofructokinase enzymes (EC 2.7.1.90), which are reversible (Phillips and Li, 1995; Costa dos Santos *et al.*, 2003; Hoffman, 2007). By measuring the forward and reverse reactions, Keen and Hoffman demonstrated that the fructose-1,6-bisphosphatase activity was 10-fold higher than the phosphofructokinase activity in *L. pneumophila* cell-free extracts (Keen and Hoffman, 1984), providing another strong indication that the gluconeogenic pathway is a key metabolic route to *L. pneumophila*. Experimental data together with genome annotation information lead us to propose the gluconeogenic pathway shown in Fig. 11.3.

Analysis of the transcriptome of *L. pneumophila* cultured in *Acanthamoeba castellanii* indicated that a number of genes required for carbohydrate metabolism are highly upregulated during the intracellular replication period (Brüggemann *et al.*, 2006). Included are genes for putative myo-inositol catabolic functions. Given the role of inositol phosphate compounds as signalling molecules in eukaryotic cells (Berridge, 1984; Strahl, 2007), the use of inositol metabolites as part of the *L. pneumophila* virulence strategy is an intriguing possibility that deserves further consideration.

Amino acids

A striking finding of the early work aimed at determining *L. pneumophila* nutritional requirements was the demonstration that this bacterium strongly prefers amino acids as carbon and energy sources (Pine *et al.*, 1979; Reeves *et al.*, 1981; Ristroph *et al.*, 1981; Tesh *et al.*, 1983). By using various radiolabelled substrates, Tesh *et al.* elegantly demonstrated that *L. pneumophila* efficiently metabolizes the amino acid glutamate, which distributes evenly in cell fractions. This is in stark contrast to the inefficient catabolism of glucose by these bacteria.

When cultured in either complex or defined media, *L. pneumophila* strain Knoxville-1 efficiently uses both glutamate and serine as energy sources, based on the ability of these amino acids to stimulate oxygen uptake (Tesh *et al.*, 1983). Additionally, threonine can be an energy source for cells grown in defined medium, whereas tyrosine is a source of energy when cells are grown in complex medium.

In addition to serving as carbon and energy sources, the amino acids arginine, threonine, methionine, serine, isoleucine, leucine, valine and cysteine appear to be absolutely required for *L. pneumophila* growth (Pine *et al.*, 1979; George *et al.*, 1980; Ristroph *et al.*, 1981; Tesh and Miller, 1981). Analysis of the *L. pneumophila* strain Philadelphia-1 genome sequence has revealed partial routes for synthesis of some of these amino acids. Nevertheless, *in vitro* growth experiments have not yet identified conditions in which these apparent auxotrophies are bypassed (Chien *et al.*, 2004).

Coenzymes, nucleic acid precursors, and metal cofactors.

Numerous studies aimed at developing an effective culture medium for *L. pneumophila* demonstrated that, aside from the absolute requirement for specific amino acids, all strains tested appear to have the biosynthetic functions for all coenzymes necessary for growth (Warren and Miller, 1979; Ristroph *et al.*, 1981). A source of purines and pyrimidines in culture media is not required, suggesting that *L. pneumophila* is fully equipped for *de novo* purine and pyrimidine biosynthesis. Nevertheless, guanine is stimulatory to the growth of several *L. pneumophila* strains and required by various other *Legionella* species (Pine *et al.*, 1986).

The trace metal requirements for *L. pneumophila* are defined as iron, copper, magnesium, cobalt, calcium, molybdenum, nickel, vanadium and zinc (Reeves *et al.*, 1981). Iron is a key metal required by *L. pneumophila* both *in vivo* and *in vitro* (Pine *et al.*, 1979; Warren and Miller, 1979; Ristroph *et al.*, 1981; Hoffman, 1984; Gebran *et al.*, 1994; James *et al.*, 1995; Byrd and Horwitz, 2000). The mechanisms by which *L. pneumophila* acquires iron are the focus of a comprehensive review (Cianciotto, 2006) and are not discussed further here.

Energy generation strategies

While amino acids are considered the primary sources of carbon and energy for *L. pneumophila*, the tricarboxylic acid (TCA) cycle is regarded as its key route to assimilate carbon and generate energy. Amino acids that serve as carbon sources are likely assimilated by *L. pneumophila* through the TCA cycle. The NADH and $FADH_2$ generated concomitantly as reducing equivalents are then oxidized via the respiratory chain and coupled to ATP generation via the proton-motive force (Hoffman, 1984).

The tricarboxylic acid cycle. Studies by Hoffman and Pine (Hoffman and Pine, 1982; Hoffman, 1984) identified nearly all of the TCA cycle enzyme activities; the exceptions are the glyoxylate cycle enzymes isocitrate lyase and malate synthase, which are deemed absent or marginal. However, measurements of oxidative metabolism of TCA cycle intermediates such as isocitrate and α-ketoglutarate indicate that these metabolites are not oxidized at appreciable rates (Warren and Miller, 1979; Ristroph *et al.*, 1981; Hoffman, 1984). On the other hand, pyruvate is an oxidizable substrate for *L. pneumophila* and thus a good energy source, as indicated by results of oxygen uptake and radiolabelling studies (Tesh *et al.*, 1983). Since no other TCA cycle intermediates besides pyruvate are appreciably oxidized, despite readily detectable TCA cycle enzyme activities, *L. pneumophila* cells may transport TCA cycle intermediates poorly, if at all (Hoffman, 1984).

L. pneumophila membrane vesicles exhibit high NADH oxidase activity, as judged by measuring respiratory chain enzyme activities (Hoffman and Pine, 1982). Genome sequence information is consistent with these data: Whereas complete TCA cycles can be inferred from the *L. pneumophila* strains Philadelphia-1, Paris and Lens, an isocitrate lyase orthologue has not been identified. Thus, the available data indicate that the glyoxylate shunt in these strains is incomplete (Cazalet *et al.*, 2004; Chien *et al.*, 2004).

Respiratory metabolism. Consistent with the conclusion from physiological and nutritional studies that *L. pneumophila* is an obligate aerobe, the genome sequences of strains Philadelphia-1, Paris and Lens indicate a complete aerobic respiratory chain (Hoffman and Pine, 1982; Hoffman, 1984; Cazalet *et al.*, 2004; Chien *et al.*, 2004). Furthermore, *L. pneumophila* strains contain detectable amounts of cytochromes *c*, *b*, *a* and *d* (Hoffman and Pine, 1982). Similarly, the terminal *c*, *a* and *d* type cytochrome oxidase activities were confirmed by photochemical reduction spectra. The major quinone in the *L. pneumophila* respiratory chain is ubiquinone (Karr *et al.*, 1982).

The high NADH oxidase activity of its membrane vesicles strongly suggests that the TCA cycle is a major energy-generating route for *L. pneumophila* (Hoffman and Pine, 1982; Hoffman, 1984). Nonetheless, preliminary measurement of H^+/O ratios indicates that coupling of NADH oxidation to ATP synthesis by *L. pneumophila* may not be very efficient, as one ATP molecule is generated per atom of oxygen consumed (Hoffman, 1984).

To our knowledge, the functional organization of the *L. pneumophila* respiratory chain has not been determined experimentally. Additional genetic and biochemical analyses are needed to further understand amino acid catabolic pathway function, flux and integration to the TCA cycle, and respiratory chain activities in *L. pneumophila*. This information will be pivotal in gaining insights as to how *L. pneumophila* cells derive energy from nutrient sources.

Fatty acids

Fatty acid profiles obtained from cell suspensions of *L. pneumophila* isolates grown on different solid media revealed very little, if any, variation in composition among different isolates (Moss *et al.*, 1977; Moss and Dees, 1979; Mayberry, 1981). The fatty acid composition of each isolate did not appear to be significantly affected by the growth medium. In contrast to several other Gram-negative bacteria, branched-chain fatty acids are highly abundant in *L. pneumophila* strains, a pattern more typical of Gram-positive genera such as *Listeria*, *Staphylococcus* and *Bacillus* and of thermophilic archaea (Moss *et al.*, 1977).

Interestingly, under the conditions tested by both Mayberry and Moss and coworkers (Moss *et al.*, 1977; Mayberry, 1981), unsaturated fatty acids, if detected, were present at very low concentrations. However, because cells for these experiments were obtained from solid agar, it is likely that heterogeneous *L. pneumophila* populations were obtained, making it difficult to assess the implications of these results to the life cycle of the organism. Analysis of the fatty acid composition of *L. pneumophila* isolated from host cells, biofilm environments, or from synchronous broth cultures may increase our knowledge of this nutritional demand in *L. pneumophila* and provide additional clues as to the fatty acid anabolic and catabolic abilities of this microbe. Regarding the latter, a few observations are worth noting:

1. Given the low pyruvate dehydrogenase activity detected in cell-free extracts of *L. pneumophila* cells (Hoffman, 1984), it has been proposed that the bulk of the acetyl-CoA entering the TCA cycle in this bacterium is derived from fatty acid catabolism (Hoffman, 2007). Consistent with this idea, genome sequence analyses reveal complete β-oxidation pathways in strains Philadelphia-1, Paris and Lens.
2. Unsaturated fatty acids have been identified as moieties of the lipopolysaccharide (LPS) component of *L. pneumophila* cells (Sonesson *et al.*, 1989), yet the sequenced

genomes of this bacterium do not appear to encode the canonical functions for synthesis of unsaturated fatty acids.

3 When added to log phase broth cultures, several classes of fatty acids trigger premature differentiation of *L. pneumophila* (R. Edwards, Z. Dalebroux and M.S. Swanson, unpublished), strongly indicating that these metabolites constitute a key regulatory component of the *L. pneumophila* life cycle.

Worth mentioning is the fact that the vegetative mature intracellular forms (MIFs) of *L. pneumophila* accumulate poly-β-hydroxybutyrate (PHB) granules as cytoplasmic inclusions (Garduño *et al.*, 2002). These polymers of the short-chain fatty acid 3-hydroxybutyric acid are utilized by a number of bacteria as carbon and energy sources in nutrient limiting environments (Steinbuchel and Schlegel, 1991; Page *et al.*, 1992; Kidwell *et al.*, 1995; Schembri *et al.*, 1995). Indeed, *L. pneumophila* utilizes PHB reserves under nutrient starvation conditions (James *et al.*, 1999; Garduño *et al.*, 2002), further highlighting the importance of fatty acid metabolites to its life cycle.

L. pneumophila nutrient acquisition strategies

L. pneumophila survives within amoeba and macrophages due to its ability to establish a unique replication vacuole that is derived from the endoplasmic reticulum and secluded from the canonical endocytic pathway (Fig. 11.3; Chapter 10) (Horwitz, 1983a; Horwitz, 1983b; Swanson and Isberg, 1995). Once within its vacuole, transmissive *L. pneumophila* cells differentiate to the replicative form (Chapter 4). Nutrient availability governs *L. pneumophila* differentiation: When nutrients are plentiful, bacteria replicate; as host cell-derived nutrients become limiting, a regulatory cascade triggers differentiation to the transmissive form (Molofsky and Swanson, 2004). Cell differentiation is a central strategy of *Legionella* pathogenesis, since mutants in one of a number of loci that encode regulators of this process are significantly attenuated (Molofsky and Swanson, 2004).

Amino acid requirements for intracellular growth

Little, if anything, is known about the nutritional environment within *L. pneumophila* replication vacuoles. The absolute requirement for specific amino acids and their preferential use as carbon and energy sources by *L. pneumophila in vitro* prompted examination of the requirements of intracellular bacteria (Wieland *et al.*, 2005). Utilizing a set of rich and defined cell culture media formulations to mimic either nutrient-rich or amino acid starvation conditions, Wieland *et al.* determined which amino acids supported proficient replication of *L. pneumophila* within an MM6 human monocytic cell line. When amino acids in the cell culture medium were limited, growth of intracellular *L. pneumophila* was significantly decreased (< 76%). Under these conditions, cysteine was the amino acid most effective in promoting *L. pneumophila* intracellular replication; its rate exceeded that of rich medium control cultures. Addition of glutamine and serine also stimulated a significant increase in growth (> 50%). Methionine, arginine, glutamate, valine, threonine and tyrosine each enhanced growth to a minimal extent, whereas leucine and isoleucine had no effect (Wieland *et al.*, 2005).

A noteworthy observation from these studies is the apparent difference in the amino acid requirements of intracellular vs. extracellular *L. pneumophila* cells. Whereas glutamate is an excellent energy source for *L. pneumophila* in broth cultures (Tesh *et al.*, 1983), it had only a minimal effect on growth in the MM6 cell culture model (Wieland *et al.*, 2005). In addition, amino acids deemed essential for growth of *L. pneumophila* in chemically defined medium, such as threonine, methionine, valine and arginine (Ristroph *et al.*, 1981), only marginally enhanced growth within monocytes. Whether each supplement had similar access to the replication vacuole or stability in the macrophage cultures was not investigated. Nevertheless, although synchronous broth culture models provide certain advantages for assessing *L. pneumophila* nutrient acquisition and utilization strategies (Hoffman, 1984; Molofsky and Swanson, 2004; Brüggemann *et al.*, 2006) (Fig. 11.4), it is important to keep in mind that the microbe's metabolism and available resources

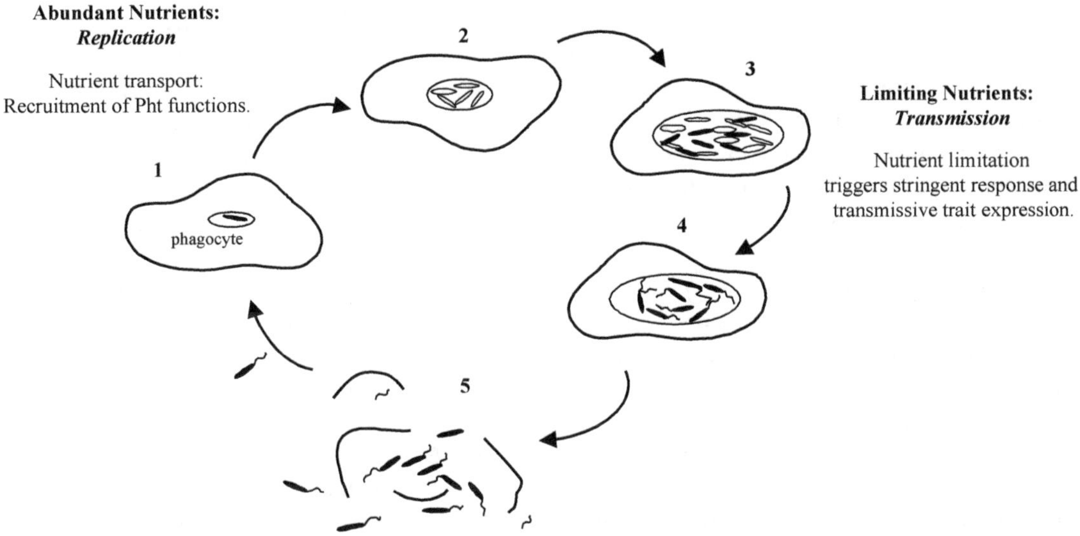

Figure 11.4 Working model for the coupling of nutrient acquisition to cell differentiation in *L. pneumophila*. A. Replicative and transmissive phases of the *L. pneumophila* lifecycle are governed by nutrient availability: 1. Transmissive *L. pneumophila* cells establish a vacuole that protects them from lysosomal degradation. 2. Utilizing high-efficiency Pht systems, *L. pneumophila* cells acquire nutrients from their host and replicate. 3. and 4. Upon deteriorating growth conditions, *L. pneumophila* halts replication, activates the stringent response and activates expression of transmissive traits. 5. Transmissive *L. pneumophila* lyses its host. Free-swimming bacteria can infect a new host cell and begin a new replication cycle.

may be quite different in macrophages, amoeba, and biofilms.

Host transport functions required for *L. pneumophila* nutrient acquisition

Host cell functions also contribute to nutrient acquisition by intracellular *L. pneumophila* (Wieland *et al.*, 2005). By examining the expression of various transporters of MM6-monocyte cells during *L. pneumophila* infection, Wieland and colleagues identified a host cell transporter, SLC1A5 that is not only up-regulated during infection, but also required for bacterial replication, regardless of whether the tissue culture medium was supplemented with amino acids. The SLC1A5 transporter is a Na^+/Cl^--coupled neutral amino acid transporter with high affinity for cysteine, glutamine, isoleucine, methionine, phenylalanine, serine, valine, tryptophan and tyrosine. Consistent with its requirement to support replication of intracellular *L. pneumophila*, the SLC1A5 transporter is thought to be located on the phagosomal membrane (Wieland *et al.*, 2005). This study provided the first evidence that intracellular *L. pneumophila* exploit a host nutrient transporter; furthermore, it provides independent evidence that amino acid acquisition and assimilation systems are also critical when the bacteria are replicating within host cells.

Nutrient transport functions in *L. pneumophila*

The *L. pneumophila* Philadelphia-1 genome is currently estimated to encode 156 transporter proteins with predicted functions ranging from nutrient transport to lipid A export (http://www.membranetransport.org/index.html). Members of the ATP-binding cassette (ABC) superfamily and the major-facilitator superfamily (MFS) each have 35 identified orthologues in the Philadelphia-1 genome, making these the predominant classes of transporters of *L. pneumophila*. Bacterial ABC transporters couple ATP hydrolysis to a variety of functions, including nutrient acquisition, toxin secretion, and drug resistance (Pao *et al.*, 1998; Schneider and Hunke, 1998; Davidson and Chen, 2004). The MFS includes families of transporters that rely on ion gradients for their activities, which range from oligosaccharide:H^+ symport, to nucleoside:H^+

symport, to drug:H$^+$ antiport functions (Pao *et al.*, 1998). The prevalence of ABC and MFS transporters in the *L. pneumophila* genome is not surprising considering that the MFS and the ABC superfamily account for approximately half of the known or predicted transport functions in most prokaryotes (Paulsen *et al.*, 1998).

Nevertheless, a striking finding from the analysis of the predicted *L. pneumophila* strain Philadelphia-1 transport functions was the identification of a unique set of 11 transporters within the putative MFS of this bacterium (Chien *et al.* 2004; Sauer *et al.*, 2005; Chen *et al.*, 2007). The predicted novel MFS members appear to be paralogous (Fig. 11.5) and to be conserved only in two phylogenetically related intracellular pathogens. The *Coxiella burnetii* and *Francisella tularensis* genomes encode 9 and 4–6 putative orthologues, respectively (Fig. 11.5). It is remarkable that within other intracellular pathogens also known to encode a large number of MFS members (e.g. *Escherichia coli*, *Salmonella enterica*), no definite orthologues of these putative MFS transporters have been identified (Chien *et al.*, 2004). Based on sequence analyses performed by Milton Saier's group and genetic data obtained in our laboratory, these unique transporters have been designated as phagosomal transporters (Phts) and classified as a distinct subfamily of the MFS (Sauer *et al.*, 2005; Saier, 2006; Chen *et al.*, 2007). Our genetic and phenotypic analyses indicate that these putative transporters equip *L. pneumophila* to acquire nutrients that are not only highly specific to its intracellular lifestyle, but also play a pivotal role in coupling the bacterium's metabolism to cell differentiation.

Phagosomal transporter A (PhtA) function is required for threonine acquisition and cell differentiation in *L. pneumophila*

The identification of PhtA as a function that *L. pneumophila* requires to acquire threonine and to replicate within macrophages offered clues to how this microbe couples its metabolic state to its life cycle (Sauer *et al.*, 2005). *L. pneumophila phtA* mutants were analysed in detail because they are defective for differentiation in broth and

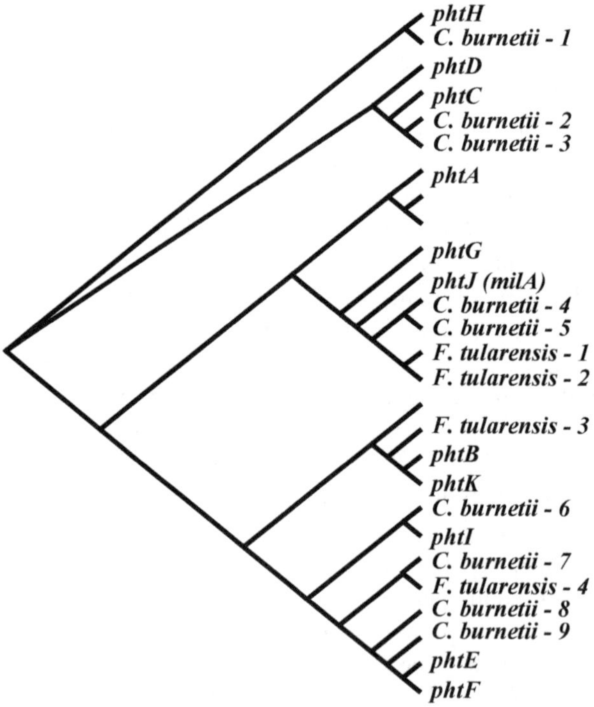

Figure 11.5 Evolutionary relationship between the Pht family of *L. pneumophila* and homologues in *C. burnetii* and *F. tularensis* (Sauer, 2006). Proposed *L. pneumophila* Pht-encoding members are indicated as designated from the genome sequence. Identified homologues in *C. burnetii* and *F. tularensis* are also indicated.

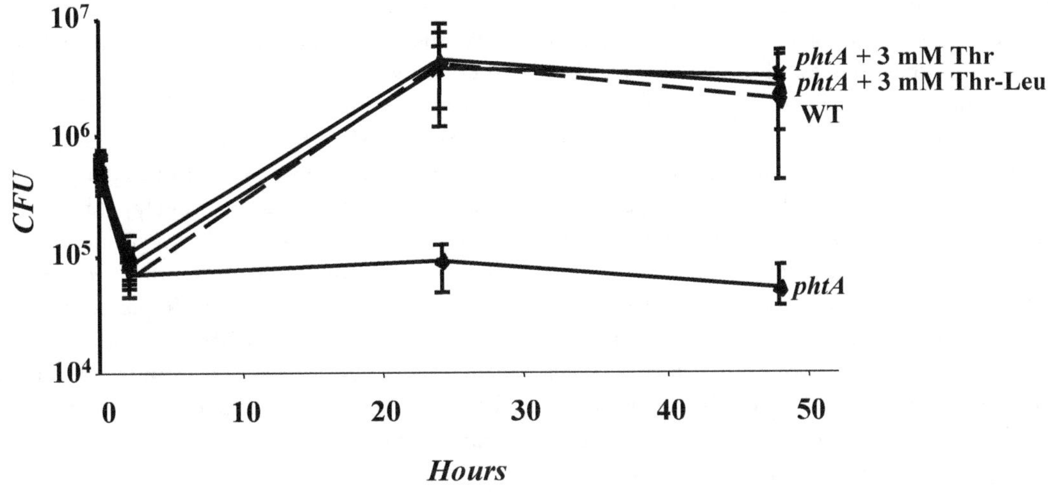

Figure 11.6 Threonine or threonine-containing dipeptides bypass PhtA function in macrophages. Macrophage culture medium was supplemented with 3 mM threonine or 3 mM of a threonine-leucine dipeptide as indicated. From Sauer *et al.* (2005).

replication in macrophages. *L. pneumophila* cells lacking PhtA infect macrophages as efficiently as the parent strain, yet they do not replicate during the >120 h period following infection. Unlike Dot/Icm type IV secretion system mutants, *phtA* mutants are competent to evade the endosomal pathway (Roy *et al.*, 1998; Sauer *et al.*, 2005). However, the intracellular *phtA* mutants fail to differentiate from the transmissive to the replicative form, as shown by their sodium-sensitivity phenotype and pattern of gene expression. Moreover, when *phtA* mutants are cultured in either broth or macrophages, their growth defect is bypassed when the medium is supplemented with excess threonine or dipeptides that contain threonine, an amino acid that is essential for growth of wild-type *L. pneumophila* (Fig. 11.6). Therefore, PhtA function is critical for *L. pneumophila* to acquire threonine, which in turns promotes differentiation to the replicative form, a process absolutely required for growth in macrophages.

Additional proposed Pht functions are required for *L. pneumophila* intracellular growth

Because of their homology to other members of the MFS and to PhtA, each ORF encoding the remaining putative members of the subfamily was given a *pht* designation, from *phtB* to *phtK*. The ORF designation was based on their genetic organization on the *L. pneumophila* genome, where *phtA* is first and *phtK* is last (Table 11.1). Initial genetic analyses of the remaining loci determined that, unlike *phtA* mutants, none of the remaining *pht* loci are required for growth in rich broth or defined medium under routine laboratory culture conditions (J.D. Sauer and M.S. Swanson, unpublished). Nonetheless, several, but not all, of these loci are essential for *L. pneumophila* to replicate within macrophages. Furthermore, phenotypic analyses have indicated that some of

Table 11.1 Phagosomal transporter (Pht) loci of *L. pneumophila* strain Philadelphia-1 (Sauer *et al.*, 2005, Sauer, 2006)

Gene designation	Accession number/locus
phtA	*YP_094583/lpg0540*
phtB	YP_094646/*lpg0610*
phtC	YP_094673/*lpg0637*
phtD	YP_094674/*lpg0638*
phtE	YP_094688/*lpg0652*
phtF	YP_094689/*lpg0653*
phtG	YP_095364/*lpg1335*
phtH	YP_095433/*lpg1404*
phtI	YP_095842/*lpg1816*
phtJ	YP_095910/*lpg1893*
phtK	YP_096523/*lpg2516*

these loci do not appear to contribute to amino acid acquisition, but instead may contribute to scavenging other classes of metabolites (M.V. Fonseca and M.S. Swanson, unpublished).

Thus, it appears that *L. pneumophila* has evolved high-affinity transport systems that support its intracellular lifestyle by allowing efficient exploitation of the host as a nutrient source. The metabolism of cysteine by *L. pneumophila* strains is also illustrative of this feature. Although cysteine is an essential amino acid to *L. pneumophila* (Pine *et al.*, 1979; Hoffman, 1984), it is not a required supplement for growth in eukaryotic cells, indicating that the microbe is equipped to scavenge this metabolite from the host (Wieland *et al.*, 2005). In support of this, studies of cysteine transport by *L. pneumophila* indicate the presence of high-affinity and low-affinity transporters for this amino acid (Ewann and Hoffman, 2006). The low-affinity transport system likely functions at high cysteine concentrations, whereas the high-affinity system is proposed to function at low cysteine concentrations. *L. pneumophila* is unable to use cystine (Pine *et al.*, 1979; Hoffman, 1984), the oxidized dipeptide form of cysteine, that *L. pneumophila* likely encounters in extracellular environments. Accordingly, Ewann and Hoffman have postulated that cysteine auxotrophy restricts replication of *L. pneumophila* to an intracellular environment where this metabolite is accessible (Ewann and Hoffman, 2006).

Conclusions

Early studies of *L. pneumophila* physiology and metabolism have greatly enabled assessment of the adaptive capabilities this bacterium has evolved to survive in both aquatic and intracellular niches. The advent of whole genome sequences, genomic microarrays, molecular genetics and bioinformatic tools has allowed data re-evaluation and incorporation of new information to current models. The nutritional environment protozoan hosts provide has likely been pivotal in shaping the biosynthetic and salvage capabilities of *L. pneumophila*. Abundance of various metabolites within these environments may account for the loss of *de novo* biosynthetic functions and acquisition of metabolic components highly specific for intracellular growth, such as the Pht systems. Indeed, our laboratory has evidence that the concentration of particular metabolites dictates when replicating bacteria differentiate to the transmissive form (R. Edwards, Z. Dalebroux and M.S. Swanson, unpublished). Thus, the ability of *L. pneumophila* to undergo drastic metabolic transformations likely enhances its versatility as it alternates between life within a biofilm community, as a planktonic microbe, or as an intracellular pathogen. Implementation of integrative genetic, biochemical and molecular approaches will further our understanding of how *L. pneumophila* utilizes its metabolic circuitry to persist in nature.

Web resources

1. BioCyc metabolic/genome database: http://biocyc.org/
2. *L. pneumophila* genome analyses: LegioList web server: http://genolist.pasteur.fr/LegioList/index.html
3. Transport database: http://www.membranetransport.org/index.html
4. Transport system classification: http://www.biology.ucsd.edu/~msaier/transport/

Acknowledgements

We would like to thank Dr Paul S. Hoffman for his numerous contributions to our knowledge of *L. pneumophila* physiology and metabolism and for his critical review of this chapter and insightful comments. Work in the Swanson laboratory is supported by Public Health Service grants R01-AI44212 and R01-AI040694 from the National Institute of Allergy and Infectious Diseases (to M.S.S.). M.V.F. is supported by the Molecular Mechanisms of Microbial Pathogenesis Training Grant T32-AI007528 from the National Institute of Allergy and Infectious Diseases of the National Institutes of Health.

References

Altschul, S.F., Gish, W., Miller, W., Myers, E.W., and Lipman, D.J. (1990). Basic local alignment search tool. J. Mol. Biol. *215*, 403–410.

Bachman, M.A., and Swanson, M.S. (2004). The LetE protein enhances expression of multiple LetA/LetS-dependent transmission traits by *Legionella pneumophila*. Infect Immun *72*, 3284–3293.

Berridge, M.J., Irvine, R.F. (1984). Inositol trisphosphate, a novel second messenger in cellular signal transduction. Nature. *312*, 315–321.

Brüggemann, H., Hagman, A., Jules, M., Sismeiro, O., Dillies, M.-A., Gouyette, G., Kunst, F., Steinert, M., Heuner, K., Coppee, J.-Y., and Buchrieser, C. (2006). Virulence strategies for infecting macrophages deduced from the *in vivo* transcriptional program of *Legionella pneumophila*. Cell. Microbiol. *8*, 1228–1240.

Byrd, T.F., and Horwitz, M.A. (2000). Aberrantly low transferrin receptor expression on human monocytes is associated with nonpermissiveness for *Legionella pneumophila* growth. J. Infect. Dis. *181*, 1394–1400.

Byrne, B. and Swanson, M.S. (1998). Expression of *Legionella pneumophila* virulence traits in response to growth conditions. Infect. Immun. *66*, 3029–3034.

Cazalet, C., Rusniok, C., Brüggemann, H., Zidane, N., Magnier, A., Ma, L., Tichit, M., Jarraud, S., Bouchier, C., Vandenesch, F., Kunst, F., Etienne, J., Glaser, P., and Buchrieser, C. (2004). Evidence in the *Legionella pneumophila* genome for exploitation of host cell functions and high genome plasticity. Nat. Genet. *36*, 1165–1173.

Chen, D., Podell, S., Sauer, J.D., Swanson, M.S., and Saier, M.H. Jr. (2007). The phagosomal nutrient transporter (Pht) family. Microbiology (in press).

Chien, M., Morozova, I., Shi, S., Sheng, H., Chen, J., Gomez, S.M., Asamani, G., Hill, K., Nuara, J., Feder, M., Rineer, J., Greenberg, J.J., Steshenko, V., Park, S.H., Zhao, B., Teplitskaya, E., Edwards, J.R., Pampou, S., Georghiou, A., Chou, I.C., Iannuccilli, W., Ulz, M.E., Kim, D.H., Geringer-Sameth, A., Goldsberry, C., Morozov, P., Fischer, S.G., Segal, G., Qu, X., Rzhetsky, A., Zhang, P., Cayanis, E., De Jong, P.J., Ju, J., Kalachikov, S., Shuman, H.A., and Russo, J.J. (2004). The genomic sequence of the accidental pathogen *Legionella pneumophila*. Science *305*, 1966–1968.

Cianciotto, N.P. (2006). Iron Acquisition by *Legionella pneumophila*. Biometals *20*, 323–331.

Costa dos Santos, A., Seixas da-Silva, W., de Meis, L., and Galina, A. (2003). Proton transport in maize tonoplasts supported by fructose-1,6-bisphosphate cleavage. Pyrophosphate-dependent phosphofructokinase as a pyrophosphate-regenerating system. Plant. Physiol. *133*, 885–892.

Davidson, A.L., and Chen, J. (2004). ATP-binding cassette transporters in bacteria. Annu. Rev. Biochem. *73*, 241–268.

Ewann, F., and Hoffman, P.S. (2006). Cysteine metabolism in *Legionella pneumophila*: characterization of an L-cystine-utilizing mutant. Appl. Environ. Microbiol. *72*, 3993–4000.

Garduño, R.A., Garduño, E., Hiltz, M., and Hoffman, P.S. (2002). Intracellular growth of *Legionella pneumophila* gives rise to a differentiated form dissimilar to stationary-phase forms. Infect. Immun. *70*, 6273–6283.

Gebran, S.J., Newton, C., Yamamoto, Y., Widen, R., Klein, T.W., and Friedman, H. (1994). Macrophage permissiveness for *Legionella pneumophila* growth modulated by iron. Infect. Immun. *62*, 564–568.

George, J.R., Pine, L., Reeves, M.W., and Harrell, W.K. (1980). Amino acid requirements of *Legionella pneumophila*. J. Clin. Microbiol. *11*, 286–291.

Green, M.L., and Karp, P.D. (2004). A Bayesian method for identifying missing enzymes in predicted metabolic pathway databases. BMC Bioinformatics 5, 76.

Grochowski, L.L., Xu, H., White, R.H. (2005). Ribose-5-phosphate biosynthesis in *Methanocaldococcus jannaschii* occurs in the absence of a pentose-phosphate pathway. J. Bacteriol. *187*, 7382–7389.

Heuner, K., Brand, B.C., Hacker, J. (1999). The expression of the flagellum of *Legionella pneumophila* is modulated by different environmental factors. FEMS Microbiol. Lett. *175*, 69–77.

Hoffman, P.S. (1984). Bacterial Physiology. In Proceedings of the 2nd. International Symposioum on *Legionella*, C. Thornsberry, A. Balows, J. C. Feeley, and W. Jakubowski, eds. (Washington, DC, American Society for Microbiology), pp. 61–67.

Hoffman, P.S. (2007). Microbial Physiology. In *Legionella pneumophila: Pathogenesis and Immunity*. P. Hoffman, T. Klein, and H. Friedman, eds., (Springer Publishing Corp.), pp. 113–128.

Hoffman, P.S., and Pine, L. (1982). Respiratory physiology and cytochrome content of *Legionella pneumophila*. Curr. Microbiol. *7*, 351–356.

Horwitz, M.A. (1983a). Formation of a novel phagosome by the Legionnaires' disease bacterium (*Legionella pneumophila*) in human monocytes. J. Exp. Med. *158*, 1319–1331.

Horwitz, M.A. (1983b). The Legionnaires' disease bacterium (*Legionella pneumophila*) inhibits phagosome-lysosome fusion in human monocytes. J. Exp. Med. *158*, 2108–2126.

James, B.W., Mauchline, W.S., Dennis, P. J., Keevil, C.W., and Wait, R. (1999). Poly-3-hydroxybutyrate in *Legionella pneumophila*, an energy source for survival in low-nutrient environments. Appl. Environ. Microbiol. *65*, 822–827.

James, B.W., Mauchline, W.S., Fitzgeorge, R.B., Dennis, P.J., and Keevil, C.W. (1995). Influence of iron-limited continuous culture on physiology and virulence of *Legionella pneumophila*. Infect. Immun. *63*, 4224–4230.

Karr, D.E., Bibb, W.F., and Moss, C.W. (1982). Isoprenoid quinones of the genus *Legionella*. J. Clin. Microbiol. *15*, 1044–1048.

Keen, M.G., and Hoffman, P.S. (1984). Metabolic pathways and nitrogen metabolism in *Legionella pneumophila*. Curr. Microbiol. *11*, 81–88.

Keseler, I.M., Collado-Vides, J., Gama-Castro, S., Ingraham, J., Paley, S., Paulsen, I.T., Peralta-Gil, M., and Karp, P.D. (2005). EcoCyc: a comprehensive database resource for *Escherichia coli*. Nucleic Acids Res. 33, D334–337.

Kidwell, J., Valentin, H.E., and Dennis, D. (1995). Regulated expression of the *Alcaligenes eutrophus pha* biosynthesis genes in *Escherichia coli*. Appl. Environ. Microbiol. *61*, 1391–1398.

Mayberry, W.R. (1981). Dihydroxy and monohydroxy fatty acids in *Legionella pneumophila*. J. Bacteriol. *147*, 373–381.

Molofsky, A.B., and Swanson, M.S. (2004). Differentiate to thrive: lessons from the *Legionella pneumophila* life cycle. Mol. Microbiol. *53*, 29–40.

Moss, C.W., and Dees, S.B. (1979). Further studies of the cellular fatty acid composition of Legionnaires' disease bacteria. J. Clin. Microbiol. *9*, 648–649.

Moss, C.W., Weaver, R.E., Dees, S.B., and Cherry, W.B. (1977). Cellular fatty acid composition of isolates from Legionnaires' disease. J. Clin. Microbiol. *6*, 140–143.

Neidhardt, F.C., Ingraham, J.L., Schaechter, M. (1990). Biosynthesis and Fueling. In Physiology of the bacterial cell: a molecular approach. (Sunderland, Massachusetts, Sinauer Associates, Inc.) pp. 133–173.

Page, W.J., Manchak, J., and Rudy, B. (1992). Formation of poly(hydroxybutyrate-co-hydroxyvalerate) by *Azotobacter vinelandii* UWD. Appl. Environ. Microbiol. *58*, 2866–2873.

Pao, S.S., Paulsen, I.T., and Saier, M.H., Jr. (1998). Major facilitator superfamily. Microbiol. Mol. Biol. Rev. *62*, 1–34.

Paulsen, I.T., Sliwinski, M.K., and Saier, M.H., Jr. (1998). Microbial genome analyses: global comparisons of transport capabilities based on phylogenies, bioenergetics and substrate specificities. J. Mol. Biol. *277*, 573–592.

Phillips, N.F., and Li, Z. (1995). Kinetic mechanism of pyrophosphate-dependent phosphofructokinase from *Giardia lamblia*. Mol. Biochem. Parasitol. *73*, 43–51.

Pine, L., Franzus, M.J., and Malcolm, G.B. (1986). Guanine is a growth factor for *Legionella* species. J. Clin. Microbiol. *23*, 163–169.

Pine, L., George, J.R., Reeves, M.W., and Harrell, W.K. (1979). Development of a chemically defined liquid medium for growth of *Legionella pneumophila*. J. Clin. Microbiol. *9*, 615–626.

Reeves, M.W., Pine, L., Hutner, S.H., George, J.R., and Harrell, W.K. (1981). Metal requirements of *Legionella pneumophila*. J. Clin. Microbiol. *13*, 688–695.

Ristroph, J.D., Hedlund, K.W., and Gowda, S. (1981). Chemically defined medium for *Legionella pneumophila* growth. J. Clin. Microbiol. *13*, 115–119.

Roy, C.R., Berger, K.H., and Isberg, R.R. (1998). *Legionella pneumophila* DotA protein is required for early phagosome trafficking decisions that occur within minutes of bacterial uptake. Mol. Microbiol. *28*, 663–674.

Saier, M.H., Jr. (2006). Transport Classification Database. http://www.membranetransport.org/index.html

Sauer, J.D. (2006), PhD Thesis, University of Michigan.

Sauer, J.D., Bachman, M.A., and Swanson, M.S. (2005). The phagosomal transporter A couples threonine acquisition to differentiation and replication of *Legionella pneumophila* in macrophages. Proc. Natl. Acad. Sci. USA *102*, 9924–9929.

Schembri, M.A., Bayly, R.C., and Davies, J.K. (1995). Phosphate concentration regulates transcription of the *Acinetobacter* polyhydroxyalkanoic acid biosynthetic genes. J. Bacteriol. *177*, 4501–4507.

Schneider, E., and Hunke, S. (1998). ATP-binding-cassette (ABC) transport systems: functional and structural aspects of the ATP-hydrolyzing subunits/domains. FEMS Microbiol. Rev. *22*, 1–20.

Sonesson, A., Jantzen, E., Bryn, K., Larsson, L., and Eng, J. (1989). Chemical composition of a lipopolysaccharide from *Legionella pneumophila*. Arch. Microbiol. *153*, 72–78.

Steinbuchel, A., and Schlegel, H.G. (1991). Physiology and molecular genetics of poly(beta-hydroxyalkanoic acid) synthesis in *Alcaligenes eutrophus*. Mol. Microbiol. *5*, 535–542.

Strahl, T., Thorner, J. (2007). Synthesis and function of membrane phosphoinositides in budding yeast, *Saccharomyces cerevisiae*. Biochim. Biophys. Acta. *1771*, 353–404.

Swanson, M.S., and Isberg, R.R. (1995). Association of *Legionella pneumophila* with the macrophage endoplasmic reticulum. Infect. Immun. *63*, 3609–3620.

Tesh, M.J., and Miller, R.D. (1981). Amino acid requirements for *Legionella pneumophila* growth. J. Clin. Microbiol. *13*, 865–869.

Tesh, M.J., Morse, S.A., and Miller, R.D. (1983). Intermediary metabolism in *Legionella pneumophila*: utilization of amino acids and other compounds as energy sources. J. Bacteriol. *154*, 1104–1109.

Tumbula, D.L., Teng, Q., Bartlett, M.G., Whitman, W.B. (1997). Ribose biosynthesis and evidence for an alternative first step in the common aromatic amino acid pathway in *Methanococcus maripaludis*. J. Bacteriol. *179*, 6010–6013.

Warren, W.J., and Miller, R.D. (1979). Growth of Legionnaires disease bacterium (*Legionella pneumophila*) in chemically defined medium. J. Clin. Microbiol. *10*, 50–55.

Weaver, R.E., and Feeley, J.C. (1979). Cultural and biochemical characterization of the Legionnaires' disease bacterium. In Legionnaires: the disease, the bacterium, and methodology, G.L. Jones, and G.A. Hebert, eds. (Atlanta, Center for Disease Control), pp. 19–25.

Weiss, E., Peacock, M.G., and Williams, J.C. (1980). Glucose and glutamate metabolism of *Legionella pneumophila*. Curr. Microbiol. *4*, 1–6.

Wieland, H., Ullrich, S., Lang, F., and Neumeister, B. (2005). Intracellular multiplication of *Legionella pneumophila* depends on host cell amino acid transporter SLC1A5. Mol. Microbiol. *55*, 1528–1537.

Dictyostelium, a Tractable Model Host Organism for *Legionella* 12

Heike Bruhn and Michael Steinert*

Abstract

The haploid amoeba *Dictyostelium* has proven to be a suitable model system for studying cellular aspects of Legionnaires' disease. In this review we describe both *Legionella* and *Dictyostelium* factors and processes that are relevant to infection. Moreover, we summarize the results of the *Dictyostelium* transcriptional host cell response to infection and discuss how genetic and molecular biology techniques available for *Dictyostelium* can further improve our knowledge of *Legionella*-host cross-talk.

Introduction

Infection models in *Legionella* research

Susceptibility to a certain infection is an innate property of a given host species and is governed by a number of different factors such as anatomy, physiology, variations in tissue or cell surface receptors, effector molecules of the innate and adaptive immune system and others. To gain insight into specific aspects of the host–pathogen interaction, the selection of an appropriate model system with an appropriate level of complexity appears to be essential. Frequently studied host models for *Legionella* research are guinea pigs, inbred mice, mammalian cell culture systems and various protozoa species. Each system provides specific advantages for the analysis of *Legionella* infection. Guinea pigs are highly susceptible to infection with *Legionella* and have been widely used since they develop pulmonary disease symptomatically and histologically similar to that observed in humans (Edelstein *et al.*, 1999; Wagner *et al.*, 2007). Moreover, they allow reliable predictions of drug efficacies. Mice are genetically well characterized; however, they are relatively resistant to *Legionella* infections. Only intratracheally inoculated A/J mice seem to be a suitable model of a replicative *L. pneumophila* lung infection (Rossier *et al.*, 2004). Accordingly, only peritoneal and bone marrow-derived macrophages from A/J mice are permissive for *Legionella* growth, whereas macrophages from other inbred strains are not (Chapter 5).

In addition to animal models, primary cell culture systems and permanent cell lines have provided important information on virulence properties of *Legionella* and corresponding host targets. In this regard, human monocytes, human alveolar macrophages, and the macrophages of guinea pigs, hamsters and rats have successfully been studied (Steinert *et al.*, 2002). At least equally important are infection studies performed with host protozoa. Ubiquitously present in aquatic habitats, *Legionella* naturally exploits numerous species of protozoa, which provide nutrients, protect the bacteria from adverse conditions and serve as a vehicle for the colonization of new habitats.

Each of the aforementioned infection models suffers from one of three drawbacks: The animal models are complex; therefore, their analysis and manipulation is time-consuming and difficult. Cell lines lack the cross-talk and other responses typical of multicellular organisms. The single celled protozoa are only poorly investigated

*Corresponding author.

and hence insufficiently understood on the molecular level. Moreover, they are not genetically tractable.

In the last years, model organisms like *Caenorhabditis elegans*, *Drosophila melanogaster* and *Dictyostelium discoideum* became increasingly attractive for pathogenicity studies because of their ease to cultivate, the short generation times and the availability of cell biological and genetic tools. Particularly, the haploid social amoeba *D. discoideum* appears to be an ideal model system for *Legionella* infection due to its life style: It is an obligate phagocyte that feeds on bacteria – just as the natural protozoan hosts of the pathogen. This phenotypic relation is corroborated by the narrow phylogenetic distance between *Dictyostelium* and *Acanthamoeba*, which predicts a similar molecular repertoire.

Model organism *Dictyostelium discoideum*

The intriguing biology of *D. discoideum* has fascinated cell biologists for decades, making it a popular model for investigating molecular cell biology. The haploid cells feed on soil bacteria and replicate by binary fission. Under conditions of starvation, approximately 10^5 cells aggregate by cAMP-relayed chemotaxis and differentiate in a complex progression into a fruiting body, which is built by spore cells and dead stalk cells (Fig. 12.1). Many fundamental processes underlying cellular procedures like cytokinesis, motility, chemotaxis and signal transduction observed in metazoans appear to have emerged in primordial precursor cells early in evolution. Therefore, these complex processes can be studied in unicellular organisms like *D. discoideum*. The life cycle of *D. discoideum* also encourages the investigation of social behaviours like cell differentiation and altruistic cell death, processes that rely on cell-cell communication and chemotactic cell sorting. Hence, it is not surprising that infection biologists have also been attracted to this organism as an experimental model (Steinert and Heuner, 2005).

Dictyostelium shares many molecular features with macrophages, the human host of *Legionella*. The cytoskeletal composition of *D. discoideum* is similar to that of mammalian cells as are the processes driven by these components, such as phagocytosis, membrane trafficking, endocytic transit and vesicle sorting. Like leukocytes, *D. discoideum* possess chemotactic capacity. Hence, *D. discoideum* represents a suitable model system to ascertain the influence of a variety of host cell factors during *Legionella* infections. Furthermore, the abundance of well-characterized tools to observe the cell biology of this protozoan provides the opportunity to analyse the pathogen's mechanisms of infection by investigating the fate of various *Legionella* mutants defective in putative pathogenicity factors.

Dictyostelium–Legionella interactions at the cellular level

Most importantly, *Legionella* is capable of infecting *D. discoideum* and using this host for

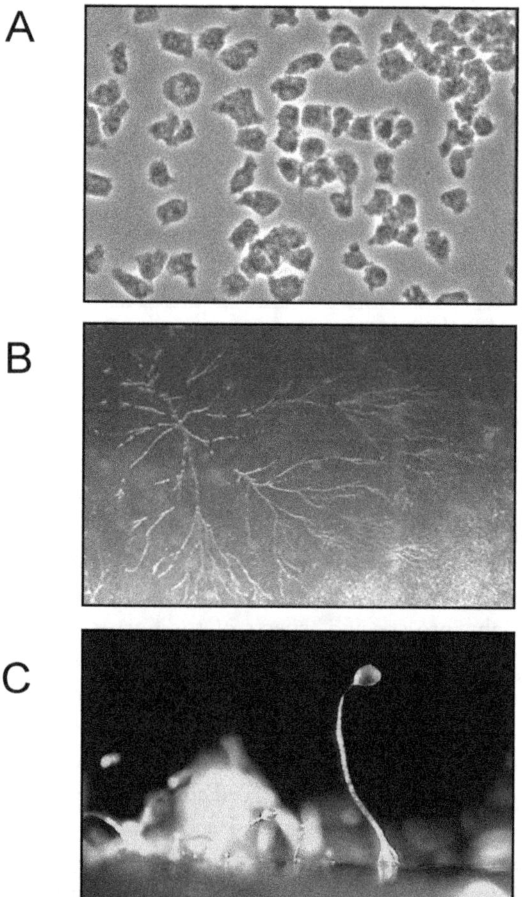

Figure 12.1 Multicellular morphogenesis of *Dictyostelium discoideum*, from single cells (A) via aggregation (B) into a multicellular fruiting body (C).

reproduction (Hägele *et al.*, 2000; Solomon *et al.*, 2000). The uptake and replication of the bacteria in *D. discoideum* resemble that which is known from amoebae and macrophages (Steinert *et al.*, 2007). The pathogen itself triggers the phagocytosis and replicates within membrane-bound vesicles that prevent fusion with lysosomes (Hägele *et al.*, 2000; Solomon *et al.*, 2000). The endoplasmic reticulum (ER) contributes to the *Legionella*-containing vacuoles (LCV) in the natural protozoan hosts *A. castellanii* and *Hartmanella* (Abu Kwaik, 1996; Bozue and Johnson, 1996) and in macrophages (Swanson and Isberg, 1995; Kagan and Roy, 2002; Robinson and Roy, 2006) as well as in *D. discoideum* (Fajardo *et al.*, 2004; Lu and Clarke, 2005). Furthermore, the growth-rates of wild-type and various *Legionella* mutants are comparable in the different hosts, demonstrating essentially similar replication mechanisms (Heuner *et al.*, 2002; Schreiner *et al.*, 2002; Skriwan *et al.*, 2002). Taken together, these prerequisites are essential and sufficient to justify the use of the non-natural host *D. discoideum* as an infection model.

Legionella factors and function

The bacterial factors provoking pathogenicity can mainly be grouped into two categories: Regulatory factors, which control the transcriptional pattern in response to the environment, and effector molecules, which mediate interactions with the host cell. The regulatory factors comprise the alternative sigma factors RpoS (Hales and Shuman, 1999) and FliA (Heuner *et al.*, 2002) as well as the two-component response regulator LetA (GacA) (Hammer *et al.*, 2002; Gal-Mor and Segal, 2003; Lynch *et al.*, 2003) which regulate the expression of proteins characteristic for the transmissive state (Molofsky and Swanson, 2004). The most prominent effector molecules are the type 4 secretion system (T4SS) Dot/Icm (Segal *et al.*, 1998; Vogel *et al.*, 1998) and several substrates secreted by it. These transferred proteins are mostly still of unknown function, but concertedly, they prevent the fusion of the *Legionella*-containing vacuole with the endosomal compartment (Chapter 10). Two additional secretion systems contributing to pathogenicity are the T2SS Lsp (Hales and Shuman, 1999) and a twin-arginine translocation secretion system (De Buck *et al.*, 2005; Chapter 8). A protein identified as macrophage infection potentiator (Mip) (Cianciotto *et al.*, 1989) is also involved in the infection of protozoan hosts (Cianciotto and Fields, 1992). It belongs to the family of FK506-binding peptidyl-prolyl-*cis*/*trans*-isomerases (Fischer *et al.*, 1992) and contributes to the dissemination of the bacteria in the tissue in concert with a yet unknown serine proteinase (Wagner *et al.*, 2007). Finally, type IV pili promote the adherence of the bacteria to cells (Stone and Abu Kwaik, 1998). Since all these factors are described in comprehensive details in particular chapters of this book, we here will concentrate on their impact on specific interactions with *D. discoideum*.

Legionella Dot/Icm mutants replicate less efficiently than wild-type bacteria in *D. discoideum* and permit *Dictyostelium* growth on bacterial lawns (Solomon *et al.*, 2000), incontrovertible evidence that the pathogen requires the secretion system to reproduce. The effect of only few of the known substrates of Dot/Icm has been investigated in *D. discoideum* thus far. Of these, SidC is the only one which provokes a specific effect in the slime mould. SidC accumulates on the LCVs by binding to phosphatidylinositol(4) phosphate (PI4-P) in a highly specific manner (Weber *et al.*, 2006). This metabolite of the phosphoinositol pathway is generated by the action of phosphatidylinositol(3) phosphate-kinase, an enzyme involved in several signal transduction pathways. Inhibitory studies revealed that the enzyme is also required for replication of *Legionella* in *Dictyostelium* (Weber *et al.*, 2006). Its activity might enrich the LCV membrane in PI4-P, which then functions as an anchor for transmitted bacterial proteins like SidC.

The only other *Legionella* protein investigated that influences the intracellular growth in *D. discoideum* is the alternative sigma factor FliA, which controls the flagellar regulon. Deletion mutants are aflagellate and are incapable of replication in *D. discoideum* (Heuner *et al.*, 2002). Although the analysis of *Legionella* mutants in *D. discoideum* is a straightforward and powerful approach, the effect of some discrete host factors is not ready at hand due to the lack of homologues in the slime mould. Prominent examples are NF-kappa B, an observation that indicates some metazoan pathways of cell responses do not exist in *D. discoideum*. A second example is caspases, enzymes that are involved in programmed cell death (Golstein *et al.*, 2003). The distinct mechanisms that orchestrate cell death in macrophages and *D. discoideum* might account for the observed cell-type specificity of the Dot/Icm substrate SdhA in the two hosts. SdhA apparently prevents host cell death in macrophages, thereby ensuring *Legionella* replication (Laguna *et al.*, 2006). The *sdhA* deletion mutants are severely impaired in macrophages but not in *D. discoideum*. For other pathogenicity factors, the effect of their deletion is more pronounced in one host than the other. For example, the mutant of the Dot/Icm substrate VipD, which possesses a phospholipase A domain and is suggested to assist in the establishment of the LCV, replicates normally in macrophages but exhibits an increased growth rate in *D. discoideum* (VanRheenen *et al.*, 2006). These results support the idea that the biochemical activity of these proteins is identical in the different hosts, but they may have evolved to operate more efficiently in specific cell types.

Dictyostelium factors and functions

The internalization of *Legionella* is actin-mediated and involves several cytoskeleton-associated proteins like myosin I (Solomon *et al.*, 2000), coronin (Solomon *et al.*, 2000; Fajardo *et al.*, 2004), comitin (Schreiner *et al.*, 2002; Skriwan *et al.*, 2002), the actin cross-linker alpha-actinin/ABP120, the F-actin fragmenting and capping protein DAip1, the cytoskeleton-membrane interface proteins LimA-D and the F-actin associated protein villidin (Fajardo *et al.*, 2004). Knock-out mutants of all these proteins show reduced uptake and/or increased intracellular growth of the bacteria. Interestingly, the only actin-binding protein which reduces bacterial internalization is profilin, a protein that sequesters G-actin (Hägele *et al.*, 2000). Knock-out mutants of profilin exhibit a generally higher rate of phagocytosis and thus a higher rate of bacterial uptake whereas phosphatidylinositol(3) kinases are apparently not required for phagocytosis of *Legionella* (Weber *et al.*, 2006), inhibitory studies revealed that the phospholipase C (PLC) pathway does contribute (Fajardo *et al.*, 2004). PLC itself is activated by heterotrimeric G-proteins, and knock-out mutants of the Gβ subunit are impaired in uptake (Solomon *et al.*, 2000; Fajardo *et al.*, 2004).

Phagocytosis and replication of the bacteria are also modulated by cytosolic calcium. The influence of two ER-localized calcium sequestering proteins has been examined, namely the luminal calreticulin and the transmembrane calnexin (Fajardo *et al.*, 2004; Lu and Clarke, 2005). According to Fajardo *et al*, both proteins extend into the phagocytic cup and accumulate permanently on the LCV, consistent with a contribution of these proteins to uptake and vacuole formation (Fajardo *et al.*, 2004). By contrast, Lu and Clarke described the LCV captured by the ER with a concurrent accumulation of calnexin about one hour into infection whereas the LCV maintains its distinct identity (Lu and Clarke, 2005). The bacterial vacuoles are transported in the cell along microtubules (Lu and Clarke, 2005). Vesicle trafficking is a pivotal process for *Legionella* reproduction. Mutants in RtoA, a protein mediating vesicle trafficking, showed severely depressed intracellular growth of *Legionella* (Li *et al.*, 2005). The *Legionella* vacuoles avoid lysosomal fusion, as demonstrated by the exclusion of lysosomal marker proteins such as DdLIMP (Hägele *et al.*, 2000). In spite of the morphological similarity of the vacuoles to autophagosomes, macroautophagy is not required for bacterial replication in *Dictyostelium* (Otto *et al.*, 2004). In mouse macrophages, a metal ion transporter protein called NRAMP1 (natural resistance associated membrane protein) decreases the susceptibility to intracellular infections. Likewise, *Dictyostelium* mutants that lack the homologous protein exhibit enhanced replication of *Legionella*; the phenotype can be rescued by NRAMP1 over-expression. The protein is located on macropinosomes, phagosomes and exocytic vesicles. In the presence of ATP, the protein acts as a proton/iron ion symporter thereby depleting the vacuole of iron, which is pivotal for intracellular pathogens. Since the protein itself does not bind ATP, the activity suggests a concerted reaction of the vesicular H^+-ATPase and NRAMP1 (Lu and Clarke, 2005; Peracino *et al.*, 2006).

The genome sequencing of *Dictyostelium* enabled the application of comprehensive strategies. The transcriptional response of the amoeba upon infection with wild-type or avirulent, DotA-deficient *L. pneumophila* and the less virulent species *L. hackeliae* was analysed using microarrays comprising approximately half of the *Dictyostelium* genes (Farbrother *et al.*, 2006). Most of the transcriptional changes occur 24 h post infection when 131 genes are differentially regulated in response to all three *Legionella* species. Genes encoding proteins that promote degradation like hydrolases, lysozymes, lipid modifying enzymes and several calcium binding proteins were down-regulated, as were genes important for the biosynthesis of proteins, whereas genes that contribute to tRNA metabolism, nucleotide biosynthesis and the cytoskeleton are apparently up-regulated, presumably to ensure the pathogen's supply of nutrients. Consistent with conclusions from mutational studies, expression of a number of amoebal genes encoding proteins primarily involved in vesicle-trafficking processes like RtoA, ARF1 and β'COP were also induced after infection.

The *Dictyostelium* tool box opens new perspectives

Due to the interest of cell biologists in particular, multifarious tools have been developed for genetic manipulation and generation of mutants in *D. discoideum* (Table 12.1). Random mutagenesis by insertion of extrachromosomal vectors into the genome can be used to create large libraries of mutant strains by applying an elegant method called restriction enzyme mediated integration (REMI) (Kuspa and Loomis, 1992; Kuspa, 2006). The cells are electroporated with a linearized plasmid in combination with a frequently cutting restriction enzyme that produces the same overhanging ends that the plasmid offers. The enzyme mediates the insertion of the plasmid into the genome at the corresponding restriction sites. Southern Blot analysis performed with a second enzyme that does not cut inside the plasmid reveals the genomic region of integration. The fragment is isolated, recircularized and amplified in *E. coli* for sequencing. Phenotypic analysis of such libraries of mutants identified a large number of genes involved in cytokinesis, motility, aggregation and later stages of development. Moreover, directed

Table 12.1 Cell biological and bioinformatic tools established for *D. discoideum*

Method	Effect	Reference
Genetic manipulation		
REMI	Random mutagenesis	Kuspa, 2006; Kuspa and Loomis, 1992
Parasexual genetics	Combination of mutations	King and Insall, 2006; King and Insall, 2003
Knock-out	Directed mutagenesis by homologues recombination	
RNAi	Gene silencing	Martens *et al.*, 2002
Phenotypic analysis: time lapse imaging		
Fluorescent proteins	Non-invasive tags visuable *in vivo*	Knetsch *et al.*, 2002; Müller-Taubenberger, 2006
DIAS	3D reconstruction and motion analysis	Soll *et al.*, 2000; Wessels *et al.*, 2006
Bioinformatic aids		
www.dictybase.org	Annotated genome sequence Respiratory of mutants	Eichinger *et al.*, 2005
www.genome.jp	Pathways of biological processes	

mutagenesis of non-essential genes can be easily achieved by homologues recombination, which is also used to verify the phenotype of random mutants obtained by REMI. The constructions of double and triple knock-out mutants is also facilitated by parasexual genetics, which refers to the formation of diploid cells by fusion of pairs of haploid cells and the subsequent spontaneous haploidization by random loss of chromosomes, (King and Insall, 2003; King and Insall, 2006). All these mutants are listed in the 'Dicty Stock Center', the central repository for *D. discoideum* strains and plasmids (http://www.dictybase.org/StockCenter/StockCenter.html). Currently, there are 1020 different strains, including natural isolates, random mutants, null mutants and GFP-labelled strains, and 290 plasmids are also recorded. Waiting to be analysed for their impact in the cell biology of the amoeba is the whole catalogue of mutant *D. discoideum* strains defective in cytoskeleton proteins, vesicle sorting, or signal transduction – exactly those processes that are of great importance to uptake and replication of *Legionella*.

Mainly three independent paths are inviting the interest of scientists keen to characterize the host side of the interaction with bacteria. The first one comprises the cellular responses to the invasion. In this respect, a great boon was the publication in 2005 of the genome sequence of *D. discoideum* (Eichinger *et al*., 2005). The six chromosomes of *D. discoideum* comprise 33.8 Mb DNA coding for approximately 13.500 genes. The genome has a higher content of simple sequence repeats and homopolymeric traits like polyglutamine in coding regions than that of any other known organism. A large number of the genes is likely to be common in metazoans and *D. discoideum* but to be absent in *S. cerevisiae*. The genome is well-annotated by now and has enabled the application of comprehensive strategies. For example, the proteome of subcellular structures and alterations thereof have become amenable, along with new approaches to analyse protein-protein interactions. Knowledge of the complete set of predicted genes also allows the use of microarrays to analyse transcriptional changes during infection. Additionally, on the basis of the sequences, the proteome of subcellular structures and alterations thereof become amenable and even points the way to protein-protein interactions.

The second major track of investigation is the visualization of the dynamic cell behaviour, organelles or even single host cell proteins *in vivo* by time-lapse imaging. Sophisticated approaches using 2D and 3D Dynamic Image Analysis Systems (Soll *et al*., 2000; Wessels *et al*., 2006) or quantitative fluorescence microscopy and fluorescence resonance energy transfer (Sekar and Periasamy, 2003; Xu *et al*., 2005) were designed to analyse the chemotactic response of the cells initiating differentiation. The application of the green fluorescent protein and variants thereof as non-invasive tags facilitates the tracking of host cell factors (Knetsch *et al*., 2002; Müller-Taubenberger, 2006). Null mutants rescued by GFP fusions ensure the functionality of the fusion. Finally, GFP fusions can be used to investigate the behaviour of cells under various conditions.

Lastly, exploiting the high frequency of homologous recombination, custom-tailored knock-out mutants of *D. discoideum* and the applicability of RNA interference (Martens *et al*., 2002) allow the characterization of the impact of single proteins in the replication pathway.

Concluding remarks

Owing to remarkable similarities in cell structure and behaviour to the natural hosts, *D. discoideum* has been proven a powerful model for *Legionella* infection. The analysis of bacterial virulence in this model reflects many mechanisms that are relevant for the infection of macrophages. Particularly the applicability of genetic manipulation of both the pathogen and the host enables the detailed characterization of their interactions by the combined analysis of attenuated bacteria mutants and the corresponding host mutants. Moreover, high-throughput approaches like functional and structural genomics and proteomics of pathogen and host will extend the knowledge of their interrelations. Additionally, in the search for yet unknown functions of putative pathogenicity factors, like the Dot/Icm substrates, the abundance of cell biological tools and the ease of handling make *D. discoideum* an optimal experimental system. Since the idea to exploit the social amoeba as an infection model and the appreciation of its versatility emerged only a few years ago, the number of studies exploiting this system will undoubtedly increase dramatically in the near future.

References

Abu Kwaik, Y. (1996). The phagosome containing *Legionella pneumophila* within the protozoan *Hartmannella vermiformis* is surrounded by the rough endoplasmic reticulum. Appl. Environ. Microbiol. *62*, 2022–2028.

Bozue, J.A., and Johnson, W. (1996). Interaction of *Legionella pneumophila* with *Acanthamoeba castellanii*: uptake by coiling phagocytosis and inhibition of phagosome-lysosome fusion. Infect. Immun. *64*, 668–673.

Cianciotto, N.P., Eisenstein, B.I., Mody, C.H., Toews, G.B., and Engleberg, N.C. (1989). A *Legionella pneumophila* gene encoding a species-specific surface protein potentiates initiation of intracellular infection. Infect. Immun. *57*, 1255–1262.

Cianciotto, N. P., and Fields, B. S. (1992). *Legionella pneumophila* mip gene potentiates intracellular infection of protozoa and human macrophages. Proc. Natl. Acad. Sci. USA *89*, 5188–5191.

De Buck, E., Maes, L., Meyen, E., Van Mellaert, L., Geukens, N., Anne, J., and Lammertyn, E. (2005). *Legionella pneumophila* Philadelphia-1 tatB and tatC affect intracellular replication and biofilm formation. Biochem. Biophys. Res. Commun. *331*, 1413–1420.

Edelstein, P.H., Edelstein, M. A., Higa, F., and Falkow, S. (1999). Discovery of virulence genes of *Legionella pneumophila* by using signature tagged mutagenesis in a guinea pig pneumonia model. Proc. Natl. Acad. Sci. USA *96*, 8190–8195.

Eichinger, L., Pachebat, J. A., Glockner, G., Rajandream, M. A., Sucgang, R., Berriman, M., Song, J., Olsen, R., Szafranski, K., Xu, Q., Tunggal, B., Kummerfeld, S., Madera, M., Konfortov, B.A., Rivero, F., Bankier, A.T., Lehmann, R., Hamlin, N., Davies, R., Gaudet, P., Fey, P., Pilcher, K., Chen, G., Saunders, D., Sodergren, E., Davis, P., Kerhornou, A., Nie, X., Hall, N., Anjard, C., Hemphill, L., Bason, N., Farbrother, P., Desany, B., Just, E., Morio, T., Rost, R., Churcher, C., Cooper, J., Haydock, S., van Driessche, N., Cronin, A., Goodhead, I., Muzny, D., Mourier, T., Pain, A., Lu, M., Harper, D., Lindsay, R., Hauser, H., James, K., Quiles, M., Madan Babu, M., Saito, T., Buchrieser, C., Wardroper, A., Felder, M., Thangavelu, M., Johnson, D., Knights, A., Loulseged, H., Mungall, K., Oliver, K., Price, C., Quail, M.A., Urushihara, H., Hernandez, J., Rabbinowitsch, E., Steffen, D., Sanders, M., Ma, J., Kohara, Y., Sharp, S., Simmonds, M., Spiegler, S., Tivey, A., Sugano, S., White, B., Walker, D., Woodward, J., Winckler, T., Tanaka, Y., Shaulsky, G., Schleicher, M., Weinstock, G., Rosenthal, A., Cox, E.C., Chisholm, R.L., Gibbs, R., Loomis, W.F., Platzer, M., Kay, R.R., Williams, J., Dear, P.H., Noegel, A.A., Barrell, B., and Kuspa, A. (2005). The genome of the social amoeba *Dictyostelium discoideum*. Nature *435*, 43–57.

Fajardo, M., Schleicher, M., Noegel, A., Bozzaro, S., Killinger, S., Heuner, K., Hacker, J., and Steinert, M. (2004). Calnexin, calreticulin and cytoskeleton-associated proteins modulate uptake and growth of *Legionella pneumophila* in *Dictyostelium discoideum*. Microbiology *150*, 2825–2835.

Farbrother, P., Wagner, C., Na, J., Tunggal, B., Morio, T., Urushihara, H., Tanaka, Y., Schleicher, M., Steinert, M., and Eichinger, L. (2006). *Dictyostelium* transcriptional host cell response upon infection with *Legionella*. Cell. Microbiol. *8*, 438–456.

Fischer, G., Bang, H., Ludwig, B., Mann, K., and Hacker, J. (1992). Mip protein of *Legionella pneumophila* exhibits peptidyl-prolyl-cis/trans isomerase (PPlase) activity. Mol. Microbiol. *6*, 1375–1383.

Gal-Mor, O., and Segal, G. (2003). The *Legionella pneumophila* GacA homologue(LetA) is involved in the regulation of icm virulence genes and is required for intracellular multiplication in *Acanthamoeba castellanii*. Microb. Pathogen. *34*, 187–194.

Golstein, P., Aubry, L., and Levraud, J.P. (2003). Cell-death alternative model organisms: why and which? Nat. Rev. Mol. Cell. Biol. *4*, 798–807.

Hägele, S., Kohler, R., Merkert, H., Schleicher, M., Hacker, J., and Steinert, M. (2000). *Dictyostelium discoideum*: a new host model system for intracellular pathogens of the genus *Legionella*. Cell. Microbiol. *2*, 165–171.

Hales, L.M., and Shuman, H.A. (1999). *Legionella pneumophila* contains a type II general secretion pathway required for growth in amoebae as well as for secretion of the Msp protease. Infect. Immun. *67*, 3662–3666.

Hales, L.M., and Shuman, H.A. (1999). The *Legionella pneumophila* rpoS gene is required for growth within *Acanthamoeba castellanii*. J. Bacteriol. *181*, 4879–4889.

Hammer, B.K., Tateda, E.S., and Swanson, M.S. (2002). A two-component regulator induces the transmission phenotype of stationary-phase *Legionella pneumophila*. Mol. Microbiol. *44*, 107–118.

Heuner, K., Dietrich, C., Skriwan, C., Steinert, M., and Hacker, J. (2002). Influence of the alternative sigma(28) factor on virulence and flagellum expression of *Legionella pneumophila*. Infect. Immun. *70*, 1604–1608.

Kagan, J.C., and Roy, C.R. (2002). *Legionella* phagosomes intercept vesicular traffic from endoplasmic reticulum exit sites. Nat. Cell. Biol. *4*, 945–954.

King, J., and Insall, R.H. (2003). Parasexual genetics of *Dictyostelium* gene disruptions: identification of a *ras* pathway using diploids. BMC Genet. *4*, 12.

King, J., and Insall, R.H. (2006). Parasexual genetics using axenic cells. Methods Mol. Biol. *346*, 125–135.

Knetsch, M.L., Tsiavaliaris, G., Zimmermann, S., Ruhl, U., and Manstein, D.J. (2002). Expression vectors for studying cytoskeletal proteins in *Dictyostelium discoideum*. J Muscle Res. Cell. Motil. *23*, 605–611.

Kuspa, A. (2006). Restriction enzyme-mediated integration (REMI) mutagenesis. Methods Mol. Biol. *346*, 201–209.

Kuspa, A., and Loomis, W.F. (1992). Tagging developmental genes in Dictyostelium by restriction enzyme-mediated integration of plasmid DNA. Proc. Natl. Acad. Sci. USA *89*, 8803–8807.

Laguna, R.K., Creasey, E.A., Li, Z., Valtz, N., and Isberg, R.R. (2006). A *Legionella pneumophila*-translocated substrate that is required for growth within macrophages and protection from host cell death. Proc. Natl. Acad. Sci, USA *103*, 18745–18750.

Li, Z., Solomon, J.M., and Isberg, R.R. (2005). *Dictyostelium discoideum* strains lacking the RtoA protein are defective for maturation of the *Legionella pneumophila* replication vacuole. Cell. Microbiol. *7*, 431–442.

Lu, H., and Clarke, M. (2005). Dynamic properties of *Legionella*-containing phagosomes in *Dictyostelium* amoebae. Cell. Microbiol. *7*, 995–1007.

Lynch, D., Fieser, N., Gloggler, K., Forsbach-Birk, V., and Marre, R. (2003). The response regulator LetA regulates the stationary-phase stress response in *Legionella pneumophila* and is required for efficient infection of *Acanthamoeba castellanii*. FEMS Microbiol. Lett. *219*, 241–248.

Martens, H., Novotny, J., Oberstrass, J., Steck, T.L., Postlethwait, P., and Nellen, W. (2002). RNAi in *Dictyostelium*: the role of RNA-directed RNA polymerases and double-stranded RNase. Mol. Biol. Cell. *13*, 445–453.

Molofsky, A.B., and Swanson, M.S. (2004). Differentiate to thrive: lessons from the *Legionella pneumophila* life cycle. Mol. Microbiol. *53*, 29–40.

Müller-Taubenberger, A. (2006). Application of fluorescent protein tags as reporters in live-cell imaging studies. Methods Mol. Biol. *346*, 229–246.

Otto, G.P., Wu, M.Y., Clarke, M., Lu, H., Anderson, O.R., Hilbi, H., Shuman, H.A., and Kessin, R.H. (2004). Macroautophagy is dispensable for intracellular replication of *Legionella pneumophila* in *Dictyostelium discoideum*. Mol. Microbiol. *51*, 63–72.

Peracino, B., Wagner, C., Balest, A., Balbo, A., Pergolizzi, B., Noegel, A.A., Steinert, M., and Bozzaro, S. (2006). Function and mechanism of action of *Dictyostelium* Nramp1 (Slc11a1) in bacterial infection. Traffic *7*, 22–38.

Robinson, C.G., and Roy, C.R. (2006). Attachment and fusion of endoplasmic reticulum with vacuoles containing *Legionella pneumophila*. Cell. Microbiol. *8*, 793–805.

Rossier, O., Starkenburg, S.R., and Cianciotto, N.P. (2004). *Legionella pneumophila* type II protein secretion promotes virulence in the A/J mouse model of Legionnaires' disease pneumonia. Infect. Immun. *72*, 310–321.

Schreiner, T., Mohrs, M.R., Blau-Wasser, R., von Krempelhuber, A., Steinert, M., Schleicher, M., and Noegel, A.A. (2002). Loss of the F-actin binding and vesicle-associated protein comitin leads to a phagocytosis defect. Eukaryot. Cell *1*, 906–914.

Segal, G., Purcell, M., and Shuman, H.A. (1998). Host cell killing and bacterial conjugation require overlapping sets of genes within a 22-kb region of the *Legionella pneumophila* genome. Proc. Natl. Acad. Sci. USA *95*, 1669–1674.

Sekar, R.B., and Periasamy, A. (2003). Fluorescence resonance energy transfer (FRET) microscopy imaging of live cell protein localizations. J. Cell. Biol. *160*, 629–633.

Skriwan, C., Fajardo, M., Hägele, S., Horn, M., Wagner, M., Michel, R., Krohne, G., Schleicher, M., Hacker, J., and Steinert, M. (2002). Various bacterial pathogens and symbionts infect the amoeba *Dictyostelium discoideum*. Int. J. Med. Microbiol. *291*, 615–624.

Soll, D.R., Voss, E., Johnson, O., and Wessels, D. (2000). Three-dimensional reconstruction and motion analysis of living, crawling cells. Scanning *22*, 249–257.

Solomon, J.M., Rupper, A., Cardelli, J.A., and Isberg, R.R. (2000). Intracellular growth of *Legionella pneumophila* in *Dictyostelium discoideum*, a system for genetic analysis of host–pathogen interactions. Infect. Immun. *68*, 2939–2947.

Steinert, M., Hentschel, U., and Hacker, J. (2002). *Legionella pneumophila*: an aquatic microbe goes astray. FEMS Microbiol. Rev. *26*, 149–162.

Steinert, M., Heuner, K., Buchrieser, C., Albert-Weissenberger, C., and Glockner, G. (2007). *Legionella pathogenicity*: Genome structure, regulatory networks and the host cell response. Int. J. Med. Microbiol., *in press*.

Stone, B. J., and Abu Kwaik, Y. (1998). Expression of multiple pili by *Legionella pneumophila*: identification and characterization of a type IV pilin gene and its role in adherence to mammalian and protozoan cells. Infect. Immun. *66*, 1768–1775.

VanRheenen, S.M., Luo, Z.Q., O'Connor, T., and Isberg, R.R. (2006). Members of a *Legionella pneumophila* family of proteins with ExoU (phospholipase A) active sites are translocated to target cells. Infect. Immun. *74*, 3597–3606.

Vogel, J.P., Andrews, H.L., Wong, S.K., and Isberg, R.R. (1998). Conjugative transfer by the virulence system of *Legionella pneumophila*. Science *279*, 873–876.

Wagner, C., Khan, A.S., Kamphausen, T., Schmausser, B., Unal, C., Lorenz, U., Fischer, G., Hacker, J., and Steinert, M. (2007). Collagen binding protein Mip enables *Legionella pneumophila* to transmigrate through a barrier of NCI-H292 lung epithelial cells and extracellular matrix. Cell. Microbiol. *9*, 450–462.

Weber, S.S., Ragaz, C., Reus, K., Nyfeler, Y., and Hilbi, H. (2006). *Legionella pneumophila* exploits PI(4)P to anchor secreted effector proteins to the replicative vacuole. PLoS Pathog. *2*, e46.

Wessels, D., Kuhl, S., and Soll, D.R. (2006). Application of 2D and 3D DIAS to motion analysis of live cells in transmission and confocal microscopy imaging. Methods. Mol. Biol. *346*, 261–279.

Index

A

D

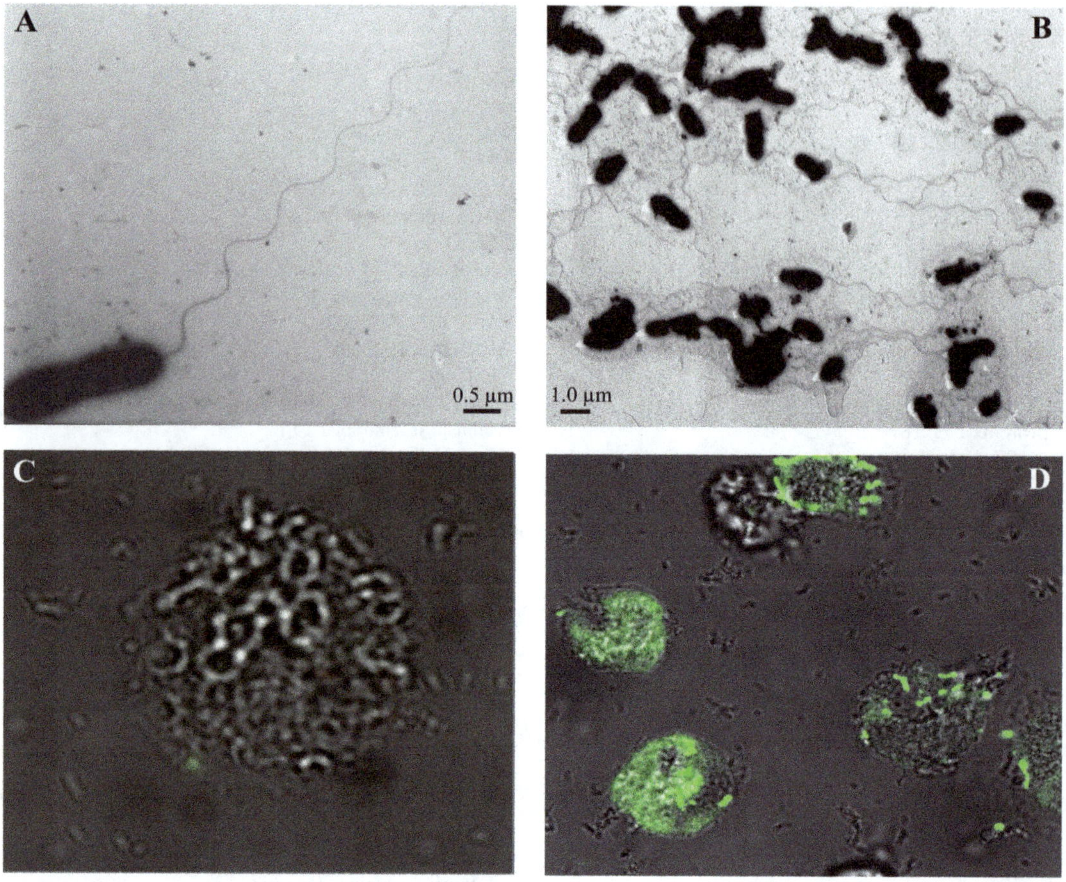

Figure 6.1 Electron microscopy (A, B) and fluorescence microscopy (C, D) of *L. pneumophila*. Micrographs A and B show the flagellation of *L. pneumophila* Corby. Bacteria were grown on BCYE agar plates at 30°C for 4 days, suspended in distilled water, transferred to Pioloform-coated copper grids and shadowed with platinium-palladium. Bacteria were examined with a transmission electron microscope at 60 kV. (C, D) Flagella expression of *L. pneumophila* Paris during infection of *Acanthamoeba castellanii* (C, after 8 h of infection; D, after 15 h of infection), visualized with a polyclonal anti-FlaA antiserum and processed for immunofluorescence (second antibody labelled with Alexa Fluor 488). Bacteria were examined with an LSM 510 confocal laser scanning microscope.

www.ingramcontent.com/pod-product-compliance
Lightning Source LLC
LaVergne TN
LVHW062230200926
848484LV00006B/1508
* 9 7 8 1 9 0 4 4 5 5 2 6 4 *